A Review of Science and Technology During the 1975 School Year

Science Year
The World Book Science Annual

1976

Field Enterprises Educational Corporation
Chicago   Frankfurt   London   Paris   Rome   Sydney   Tokyo   Toronto

The publishers of *Science Year* gratefully
acknowledge the following for permission to use
copyrighted illustrations. A full listing of illustration
acknowledgments appears on pages 430 and 431.

 78 Photographs: Dobelle, Mladejovsky and
     Girvin, from *Science*. Copyright 1974 by
     the American Association for the
     Advancement of Science

131 Copyright Editions Pierre Charron—Tanguy
     de Remur Paris

232 Larry Lahren and Robson Bonnichsen, from
     Science. Copyright 1975 by the American
     Association for the Advancement of Science

232 © Times Newspapers Ltd., 1974

234 U.S. Forest Service, from *Science*. Copyright 1974 by the
     American Association for the Advancement of Science

248 John Brandt, Fred V. Lucas, Sr., Arlene
     Martin, Marie L. Vorbeck, University of
     Missouri School of Medicine. Reprinted
     with permission from Biochemical Biophysical
     Research Communications. © Academic Press, Inc.

295 Peck-Sun Lin, Donald F. Hoelzl Wallach, Tufts-
     New England Medical Center, from *Science*. Copyright 1974
     by the American Association for the Advancement
     of Science

333 David Premack, University of California,
     Santa Barbara, from *Science*. Copyright 1975 by the American
     Association for the Advancement of Science

358 Baron Hugo van Lawick,
     © National Geographic Society

358 Robert F. Sisson,
     © National Geographic Society

The Cover: Man meets manatee in its home territory.

Preface

Long-time readers of *Science Year* may have noticed that the faces on the Advisory Board page have been changing. With the 1974 edition, *Science Year* began a program of rotation of Advisory Board members to take better advantage of the large community of scientists and engineers available to guide us. This has meant, of course, that the long and faithful service of the original board members, some of whom had been with us for nine years, has come to an end.

The *Science Year* Advisory Board helps to keep the staff informed of the most important and interesting work going on in the research laboratories today. They do this at a two-day meeting held each year in early October. First, the board members critique the latest edition, with candid, constructive criticism. Then they discuss trends in science and technology, and suggest topics for future articles.

We will miss our retired advisers. They always assured us a fascinating, information-packed meeting. Caltech professor Harrison Brown, whose work has taken him all over the world, would tell tales of science on six continents. Physicist Alvin Weinberg would constantly remind us to cover that science thoroughly. Environmentalist Barry Commoner's provocative comments started many a lively discussion and taught us an important lesson—that scientists do not always agree. And, when the talk turned a bit pompous and pedantic, Dr. Walsh McDermott put things back on the track with his ready wit.

The new board members are continuing in this vein. And they agree that a periodic turnover is a healthy thing for *Science Year*. It should make the book an even better resource and source of reading pleasure about the excitement of science.   [Arthur G. Tressler]

Contents

Staff

Editorial Director
William H. Nault

Editorial

Executive Editor
Arthur G. Tressler

Managing Editor
Michael Reed

Chief Copy Editor
Joseph P. Spohn

Senior Editors
Robert K. Johnson, Edward G. Nash,
Kathryn Sederberg, Darlene Stille,
Foster P. Stockwell

Assistant Copy Editor
Irene B. Keller

Editorial Assistant
Deborah Ellis

Art

Executive Art Director
William Dobias

Art Director
Alfred de Simone

Senior Artists
Roberta Dimmer
Gumé Nuñez

Artists
Wilma Stevens
Stanley Schrero

Photography Director
Fred C. Eckhardt, Jr.

Photo Editing Director
Ann Eriksen

Senior Photographs Editors
Blanche Cohen,
Leslie J. Cohn,
Marilyn Gartman,
John S. Marshall

Research and Services

Director of Educational Services
John Sternig

Director of Editorial Services
Carl A. Tamminen

Head, Cartographic Services
Joseph J. Stack

Senior Index Editor
Pamela Williams

Pre-Press Services

Director
Richard A. Atwood

Manager, Pre-Press
John Babrick

Assistant Manager, Pre-Press
Marguerite DuMais

Manager, Art Production
Alfred J. Mozdzen

Assistant Manager, Art Production
Barbara J. McDonald

Manufacturing

Executive Director
Philip B. Hall

Production Manager
Jerry R. Higdon

Manager, Research and Development
Henry Koval

Editorial Advisory Board

Contributors

Adelman, George, M.S.
Managing Editor
Neurosciences Research Program
Massachusetts Institute of Technology
Neurology

Alderman, Michael H., M.D.
Assistant Professor of Public Health
Cornell University Medical College
Medicine, Internal
Public Health

Araujo, Paul E., Ph.D.
Assistant Professor
Department of Food Science
University of Florida
Nutrition

Auerbach, Stanley I., Ph.D.
Director, Environmental Sciences Division
Holifield National Laboratory
Ecology

Avery, Mary Ellen, M.D., Sc.D.
Professor of Pediatrics
Harvard Medical School
The Breath of Life

Bell, William J., Ph.D.
Associate Professor of Entomology
University of Kansas
Zoology

Belton, Michael J. S., Ph.D.
Astronomer, Kitt Peak
National Observatory
Astronomy, Planetary

Brill, Winston J., Ph.D.
Professor of Bacteriology
University of Wisconsin
The Nitrogen Fix

Bromley, D. Allan, Ph.D.
Henry Ford II Professor and Chairman
Department of Physics
Yale University
Physics, Nuclear

Chatfield, Eric J., Ph.D.
Head, Electron Optical Laboratory
Ontario Research Foundation
A New Dimension in the Microscope

Chesher, Richard H., Ph.D.
President
Marine Research Foundation
Oceanography

Chiller, Jacques M., Ph.D.
Associate Professor
National Jewish Hospital
Immunology

Copeland, John A., Ph.D.
Member of Technical Staff
Bell Telephone Laboratories
Communications

Cromie, William J., B.S.
Executive Secretary
Council for the Advancement
of Science Writing
Earthquake Early Warning

Deffeyes, Kenneth S., Ph.D.
Associate Professor of Geology
Princeton University
Geoscience, Geology

Dorozynski, Alexander
Free-Lance Science Writer
Journey to the Birthplace of Continents

Drake, Charles L., Ph.D.
Professor of Earth Sciences
Dartmouth College
Geoscience, Geophysics

Dupree, A. Hunter, Ph.D.
George L. Littlefield Professor of History
Brown University
American Science—The First 200 Years

Eberhart, Jonathan
Space Sciences Editor
Science News
Space Exploration

Eigner, Joseph, Ph.D.
Chief, Hazardous Waste Project
Missouri Department of Natural Resources
Close-Up, Environment

Eisenberg, Richard, Ph.D.
Associate Professor of Chemistry
University of Rochester
Chemistry, Structural

Ensign, Jerald C., Ph.D.
Professor of Bacteriology
University of Wisconsin
Microbiology

Gallo, Robert C.
Chief, Laboratory of Tumor Cell Biology
National Cancer Institute
Close-Up, Medicine, Internal

Gerhardt, John R., M.S.
Technical Editor
American Meteorological Society
Meteorology

Giacconi, Riccardo, Ph.D.
Associate Director
Center for Astrophysics
Smithsonian Astrophysical Observatory
Astronomy, High Energy

Goldhaber, Paul, D.D.S.
Dean, Harvard School
of Dental Medicine
Medicine, Dentistry

Gray, Ernest P., Ph.D.
Principal Staff Member
Applied Physics Laboratory
Johns Hopkins University
Physics, Plasma

Griffin, James B., Ph.D.
Director, Museum of Anthropology and
Chairman, Department of Anthropology
University of Michigan
Archaeology, New World

Gump, Frank E., M.D.
Associate Professor of Surgery
College of Physicians and Surgeons
Columbia University
Medicine, Surgery

Hartl, Daniel L., Ph.D.
Associate Professor of Biological Sciences
Purdue University
Genetics

Hawthorne, M. Frederick, Ph.D.
Professor of Chemistry
University of California at Los Angeles
Chemistry, Synthesis

Hayes, Arthur H., Jr., M.D.
Associate Professor of Medicine and
Pharmacology
Milton S. Hershey Medical Center
Drugs

Henahan, John F., B.S.
Free-Lance Science Writer
James Gunn

Herbert, Victor, M.D.
Clinical Professor of Pathology and
Medicine
Columbia University
Close-Up, Nutrition

Irwin, Howard S., Ph.D.
President
New York Botanical Garden
Botany

Johnson, Richard T., M.D.
Professor of Neurology and Microbiology
The Johns Hopkins University
School of Medicine
Agents of Slow Death

Kates, Robert W., Ph.D.
Professor of Geography
Clark University
Dealing with Disaster

Kessler, Karl G., Ph.D.
Chief, Optical Physics Division
National Bureau of Standards
Physics, Atomic and Molecular

Ketterson, John B., Ph.D.
Professor of Physics
Northwestern University
Physics, Solid State

Lee, Richard B., Ph.D.
Associate Professor of Anthropology
University of Toronto
The IKung's New Culture

Lewin, Roger, Ph.D.
Science Editor
New Scientist
The Leakeys

Maglio, Vincent J., Ph.D.
Research Associate
Department of Geology
Princeton University
Geoscience, Paleontology

Maran, Stephen P., Ph.D.
Head, Advanced Systems and
Ground Observations Branch
Goddard Space Flight Center
Astronomy, Stellar

March, Robert H., Ph.D.
Professor of Physics
University of Wisconsin
A Subatomic Surprise
Physics, Elementary Particles

Merbs, Charles F., Ph.D.
Professor and Chairman
Department of Anthropology
Arizona State University
Anthropology

Nelson, David L., Ph.D.
Assistant Professor of Biochemistry
University of Wisconsin
Biochemistry

Norman, Colin, B.Sc.
Washington Correspondent
Nature
Science Support

Novick, Sheldon
Editor, *Environment*
Environment

O'Neill, Gerard K., Ph.D.
Professor of Physics
Princeton University
Settlers in Space

Price, Frederick C., B.S.
Managing Editor
Chemical Engineering
Chemical Technology

Richardson, Herbert H., Sc.D.
Professor and Head of
Mechanical Engineering Department
Massachusetts Institute of Technology
Transportation

Rodden, Judith, M.Litt.
Research Archaeologist
Archaeology, Old World

Shank, Russell, D.L.S.
Director of Libraries
Smithsonian Institution
Books of Science

Shepley, Lawrence C., Ph.D.
Associate Professor of Physics
University of Texas at Austin
Close-Up, Astronomy, Cosmology

Silk, Joseph, Ph.D.
Associate Professor of Astronomy
University of California at Berkeley
Astronomy, Cosmology

Snider, Arthur J.,
Science Editor
Chicago Daily News
New "Eyes" for the Blind

Strong, Ian B., Ph.D.
Senior Staff Member
Los Alamos Scientific Laboratory
The Case of the Baffling Bursts

Tilton, George R., Ph.D.
Professor of Geochemistry
University of California at Santa Barbara
Geoscience, Geochemistry

Vale, Wylie W., Ph.D.
Associate Research Professor
The Salk Institute for Biological Studies
Messengers from the Brain

Veverka, Joseph, Ph.D.
Assistant Professor of Astronomy
Cornell University
Solar Systems in Miniature

Vietmeyer, Noel D., Ph.D.
Professional Associate
National Academy of Sciences
The Menaced Mermaid

Weber, Samuel, B.S.E.E.
Executive Editor
Electronics Magazine
Electronics

Wick, Gerald L., Ph.D.
Assistant Director, Assistant Research
Physicist
Institute of Marine Resources
University of California at San Diego
Close-Up, Meteorology

Wittwer, Sylvan H., Ph.D.
Director, Agricultural Experimental Station
Michigan State University
Agriculture

Zare, Richard N., Ph.D.
Professor of Chemistry
Columbia University
Chemistry, Dynamics
Close-Up, Chemistry, Dynamics

Zuckerman, Harriet, Ph.D.
Associate Professor of Sociology
Columbia University
Close-Up, Science Support

Contributors not listed on
these pages are members of the
Science Year editorial staff.

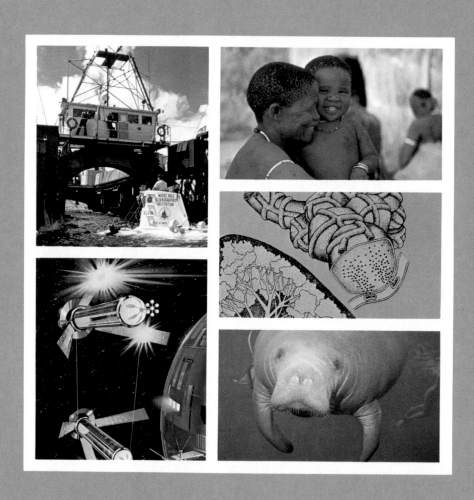

Special Reports

The Special Reports and an exclusive *Science Year* Special Feature
give in-depth treatment to the major advances in science. The subjects
were chosen for their current importance and lasting interest.

The Menaced Mermaid

By Noel D. Vietmeyer

Munching through weed-choked waterways, the homely, harmless manatee could prove a useful friend, if we can save it from its only enemy—man

One sweltering day in February, 1974, as I crouched beside a pond in the Botanic Gardens in Georgetown, Guyana, the brown water heaved suddenly, and a gray face with a broad snout and two beady, black eyes broke through the surface. A small Guyanese girl plucked a handful of grass from the lawn and held it out. The animal's mouth opened and its bristle-covered lips closed over the grass in the girl's palm. Then, with a roll of its great ashen back and a sweep of its broad, curved tail, the creature was gone.

That was my introduction to the manatee, or sea cow, a shy, little-known animal that, in centuries past, inspired sailors' tales of mermaids. Unlikely enough as a mermaid, the manatee could be of great help to mankind if science can prevent its extinction. I had traveled some 2,500 miles from Washington, D.C., to Georgetown, a seaside city on the northeastern shoulder of South America, to confer with other scientists about manatees. Under the sponsorship of the Guyanese National Science Research Council, the United States National Academy of Sciences (NAS), and the Canadian International Development Research Centre, 43 scientists from Brazil, Canada, Colombia, Great Britain, Guyana, Trinidad, the United States, and Venezuela were meeting in Georgetown to study Guyana's experience in using manatees to clear weeds from its waterways. We hoped to determine how the animals could be put to work elsewhere in the world. Our conclusions, although hopeful, were colored by a realistic assessment of the manatee's future: "The manatee offers a most attractive possibility for the biological control of aquatic weeds—on a small scale it has been demonstrated convincingly," said Donald S. Farner, chairman of the U.S. delegation, in his report. "Guyana has something to show the world . . . but, to capitalize on it, we face a race against time, for the manatee is fast becoming extinct."

Few people in temperate climates realize that aquatic weeds have become a global environmental plague. With growing numbers of dams and irrigation projects in developing tropical and subtropical countries, the menace of these weeds is intensifying. They block hydroelectricity production, clog canals and irrigation pumps, hinder small-boat traffic, and cause floods.

The weeds are also a rich breeding ground for tropical disease. They provide shelter for the larvae of mosquitoes which carry parasites causing malaria, filariasis, and encephalitis. Snails that spread schistosomiasis, one of the most widespread and debilitating tropical afflictions, find food and protection among the weeds. Weed pollution can also drastically reduce fishing.

The world's most troublesome aquatic weed is the water hyacinth, which was carried from Brazil to Africa early in the 20th century, reputedly by a missionary entranced by its purple flowers and green rosette of floating leaves. It spread quickly in its new home, densely infesting the Nile River in Sudan and Egypt. First reported on the Congo River in 1952, the water hyacinth spread 1,000 miles (1,600 kilometers) upriver to Stanleyville, in the Belgian Congo (now Kisangani, Zaire), in less than three years and, there, it soon began hindering river travel and trade.

"In the vicinity of my father's sawmill, whole villages had to be abandoned shortly after the water hyacinth arrived," Bernard Tursch, a Belgian biologist who grew up in the Belgian Congo, told me. "The African inhabitants couldn't get their canoes through the shore-to-shore tangle of plants, and they were deprived of their essential food supplies and markets."

Large irrigation projects in India's Rajasthan Desert have been rendered useless by the plants. Peasant farmers, lured to that harsh and arid land by the promise of water, face ruin and starvation when their parched crops shrivel and die for want of the irrigation water that never arrives.

The author:
Noel D. Vietmeyer, a professional associate of the U.S. National Academy of Sciences, helps emerging nations find useful new plant and animal resources.

A manatee cow and her
calf glide through
clear tropical water.
Their affectionate bond
will last several years.

Farms in the wet lowlands of Bangladesh are ruined, too, when floodwaters float huge hyacinth "rafts," weighing up to 150 tons per acre, over the rice paddies. As the floods recede, the weeds settle on the germinating rice, crushing and killing it. Panama Canal engineers have estimated that the canal would become impassable within about three years without continuous and expensive mechanical and chemical weed-control measures.

The water hyacinth's beauty also brought it to the United States. Charmed by the flower, Japanese exhibitors at the Cotton States Exposition in New Orleans in 1884 imported it from Venezuela and gave free samples to visitors to their booth. It seemed ideal to decorate plantation fishponds but, once established, the plant quickly overran the South. Only 13 years later, Florida's lumber industry petitioned Congress for disaster relief because the plants blocked rivers so completely that logs could no longer be floated to sawmills. Water hyacinths still pose an economic problem in Florida and Louisiana; each state has at least 200,000 acres (81,000 hectares) of weed-covered lakes, rivers, and canals.

Unfortunately, there is no easy way to reduce these weed infestations. All methods are unsatisfactory in some respect. There are machines that can harvest or destroy aquatic plants, but they are expen-

sive and difficult to operate in remote areas. Herbicides that kill aquatic weeds also harm the environment and, in remote areas, they are also expensive, difficult to apply, and often ineffective. For example, masses of floating plants hid ship channels and camouflaged navigation buoys in the Congo River in 1957, slowing ships carrying copper and uranium ore to Europe and North America. The Belgian Congo government spent a million dollars trying to eradicate the weeds. Herbicides were sprayed from boats, planes, and helicopters along the infested stretch of river. But the weeds, which can double in number in only two weeks, made a comeback, and Belgian biologist J. Lebrun estimated that within a year 143 short tons (130 metric tons) of water hyacinth were floating down the Congo every hour.

Some form of natural control would be ideal. One or two insect species will eat aquatic weeds and spare useful plants, such as rice, but their control potential is still unknown. Guyana has used manatees to clear the weeds from canals and ponds since the 1800s. By the late 1950s, about 100 manatees were performing this service there, and some are still at it 20 years after they were captured in Guyana's rivers and dumped into the canals.

"In 1951, we built a canal 2,000 feet long and 40 feet wide [610 by 12 meters], for the city's water," Jim Richardson, engineer at the

Tropical waterways clogged with
weeds, *top left,* could look like
one in Georgetown, Guyana,
above, which is kept weed-free
by its resident manatee. The
animal harvests the weeds by
guiding them to its mouth
with its stubby flippers.

Georgetown waterworks, recalled. "But six weeks later, we couldn't get enough water through to supply the city. Water flow was blocked by weeds rooted in the earth bottom. So we purchased two manatees from a village patriarch—who sent his family into a nearby river to haul them out by hand—and added them to the canal.

"In the 23 years since then," he said, "the manatees have kept the weeds cropped to an underwater lawn just an inch or two [25 to 50 millimeters] high. There has been no obstruction, and the city has been getting all the water it needs."

According to W. Herbert L. Allsopp, who pioneered much of Guyana's manatee program, two medium-sized animals can clear a weed-infested canal 4,800 feet long and 22 feet wide (1,500 by 7 meters) in 17 weeks. And Lawrence G. Charles, chief engineer at the Ministry of Works and Hydraulics, gave the conference an example of the economic benefit of using manatees. "A small canal in western Guyana cost us $750 annually to clear the weeds until two manatees were introduced in 1956," he pointed out. "Since then, we have not spent a penny on clearing that canal."

Charles organized a demonstration so the visiting scientists could see for themselves. A canal, half cleared for canoe travel, the rest overrun with weeds, was selected in the Garden of Eden, a district near Georgetown. A manatee cow and its calf were placed in the canal one afternoon. The next morning, we found that the pair had cleared out a stretch of canal about 50 yards (46 meters) long, right up to the bank. It was obvious that manatees are systematic grazers. Rather than a piecemeal attack, the pair started at one point and moved along the canal clearing out all plants in their path. I remember thinking that manatees could be called "aquatic lawn mowers."

The canal demonstration was impressive, but the Guyanese experience has not shown whether manatees can clear large lakes or reservoirs, or how much vegetation they can remove under varying conditions. This is largely because many of those who started the manatee program left Guyana during the political turmoil of the early 1960s, and the pioneering work was not systematically continued. The government then began using herbicides for aquatic-weed control. Now, like many other nations, Guyana has become concerned about the effect of these chemicals on domestic water that is taken from the rivers and canals that crisscross the country. This growing disenchantment with chemical weed control led to the manatee conference.

We know little about manatees. We know that they are gray, barrel-shaped animals with flippers and a spade-shaped horizontal tail—not vertical like a fish's tail. They grow to as much as 14 feet (4 meters) long, and weigh about 1,000 pounds (450 kilograms). They breathe air, are warm-blooded, and bear live young that are suckled on milk.

But almost all of the internal physiology of manatees is unknown. We know nothing about their brain, or how their digestive, nervous, and reproductive systems work. We are not sure of their gestation

A Bony Link to the Sea Cow's Kin

Manatees evolved from land mammals millions of years ago. Flippers retain the five-fingered structure, but the hind limbs are gone and the pelvis is reduced to two small bones.

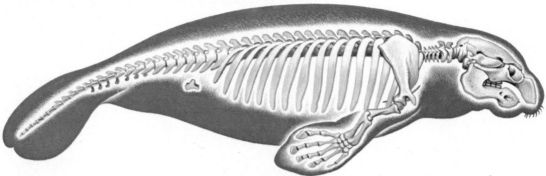

Manatees, dugongs, and elephants have similar jaws and teeth. Manatees have only molars. Some dugongs have tusks. All three replace worn-out teeth.

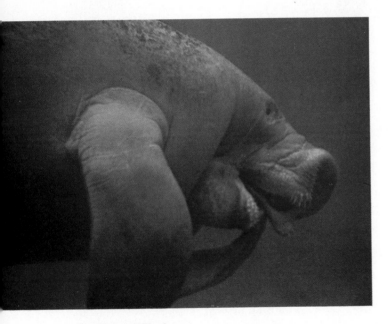

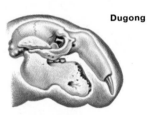

Dugong

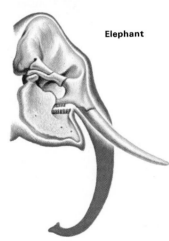

Elephant

Feasting on Weeds

The manatee's fleshy lips draw plants into its mouth, *above*. Then it grinds the food with molars and a rough, horny palate, *below*.

The sons and grandsons
of Datakaran Jeetlall
wrestle a stubborn
manatee onto a canal
bank so scientists
can examine it.

period or even how long they live, though authorities suspect they may live more than 30 years. We do not know how often they breed or if they ever have more than one offspring at a time. We have no idea how many there are in the world. Even their classification is disputed, though most biologists differentiate three species of manatees.

Manatees are mammals, not fish. But they are not related to the more famous marine mammals such as whales, dolphins, and seals, all of which eat fish and other marine animals. Manatees eat only plants. Their closest relative, the dugong, lives in the coastal waters of countries surrounding the Indian Ocean. Oddly enough, their next nearest living relative is the elephant.

Although biologists disagree on the specific details of the origins of modern mammals, one widely held view is that three kinds of small, furry mammals lived in shallow coastal marshes about 100 million years ago. Down through the ages, they gave rise to various land-

based and aquatic mammals. One of the three lines led to cows and toothed whales; one to bears, dogs, and baleen whales; the third, to elephants, manatees, and dugongs.

With water to support much of their weight, the aquatic forms ballooned to the blimplike shape of the whale and manatee. Their hind legs, no longer needed in the water, shrank, and, over the eons, disappeared entirely. This streamlining was an advantage for an animal that moves through water. To make swimming even easier, the "fingers" of the forefeet grew together to form stubby flippers. Even today, the Caribbean and the African species of manatee still have "fingernails" at the tip of their flippers and X rays clearly show jointed finger bones. A tail was needed for swimming, and it swelled out to a broad blade attached to the manatee's body by a large, powerful bundle of muscle. The tail was not vital to survival on land, so it became the elephant's tiny "fly swatter."

Manatees live only in tropical waters. In water colder than 60°F. (15°C), they quickly get pneumonia and die. They are found in more than 40 countries—around the Caribbean Sea, in the headwaters of the Amazon River, and along the west coast of Africa from Senegal south to Angola. About 1,200 manatees lead a marginal existence in Florida. Every few years, the water temperature in most parts of the state becomes too cold for them to survive, so they seek out warm springs for sanctuary or they migrate down the Florida coast to the warmth of the Everglades.

Although manatees spend most of their lives submerged, their nostrils closed against the water by a flap of skin, they must rise to the surface every few minutes, open their nostrils, and breathe. They prefer fresh water, but they are also at home in the sea and routinely swim up and down coasts looking for rivers with better grazing. They will eat almost any water plant, though they prefer succulent, submerged plants to fibrous rushes and grasses. They can consume over 100 pounds (45 kilograms) a day.

Despite having no front teeth and two stubby flippers that barely reach its mouth, the manatee has very little trouble eating. It uses its fleshy, bristle-covered lips to rip plants from the bottom (or from a floating weed mat) and to pass them back to its grinding molars. New molars appear regularly at the back of the jaw. With each new one, the whole row moves forward and the front molar falls out. In this curious way, a manatee may produce 60 new teeth during its lifetime.

Because I and several other scientists at the conference knew so little about manatees, our Guyanese hosts decided to capture one so we could examine it. We were bused to the Garden of Eden, where we found Datakaran Jeetlall, who had captured 80 manatees for the weed-control projects in the 1950s, directing a team consisting of his sons and grandsons. The men walked chest-deep in the chocolate-brown water of the canal. Ahead of them, we could see the ripple created by the submerged manatee that they were herding. Finally, at a low-banked place with soft, cushioning grasses, they cast a rope around the bulky animal's tail and heaved it up onto the bank.

For 15 minutes, our 1,000-pound manatee remained quite still even though the workers threw buckets of water over it. Its coarse skin felt like cold, hard rubber. Its pea-sized eyes were closed, and its nostrils opened for air every few minutes. Then, an inquisitive physiologist, Emmanuel C. Amoroso of Cambridge, England, asked the workers to roll it over so he could determine its sex. At that, it rebelled. Curving into a crescent shape, it recoiled with a powerful snap that pitched workers holding on at both ends into the air. After 10 attempts, they finally managed to roll the manatee onto its back. Teats were clearly visible—one at the base of each flipper. She was christened Eve.

Later, the workers, who had known its sex all along, told us that Eve had a calf in the canal. A manatee calf suckles from the fist-sized teat while swimming stroke for stroke with its mother. When the mother surfaces to breathe, she takes care that the suckling calf's nostrils break the surface too, and because of the calf's smaller lung capacity, she surfaces more frequently.

A newborn calf may be 4 feet (1.22 meters) long and weigh 75 pounds (34 kilograms). Calf and mother have a close, affectionate bond. They spend much time nuzzling each other and chirp incessantly to maintain contact, especially in murky waters. Calves begin to eat vegetation after only a few months, but they still remain with their mother for several years.

Manatees are sluggish, docile, and harmless. They never attack humans or wildlife. Unlike other animals, they have no aggressive instincts. They share their weeds with each other and their waterways with fish, alligators, crocodiles, and electric eels. However, they shun

Scars on the back of a Florida manatee are souvenirs of encounters with boat propellers. Most U.S. manatees bear similar marks.

Where the Sea Cows Graze

■ Manatee   ■ Dugong   ■ Steller's Sea Cow (extinct)

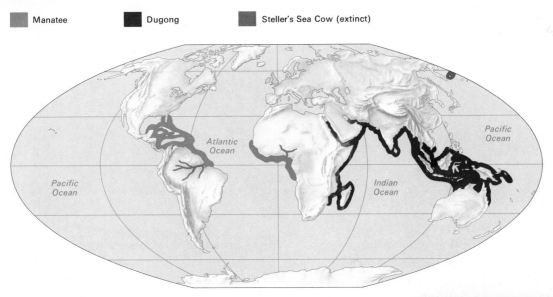

Workers at the Miami Seaquarium use a sturdy sling to send a manatee off to a weed-control test in a Florida canal.

contact with humans, preferring remote swamps and estuaries. In captivity, however, with good treatment, they have developed affectionate relationships with people, and have even learned a few simple tricks, such as swimming through hoops. Few, if any, wild animals of comparable size are so docile and easy to control. Workers at San Francisco's Steinhart Aquarium once kept a newly arrived and very young Amazonian manatee out of water to treat a harpoon wound. Lying on a canvas sling–covered with damp sacks to keep its skin moist to prevent cracking and peeling–it happily munched lettuce for two weeks. That manatee is still at the Steinhart–one of only a handful living in zoos throughout the world.

In Columbus' time, manatees could be found throughout the Caribbean Sea. The explorer's log for Jan. 9, 1493, three months after he reached the New World, reads: "On the previous day, when the Admiral went to the Rio del Oro, he saw three mermaids, which rose well out of the sea." We know now that he actually saw manatees. He saw them again in 1502 during his fourth voyage. His 13-year-old son, Ferdinand, who was on board, later speculated that they might be "not fishes but real calves," because their meat looks and tastes like veal, they feed only on grass, and "inside they have nothing like a fish." Such observations led to the manatee's alternate name, sea cow.

Settlers following Columbus to the Caribbean found manatees in abundance. The Roman Catholic Church conveniently classified the animal as a fish, and manatee meat was relished on Fridays and days of religious abstinence when other meats were forbidden. In the 1600s, French colonists copied Indian methods of preserving meat, including manatee, by drying it over a smoky fire to form strips of *boucan* (now

Moving through water with ponderous grace, the gentle sea cow faces an uncertain future.

often called jerky). These colonists became known as *boucaniers*; later Anglicized to buccaneers. Some became renegades, and boucaned manatee meat went along on many pirate raids. In 1670, the English pirate Sir Henry Morgan took it with him when he crossed the Isthmus of Panama to attack Spanish settlements on the Pacific shore. Expeditions were sent from as far as Holland to collect manatee meat, and the manatee population began a dismal decline.

Manatees produce clear, pleasantly flavored blubber oil and their hides can be processed into tough leather. Even their ribs can be used, as a substitute for ivory. As a result, though it is protected by law in most countries, illegal hunting of the manatee continues today.

Paulo de Almeida Machado, Brazil's minister of health, described to conference participants the slaughter of 8,000 manatees in the Amazon during a two-year period in the early 1900s. It was also reported that a pack train carrying the meat of over 200 manatees was seen near the headwaters of the Amazon River in eastern Peru in the mid-1960s. "Manatee is much tastier and more tender than beef," a Georgetown taxi driver told me. "It has little fat and looks like veal."

Realidade, a Brazilian magazine, published a color-illustrated feature story in 1971 showing hunters capturing manatees and cruelly hammering wooden plugs into their nostrils to kill them. Because they cannot breathe through their mouth, they suffocated.

In other ways, too, man proves unworthy of these gentle giants. For example, powerboats, driven thoughtlessly at high speed, have killed many, maimed some, and scarred almost every manatee left in Florida. Many can be identified by the individual pattern of white propeller scars slashed into their back.

Man's excesses have now brought manatees to the brink of extinction. Except for some species of seal, the manatee and the dugong are the most endangered aquatic mammals. They may soon share the fate of Steller's sea cow, a gigantic—23 feet (7 meters) long—Arctic manatee first seen by a biologist off the Alaskan coast in 1741 and hunted out of existence by 1760.

While manatee numbers decrease, the aquatic-weed threat grows. Manatees could be put to use clearing waterways, but there are just not enough left. To preserve them and increase their numbers, we must learn to breed and rear them. On May 3, 1975, the first manatee ever conceived in captivity was born at the Miami Seaquarium. Studies of the parents' mating and fertility cycles and the 35-pound (16-kilogram) calf's development may yield valuable information.

To stimulate this vitally needed manatee research, the Georgetown conference called for the establishment of an international center where researchers can promote manatee survival and domestication. Researchers from all nations that have or need manatees could work there. It would have field programs to help countries inventory and preserve their manatees and make use of research findings.

The proposal for a manatee laboratory seemed like a wishful dream when the conference ended. But a deeply committed group of conference participants is making a start. Through their efforts, the government of Guyana has already provided 100 acres (40 hectares) of land for a laboratory and has offered to dig enough canals and ponds to hold 50 manatees. Legislation establishing the laboratory, the International Centre for Research on Manatees, as an independent international body has been drafted by the Guyanese government, and the NAS has agreed to help in preparing its scientific program. The Canadian International Development Research Centre in Ottawa is coordinating all the activities.

But it will take time to halt the momentum that is carrying manatees to extinction. Donald Farner cautions that, "It may take a decade before the first manatee farm is turning out enough calves for herds to be established." Meanwhile, the laboratory will provide a focus for basic research on the animal. It will produce breeding stock to put manatees back into regions where they are now extinct, and to introduce them to new countries where they can be used. Perhaps, in the future, manatee farmers will harvest aquatic weeds as a resource to feed their stock—just as ranchers raise alfalfa to feed cattle—converting the weeds to meat, oils, and hides. Tropical nations would then have a valuable new resource and Columbus' mermaid, the manatee, would be spared extinction.

Settlers In Space

By Gerard K. O'Neill

Self-sufficient colonies, built with today's technology, could allow millions to live in space within a century

"Jan. 15, 1996: Jennie and I have been in Transfer Station One, in low orbit around the Earth, for 24 hours. We have decided to start this journal while our impressions are still fresh.

"There were 150 of us on the shuttle from Earth to the transfer station. The video panels gave us a good view of our lift-off, but it was quite different from watching a rocket lift off on a regular television show. This time, we knew we were on top of all those fireworks. Lying in narrow bunks, however, we weren't bothered much by the acceleration and, after 10 minutes, the engines shut down. A few more minutes of coasting in zero gravity, and we docked onto Station One. As we walked down a ramp leading to the outer rim of the rotating station, gravity built up to normal.

"They have a nice mobile hanging from the ceiling of the Orbiter Hotel lobby. The Earth and

Moon are two points of an equal-sided triangle and a small green light at the third point marks L5, where we are headed. Dots of white light that move around the green point in a big kidney-shaped orbit represent all the colonies, including Colony Two, our final destination.

"Every six hours another shuttle arrives. The hotel is about two-thirds full now, and there's quite an international mix among the new colonists. Jennie and I are starting to try our Russian and Japanese, even though we are pretty shy about it still.

"Jan. 17, 1996: Yesterday we boarded the *Konstantin E. Tsiolkovsky*, named after the Russian space pioneer, for the eight-day trip to L5. It's a huge ship, holding about 2,000 passengers. It runs on three separate schedules, eight hours apart, that match the Moscow, Cape Canaveral, and Western Pacific time zones on Earth. One nice custom they have is seating us by place cards at breakfast, with a couple from another time zone having dinner. The Japanese couple we met this morning are in power-plant construction, just as we are, so we were able to get into some shoptalk.

"Jan. 20, 1996: There was lots of excitement early today when we passed the *Tsiolkovsky*'s counterpart, the *Robert H. Goddard*, on its way back to Station One. The *Goddard* was in view for more than an hour, and our crew gave us some nice telescope views on the video screens. It was a pretty sight. At the end of the engine, a bright searchlight showed up the puffs of rock dust as they came shooting out. They looked like the frames of a movie flickering when the projector is slowed down, and we felt we could almost count them. The engine was a long, thin, white line of flashing lights. We couldn't see the guy wires that keep it all straight. Then there was the main passenger section, a big ball with a few windows. And beyond that was the solar power station, with a big dish collector for the sunlight, and the waste-heat radiators glowing a dull red.

"Jan. 24, 1996: We docked at Colony Two this morning and got a good view of it on the video screen in our room as we came in. It's quite impressive. Two parallel cylinders rotate slowly in space, about 10 kilometers [6 miles] apart. Each one is about 700 meters [0.4 mile] in diameter and 3,400 meters [2.1 miles] long. Three big mirrors rotate with each cylinder, sweeping over a wide area as they gather solar energy. As we docked, we passed between the colony and the Sun for a few minutes and we could look right into the mirrors and see the reflected valley areas inside, green with crops.

"The colonists live in terraced apartments in the end caps of the cylinders. Jennie and I have one of the smaller apartments, but it still has two large rooms, plus kitchen and bath, and a nice garden. The climate, controlled by opening and closing mirrors, resembles springtime in Hawaii, so there are masses of flowers in the garden and we picked a fresh papaya for lunch."

"Jan. 25, 1996: Edward has gone off to the power-plant construction factory, so I'll take over the writing for a while.

The author:
Gerard K. O'Neill, professor of physics at Princeton University, has been developing the space-colony concept since 1969.

"Sunning ourselves in the garden yesterday, we could look up and see the gardens of the apartments on the opposite side of the cylinder. For some reason, it doesn't seem as strange to have trees growing straight down as it does to have them coming out sideways, as they do from the gardens a quarter circle away. Most people seem to have little gauze sunshades over part of their grass, so they can sunbathe and not be seen from the land above them. Everyone spends a fair amount of time outside, and those 'colony tans' on the old residents make us feel pale and white.

"Looking out and across, we can see the big disk that seals us off from the main part of the cylinder, where the crops are grown. It is about 50 centimeters [20 inches] thick, to block the cosmic rays. The disk is gently curved like a dish and, on our side, it is steeply terraced and planted with tropical flowers, a mass of green and bright colors. Early each morning there is a rainshower, so everything is fresh when we wake up, and the air has a nice clean smell of rain and flowers. Our colony is not quite big enough to have weather of its own, though, so the 'rain' comes from spray pipes that are just a bit too high and too thin to be seen.

"After Edward gets home, we are going down to the little beach for a swim. Maybe tomorrow we'll walk up the garden paths, past the shops and restaurants, and into the park. They have zero-gravity clubrooms there where we can try flying under pedal-power and slow-motion diving in low gravity – that pool must be fun."

■   ■   ■   ■

Those journal entries may sound like science fiction about life centuries from now, but every detail described is practical in terms of 1975 technology. We require no new breakthroughs, no superstrength materials, not even any new inventions to carry through a space-colonization program that could have as many as 1 million people living and working in space by the year 2000.

The concept of such a program grew out of an informal seminar that I held in 1969 for a few of my first-year physics students at Princeton University in New Jersey. When we studied whether a planetary surface was really the best place for an expanding industrial civilization, we found some surprising answers. For example, enclosing atmospheres in cylinders was more than 1 billion times more efficient than trying to hold an atmosphere by gravity, as a planet does.

The seminar lasted only a few weeks, but I continued to work on details of the space-community concept in my spare time. By 1972, I was lecturing about space colonies at other universities, and in 1974, the first small, informal conference on the topic was held at Princeton. Suddenly, the concept became front-page news, and hundreds of letters began to pour in from interested and enthusiastic people who wanted to help make the idea a reality. Magazine and newspaper

Where Colonies Roam

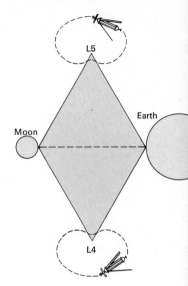

The colonies can move in huge kidney-shaped orbits around L4 and L5, points in space where the gravities and centrifugal forces of the Earth and Moon are balanced.

At a lunar mining town, a solar-powered mass driver hurls materials mined from the Moon to the space-colony construction site.

articles have appeared in many countries, formal conferences are being held, and the National Aeronautics and Space Administration has awarded a grant to finance further study at Princeton.

Although the concept of space colonization can be traced back more than 75 years, a serious, consistent plan, based on available technology and calculable costs, could never have been formed before the Apollo astronauts landed on the lunar surface. Within just a few years, the Apollo program greatly advanced rocket technology. More, it proved that people could live far from the surface of our planet. Above all, it gave us samples of the lunar surface to analyze. We know that the lunar surface is a rich mine of aluminum, iron, silicon, oxygen, titanium, calcium, and other useful elements. And we know that Earth plants can grow in lunar soil, if they are given water and fertilizer.

Up to now, most space activities have been very expensive and the only return has been scientific information. But colonizing the space frontier offers an opportunity to improve the material well-being of all humankind. In the developed nations of the world, the industrial revolution has been the means of achieving decent living standards. Now, just when many of the world's peoples are at the threshold of industrialization, their hopes for progress are being dashed by shortages of energy and materials. Can the industrial revolution, which has been halted at so crucial a moment by our Earth-bound limits, possibly be continued in the vastness of space?

Space has abundant supplies of the two key requirements for continuing industrial progress—energy and materials. A colony in space could have a clean, unlimited, absolutely reliable energy source—the Sun. Averaged over a year, a given area in space near the Earth intercepts about 10 times as much total energy as an equal area in the sunniest part of Arizona or New Mexico, so that even today's solar technology would be effective and economical in the space environment. In space, sunlight is available 24 hours a day, even in midwinter, and it is never obscured by clouds or rain. With no atmosphere to penetrate, the intensity is about 50 per cent higher than it is on Earth when the Sun is directly overhead.

Abundant materials for industry are available from the Moon and, later, from the asteroids, those chunks of rock that orbit the Sun in a zone between Mars and Jupiter. We think the asteroids, like the Moon, contain iron, calcium, oxygen, and aluminum. They also have three other essential elements that are rare on the Moon but which would be needed for any long-term development of space—carbon, nitrogen, and hydrogen. The asteroids and the Moon share one great asset—no one owns them, and there are no living creatures on them whose needs and wishes must be balanced against exploitation.

The timetable on which space colonization might take place depends on many variable factors. Assuming we start now, the dates I use as rough guidelines are the earliest possible dates, based on current technological capabilities. But many of the details of colony design may be different from those I outline. We are not yet at the stage of final design, but we can demonstrate that at least one workable solution exists for each problem so far anticipated. If experience is any guide, once the door has been opened, technology will accelerate. The possibilities based on 1975 technology will seem cautious, conservative, and unimaginative when compared with reality.

The first colony will probably be built about 384,000 kilometers (240,000 miles) from the Earth in orbit around a point in space known as L5, the third point of an equilateral triangle formed with the Earth and Moon. At L5, the gravities and centrifugal forces of the Earth and Moon cancel out each other so that any object placed nearby would orbit the point forever without drifting away. The stable orbit would be kidney-shaped, because of the influence of the Sun, and it would be large enough to contain several thousand colonies.

Before we can build the first colony, we will need careful studies on such subjects as architecture for space, growing crops in constant bright sunshine and an enriched, low-pressure oxygen atmosphere, and adapting chemical and industrial processes to take advantage of low-cost energy and zero gravity. In the late 1970s, we may begin building pilot plants, special-purpose machines, and prototypes of two essential devices—an unmanned freight rocket that can carry tens of tons of material and a Transport Linear Accelerator, or mass driver, that can catapult raw materials mined from the surface of the Moon to the L5 region where the colonies will be built.

Freight rockets may start carrying the first loads from Earth to L5 and the Moon in the early 1980s. The loads to L5 will include the parts of a space station big enough to house a work force of 2,000 persons, solar electric power plants, and industrial processing plants. To the Moon, we must bring the unassembled mass driver, its solar-powered generator plant, surface-mining machinery, and materials to build living quarters for a self-sustaining outpost of some 200 persons.

It's All Done With Mirrors
Large, flat mirrors reflect the Sun's rays into a colony through long windows, *below.* They provide sunlight for intensive farming in the central land areas, *opposite page.*

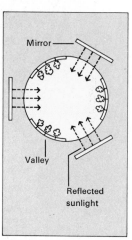

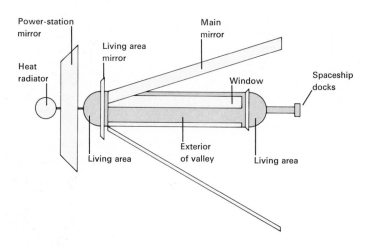

The mass driver can be assembled by sections inside an aluminum tunnel covered with lunar soil to shield the workers from cosmic rays. The mass driver will be a long, slim guideway, stretching about 10 kilometers (6.2 miles). Small buckets, each carrying a chunk of compacted lunar soil, will be continuously accelerated along this guideway. The interaction of permanent magnets in each bucket with the aluminum surface of the guideway will produce a lifting force that will allow a bucket to zip along the guideway without touching it. By the time a bucket reaches the end of the guideway, it will be traveling 2,400 meters (1.5 miles) per second, faster than a high-speed rifle bullet. The bucket will then slow down slightly, releasing its load, which will continue flying out into space at a speed fast enough to escape the Moon's low gravity. After releasing its payload, the bucket will circle back to its starting point to pick up another load.

By the time a load reaches free space, it will have slowed to about 5 or 6 kilometers (3 or 4 miles) an hour. The crew of a space tug, waiting near L5, may use a large net to collect the lunar payloads. After two weeks, when the lunar night shuts down the mass driver's solar power station, the tug can dump its load at the colony construction site near L5. When the mass driver starts up with the arrival of the next lunar day two weeks later, the space tug can be back in position to receive another load from the Moon.

Space colonization depends heavily on the success of a device such as the mass driver. Current estimates set the cost of moving lunar material to L5 at about $6 per kilogram (2.2 pounds), about 150 times less than the estimated cost of lifting the same load from Earth. That cost saving is essential if space colonies are to be built at a price we can afford. Ultimately, costs may drop much lower.

The initial space station assembled at L5 will contain living quarters, greenhouse areas, and a large spherical working area where the first colony will be assembled. A processing plant will turn lunar ores into aluminum for the colony structure, glass for its windows, oxygen for its atmosphere, and industrial slag to be crushed into sand and soil.

Slowly, the pieces of the colony structure will begin to take shape. Each time one of the large curved sections is finished, the working area will be temporarily depressurized and opened. The new section will be floated out and welded to the growing shell by remotely directed machines using solar heat.

The first colony may take the form of two parallel cylinders, each 1,000 meters (1,100 yards) long and 200 meters (220 yards) in diameter, with round caps at the ends. Each cylinder will have three land areas running lengthwise on its inner surface, alternating with window areas of the same width. Multistrand aluminum cables, rather like bridge cables, will run lengthwise along each cylinder, and will also circle it at intervals of a few meters or yards to give it strength. An aluminum lattice will strengthen the window areas, dividing the glass into small panels. Large, flat mirrors, attached to each cylinder by cables, will rotate with the cylinder, reflecting the Sun's rays into the land areas. At one end of each cylinder, a round aluminum mirror will concentrate the Sun's rays for the colony's power station. A turbine, run by high-pressure, solar-heated helium, will produce electricity. A spidery network of lightweight beams will keep the cylinders parallel and pointing toward the Sun.

It will take several years to finish the cylinders and make them airtight. The store of liquid oxygen accumulated during the industrial processing of oxide ores can then be warmed by solar heat, releasing gaseous oxygen that will slowly build up inside each cylinder. Motors will begin to turn the cylinders in opposite directions until they are eventually rotating three times per minute, providing normal Earth gravity for people standing on the land areas inside. This simulated gravity will gradually decline as one moves toward the axis, reaching the weightlessness of zero gravity at the center.

Once the colony is finished, perhaps as early as 1988, colonists like Edward and Jennie can begin traveling to their new homes. A fleet of shuttle vehicles, each carrying from 50 to 150 passengers, will begin taking them from Earth to an orbit about 160 kilometers (100 miles) from the Earth where they will meet the space liner that will carry them to L5. The space liner will be as different from the shuttle as a steamship is from a city bus. It will be large, with a loaded mass of about 10,000 metric tons (11,000 short tons), and it may hold up to 6,000 passengers. It will probably be spherical in shape, with a long, thin tail which is its reaction engine.

Building Colony One may cost several tens of billions in 1972 dollars; for comparison, the Apollo program cost about $33 billion in 1972 dollars. If space colonization were only another scientific program, like Apollo, it certainly would not receive wide support. Fortunately, it should pay its own way within a very few years. This is because production at L5 of any product to be used in space will save the cost—$950 per kilogram ($430 per pound)—of shipping the same product from Earth. One worker at L5, producing 20 metric tons (22 short tons) of finished products each year, can turn out an annual value of $19 million simply because of this saving on shipping costs.

A Skyline of the Future

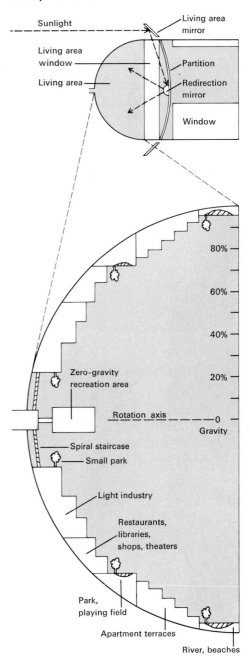

A circular mirror, *top,* reflects sunlight into the end caps of the colony, where concentric terraces hold apartments, shops, and parks, *above and overleaf.* A foliage-covered partition separates the inhabitants from the farming area and protects them against cosmic rays.

Colony One will hold about 10,000 persons. I hope that the early residents will include a wide spectrum of settlers from all countries—not only men and women in their working years, but also children and elderly persons. Perhaps they will already have worked and studied together during a preparation program on Earth so that they will have built up friendships, mutual confidence, and trust. Edward, Jennie, and their friends will find difficult conditions, but not nearly as hard as those faced by early colonists of the New World.

Agriculture, in the main land areas of the cylinders, will be intensive and highly mechanized. By controlling the solar mirrors, there can be long summer days throughout the year, with never a cloud or storm. Temperatures can stay always near 95°F. (35°C) so that corn, sweet potatoes, sorghum, and other fast-growing crops can be harvested four times a year. Chickens, turkeys, and pigs, raised on cuttings and grains, will complete the colonists' varied and nutritious diet. Although beef cattle may be too inefficient in converting grain to meat to earn space in this first colony, there may be a small stock of dairy cows—children will still need their milk. There could be insects for the birds to eat, but perhaps we can do without mosquitoes or cockroaches. We can also exclude other pests such as mice and rats.

The first colony will need about 50,000 metric tons (55,000 short tons) of water to irrigate its croplands and gardens and to provide such luxuries as swimming pools and small rivers. We could not afford to bring so much from Earth, but fortunately we do not have to. Water is 89 per cent oxygen by weight and only 11 per cent hydrogen. Plenty of oxygen will be available at L5 as a by-product of processing the lunar ores, so that only liquid hydrogen need be brought from Earth.

The living areas of Colony One will be in the end caps of the cylinders, set off from the hot, humid agricultural areas. Homes may be terraced apartments, each with its own garden. Space clothes may be quite light, because the environment can be free of harsh extremes.

Because all manufacturing that involves large objects can be done most easily in zero gravity, almost all large-scale industry will be located outside the spinning cylinders. One such industrial capsule may hold Colony One's electric power station. The early stages of ore processing, requiring the crushing of rock into powder, can also be done in space, where no noise can cross the vacuum to the colony.

Even while Colony One is being built, its designers will be planning Colony Two. Our present guess would put Colony Two near 700 meters (0.4 mile) in diameter and about 3,400 meters (2.1 miles) in length. Such a colony could support 100,000 persons, perhaps in four villages of about 25,000 each, at the cylinder end caps. Costs should taper off rapidly as space-based industry grows and each colony builds further "daughter" colonies.

A small fraction of Colony One's work force may engage in scientific work. They could build large optical telescopes, free from atmospheric interference, with a resolution fine enough to pick out

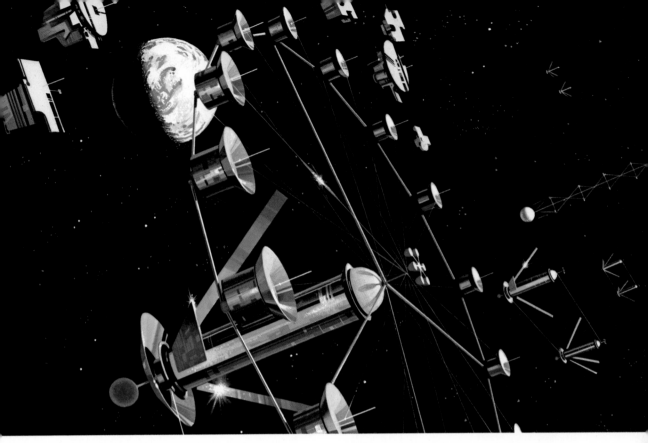

planets around the nearest stars. Or they might assemble the delicate webs of enormous radio telescopes that could stretch across vast areas of space, unhampered by the restrictions of gravity. Large, low-thrust exploration ships, built at the colony, could visit the asteroids and outer planets of the solar system. A larger group of technicians, working to satisfy Earth's urgent need for energy, might build large solar power stations that could beam microwave energy down to Earth.

One of the diversions of space-colony planning is imagining the new options that the colonists will have. The first is one that I call seasonal counterpoint. The two cylinders that form one colony will be entirely independent in determining their day-night cycle, climate, and seasonal phase. One cylinder could have a New England climate, with brisk winters and crisp autumns, to turn leaves to rainbow colors. Its twin might have the perpetual springtime of Hawaii, with moist air, palm trees, and beaches for warm, lazy swimming.

Many of the recreations will be those of a small, wealthy resort community on Earth—good restaurants, movie theaters, libraries, and perhaps small discothèques. Yet some things will be very different. There will be no cars and no smog; colonists will travel on foot or on bicycles in Colony One and perhaps use electric runabouts in the larger colonies. Some favorite colony sports would be impossible on Earth—low-gravity swimming and diving, and surfing on waves that break as slowly as in a dream; man-powered flight near the cylinder

At a large colony of the future, crops are raised outside the cylinder in a ring of agricultural modules tied together by guy wires and hollow tubes. Industrial capsules drift nearby. Other pairs of colonies float in the distance.

axes where gravity will approach zero; or mountain climbing at the end caps, where the gravity will become weaker as the climber mounts higher until at the top he will weigh nothing at all.

But along with the conveniences and pleasures, the colonies will also have their hazards. With no atmospheric shield outside, the window areas of Colony One will be exposed to a constant rain of meteoroids. Nearly all of these will be of microscopic size, far tinier than a grain of sand. But on rare occasions, a larger one will break a window panel. When this happens, there need be no panic, because it would take several days for Colony One to lose all its atmosphere. For larger colonies, the leak-down time will be even longer—up to a year on Colony Two. Within minutes of a break, workers can apply a crude patch to the hole and a permanent repair can be made at leisure. Even if it takes an hour to apply the initial patch, the inside pressure will drop only as much as it does on Earth when we climb a small hill—not even enough to affect our eardrums. Oxygen losses can be made up from the oxygen released during industrial processing. Losses in water vapor will be more expensive to replace because additional hydrogen must be brought from Earth. Still, an occasional small loss will not be difficult to make up.

Cosmic rays will be another danger. On Earth, we are shielded from cosmic rays by an atmosphere equivalent in depth to about 10 meters (11 yards) of water. Only colonies that are several kilometers in diameter will have enough atmosphere to provide comparable protection. The early colonists must accept a somewhat higher exposure to the least damaging kind of cosmic rays, those particles that move so fast they do not leave much energy behind if they pass through our bodies. But even with no protection against these fast cosmic rays, colonists will be exposed to radiation levels that are much lower than those regarded as safe for atomic energy workers.

However, the heavy primaries—nuclei of such elements as carbon and iron arriving from distant stars—will be another matter. An iron nucleus traveling at high speed can kill a human cell, and several important kinds of cells do not regenerate. It would take material about 50 centimeters (20 inches) thick to shield completely against these particles. We could not afford so much shielding for all of Colony One, but we could provide it around the living areas at the end caps.

Another potential hazard would be a break in the structure that ties together two parallel cylinders, causing the cylinder axes to drift away from pointing toward the Sun. However, they would shift only about 1 degree per day, so it would take several days before the change in orientation would affect industry or agriculture. Unless someone is so careless as to ram the structure with a spaceship, there seems no reason why it should break; but if it does, the colonists would have ample time to repair it and correct the alignment with the Sun.

The earliest possible schedule would see Colony One completed by 1988, Colony Two by 1994, and 10 more colonies the size of Colony

Two by the year 2000. By the year 2020, the rate of colony construction could be so great that emigration to better opportunities for jobs and living could begin to depopulate the Earth. As time goes on, larger colonies will almost certainly be built. The biggest may be from 6 to 15 kilometers (4 to 10 miles) in diameter and from 25 to 60 kilometers (15 to 40 miles) long. However, people may prefer to live in smaller colonies where government, transportation, and life styles might be simpler and more manageable.

Quite early in colonial history, some families may choose to cut loose from the colonies and homestead the asteroid belt on their own. A colonist living at L5 could, with comparative ease and modest cost, assemble a spacecraft quite capable of deep-space operations, where high thrusts are not required, nothing happens quickly, and navigation can be done at leisure. The same spirit that fired the pioneers who settled the American West may drive individuals of unusual energy and competence to take off to build their own small worlds.

At some point, it will become commonplace to build new colonies near asteroids. The 10 largest asteroids would provide enough material to continue building new colonies for more than 2,000 years. Colonies more distant from the Sun than the Earth could easily set up concentrating mirrors to obtain the same intensity of sunshine that exists near Earth.

The possible long-term effects of space colonization on the Earth and on the human race raise many questions. Would it be possible that the provision of low-cost energy for the Earth, probably the earliest task of the colonies, could save millions of human lives by accelerating the spread of intensive, high-yield agriculture? What would be the effect of the enormous diversity of social systems, governments, and life styles likely to develop in the colonies? Psychological and sociological results are difficult to predict, but the effect on the colonists might be a healthy one. Could the opening of a doorway into another set of worlds provide hope for many people trapped in an environment growing ever more crowded and lacking in resources?

For the Earth itself, space colonization may offer the opportunity to return to a natural environment that will be greener, less crowded, and less burdened with industry. Earth's remaining inhabitants may enjoy remolding it into the open, pastoral state that existed before the industrial age. It may ultimately become a sort of worldwide park, which the space colonists may visit as tourists. To them, life on a planet may seem as strange as life in space will at first seem to us.

Eventually, many millions of colonies may be scattered over the solar system, and at least a few explorers may begin to venture out to our nearest neighbor stars. After 1,000 years of living in space, our descendants may find those seemingly unimaginable voyages as natural as we find journeys to other continents. How lucky we are to be the generation that has the power to take the first giant step by which humankind will climb to the stars!

The Breath Of Life

By Mary Ellen Avery

Learning how to keep the first cry of a premature baby from becoming its last gasp may lead to the elimination of the leading killer of the newborn

Ryan Theodore Shaf was born in Chicago on Feb. 8, 1975, the third of premature quintuplets. Almost immediately, the doctors in the delivery room realized that he was having serious problems. His temperature was low and he needed extra oxygen to help with breathing. Unlike most infants, he did not ease into a normal pattern of breathing, but continued to fight for each breath. His tiny chest heaved up and down as he struggled to draw in enough air. Then he began breathing in a peculiar grunting fashion, trying to stretch out each inhalation to bring in more air and breathing out in quick explosive exhalations that would not allow all the air to escape. Ryan was a victim of hyaline membrane disease, a common problem among babies who are born prematurely.

More than 12,000 tiny victims die each year in the United States from this disease, the leading killer of premature infants. About 25,000 more who might once have died now recover with the help of special care in the critical first few days of life. Ryan Shaf might also have survived had he not died of an unrelated problem two days later. His two brothers and two sisters had no problems and were allowed to go home about two months later.

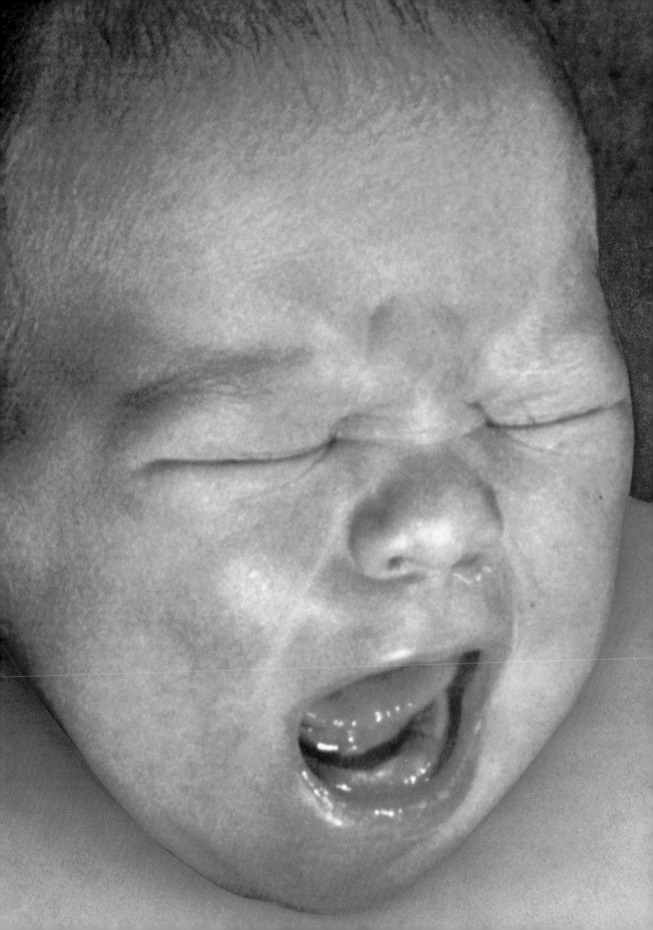

Anatomy of a Breath
Inhaled air travels from the trachea, through bronchioles, to alveoli, where oxygen is exchanged for carbon dioxide in the blood, *opposite page*. Many open alveoli characterize normal lung tissue, magnified 180 times, *bottom left*. Collapsed alveoli that hinder air flow are typical of a hyaline lung, *bottom right*. For a 3-D view, see page 135.

In Ryan's case, birth was an especially difficult experience. But even in the best of circumstances, being born is traumatic. For nine months, the developing infant is safe and warm, floating tranquilly in the insulating fluid within its mother's uterus and receiving oxygen and nourishment effortlessly through the umbilical cord. Then it is thrust out through the birth canal into a strange new world that is comparatively cold and cruel. The temperature drops by some 20°F. (about 11°C), air rushes across the baby's naked skin, bright lights and noise assault its senses. Then the umbilical cord is cut, depriving the infant of oxygen and food.

Shocked by these abrupt changes, and stimulated by changes in blood chemistry, the infant opens its mouth in a great gasp—the sign that it has taken its first breath.

That first breath requires great effort, a force from 10 to 15 times greater than is needed for normal breathing. Before breathing begins, the infant's lungs are filled with a liquid that must be expelled so that air can enter. Fortunately, the newborn child's first breath and those that immediately follow are extremely powerful, drawing air into the lungs and displacing the liquid, most of which is absorbed into the infant's circulatory system. In a normal child, breathing continues with an easy, regular rhythm once the lungs are open. But in infants suffering from hyaline membrane disease, the immature lungs are unable to stay open. Each succeeding breath is as difficult and demanding as the first. Without help, the infants eventually suffocate.

Hyaline membrane disease, sometimes called respiratory distress syndrome, gets its name from the hyaline substance that fills some of the tiny air sacs in the lung. The substance is actually not a membrane, but a fluid. Apparently, as the infant struggles to draw in air, he stretches and injures his immature lungs, allowing fluid to leak out of the cells and coat the air spaces.

Perhaps the lung can be best understood if you think of it as a spongy mass of moist tissue. From the point where the windpipe and the lung meet, progressively smaller passageways, coated with liquid, branch off. At the ends are clusters of *alveoli* (tiny air sacs)—some 300-million in an adult; about one-tenth as many in a newborn infant. Through the ultrathin walls of the alveoli, oxygen passes from the lung into the blood, and carbon dioxide is removed.

Because of the large quantity of liquid that bathes the tissues of the lungs, surface tension—the strong attraction that draws together the molecules on the surface of a liquid—is a major force contributing to the contraction of the lungs. The surface tension is strong enough to

The author:
Mary Ellen Avery, professor of pediatrics at Harvard Medical School, specializes in the study of infant respiratory problems.

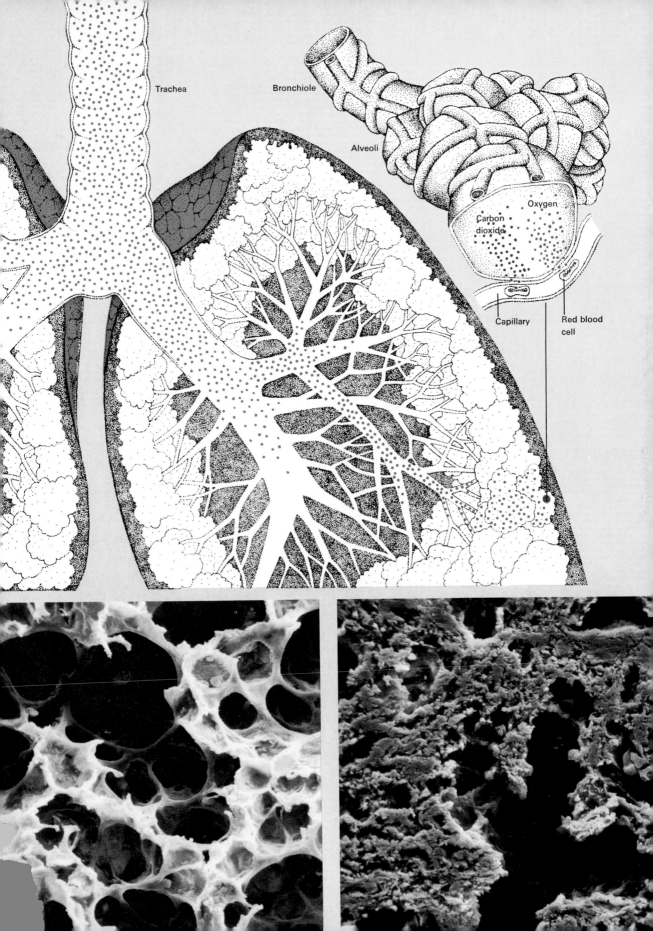

Trachea

Bronchiole

Alveoli

Oxygen

Carbon
dioxide

Capillary

Red blood
cell

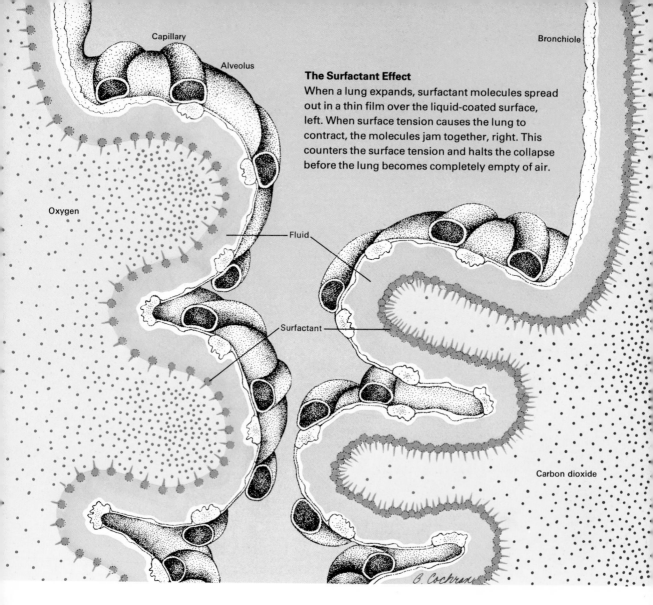

Capillary

Alveolus

Bronchiole

The Surfactant Effect
When a lung expands, surfactant molecules spread
out in a thin film over the liquid-coated surface,
left. When surface tension causes the lung to
contract, the molecules jam together, right. This
counters the surface tension and halts the collapse
before the lung becomes completely empty of air.

Oxygen

Fluid

Surfactant

Carbon dioxide

C. Cochran

bring about total collapse of the lungs at the end of each breath, unless
it is counteracted by some other factor. After the first breath, and for
the rest of his life, the normal person retains a certain amount of air in
his lungs at all times. Indeed, each average breath in an adult ex-
changes only about one-seventh of the total air in the lungs.

This ability to keep the lungs partly expanded depends on the presence of pulmonary surfactant, a material secreted by certain specialized cells that line the alveoli. The surfactant, composed of *hydrophilic*
(attracted to water) and *hydrophobic* (repelled by water) molecules,
forms a thin film on the surface of the liquid where it meets the air.
This film acts somewhat like a detergent in breaking up the surface
tension. As the lungs expand with each inhalation, the total surface
area expands and the surfactant molecules spread out more and more
thinly. When this happens, the surfactant's counterbalance to surface

tension becomes weaker. As a result, the surface tension takes over to limit lung expansion and the lungs start to collapse as air is exhaled. Then, as the lungs shrink, the amount of surfactant per surface area unit increases. The increased concentration of surfactant once again reduces surface tension and brings the lungs' recoil to a halt before they completely collapse and empty of air.

The lungs of most infants born after the 37th week of the normal 40-week gestation period have matured enough to produce the needed surfactant. But Ryan Shaf, like many other premature babies, was born before his lungs had reached this stage. Not only does this make the mechanical act of breathing difficult, but also the oxygen in the inhaled air is spread unevenly among the alveolar air sacs. The larger alveoli receive more of the air, and some of the smaller and more distant sacs get no air at all. In the normal lung, all of the alveoli participate in the exchange of oxygen and carbon dioxide with the blood. But in the immature lung, many blood vessels carry blood past alveoli that have not received any air. As a consequence, that blood is returned to the heart still laden with carbon dioxide and without any fresh oxygen to be transported throughout the body.

The more premature an infant is, the greater is the risk that it will suffer from hyaline membrane disease. Symptoms of the disease usually appear in the first minutes after birth. Recovery, if it occurs, generally takes place in a few days, when the infant's lungs mature enough to produce surfactant. A few premature babies show no symptoms of trouble until a few hours after birth. Apparently, they are born with enough surfactant to enable them to breathe normally for a short time, but they cannot manufacture new quantities fast enough to keep up with the rate at which it needs to be replaced.

If the obstetrician knows that, for one reason or another, a baby is likely to be delivered prematurely, there are several tests that he can

The lungs of a normal infant show up clear and dark in an X ray, *left,* while the lungs of an infant who has hyaline membrane disease are cloudy because they are filled with fluid, *right.*

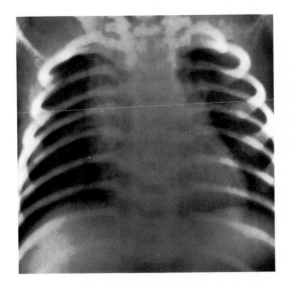
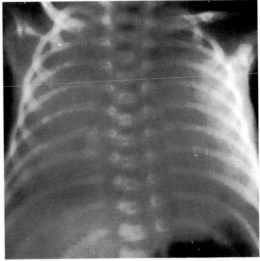

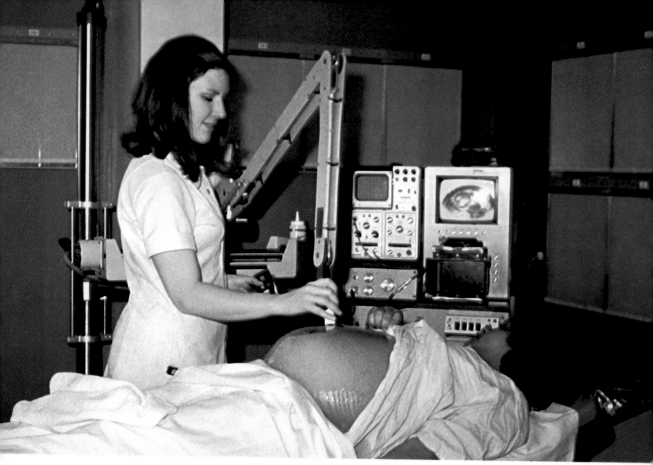

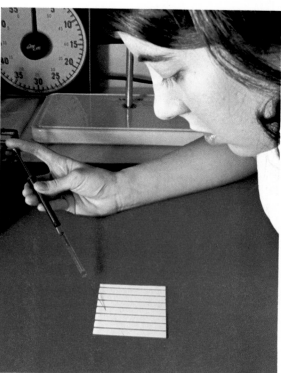

Testing for hyaline membrane disease includes locating the placenta with ultrasound, *above,* and testing amniotic fluid, *left.* Relative sizes of spots made by two components of the fluid, *below,* tell if the infant will develop the disease.

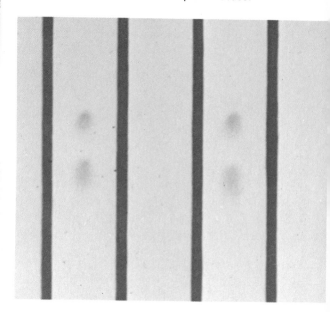

make to determine whether the infant is likely to have hyaline membrane disease. He can detect the surfactant secreted by the unborn infant's lungs by analyzing the amniotic fluid that surrounds the baby inside its mother's uterus. First he locates the placenta, the organ that keeps the baby attached to the mother's uterus, by using ultrasound techniques. In this procedure, considered much safer than X-raying, high-frequency sound waves are beamed at the expectant mother's abdomen. A picture is obtained by recording the echoes bounced back by the internal tissues and fluids, each of which reflects the ultrasound waves differently. Having located the placenta, the physician can then avoid puncturing it when he inserts a needle into the uterus and draws off some of the amniotic fluid for testing. There is no reason to use such a serious procedure for every mother, but it is valuable when an obstetrician plans to deliver a premature baby. Examples would be mothers with heart disease or diabetes.

Once the amniotic fluid is obtained, there are several ways to detect the presence of surfactant. The most widely used test, first proposed in 1971 by pediatrician Louis Gluck at the University of California at San Diego, measures the concentrations of lecithin and sphingomyelin –two surfactant-related substances–in the amniotic fluid. The concentration of sphingomyelin does not change significantly at any time during pregnancy. But that of lecithin jumps greatly at about the 36th or 37th week, indicating that the lung is mature enough to produce sufficient surfactant. If the concentration of lecithin is more than twice that of sphingomyelin, the infant will probably be able to breathe normally at birth. However, if there is less than 1½ times as much lecithin as sphingomyelin, the baby is likely to have hyaline membrane disease. This test was used for the Shaf quintuplets and accurately pinpointed the third-born baby–Ryan–as a borderline case, while the other four were expected to have no trouble.

Doctors may also use the "bubble test," devised in 1972 by physiologist John A. Clements and his colleagues at the University of California at San Francisco. In this test, the amniotic fluid is mixed with ethyl alcohol and shaken in a test tube. If bubbles persist at the surface for several minutes, it indicates that sufficient surfactant is present. This simple, speedy test has proved highly accurate in predicting whether an infant will be free of breathing problems.

These tests are useful only if the absence of surfactant can be detected at least 24 hours before delivery. Unfortunately for babies like Ryan Shaf who are born unexpectedly, this cannot always be done. But in the case of a Caesarean delivery, which can be planned ahead of time, it is possible to delay delivery and use the extra time to speed up lung development to produce the missing surfactant.

The first success in preventing hyaline membrane disease in humans by stimulating lung development was reported only in 1972. But the search for ways to speed up cell maturation started more than 20 years ago, and involved many individuals in different fields.

One of the first to study the hormonal factors that influence cell maturation was Florence Moog, professor of biology at Washington University in St. Louis. She discovered in 1953 that administering cortisol—one of the glucocorticoid hormones from the adrenal gland— hastened the development of the cells of the intestinal lining in young rats. Fifteen years later, pediatrician Sue Buckingham at Columbia University in New York City suggested that cortisol might also stimulate lung cells to develop faster.

At about the same time halfway around the world, Mont Liggins, a New Zealand obstetrician, was investigating the factors that trigger labor in pregnant sheep. He noted that the fetal adrenal gland appeared to influence the timing of labor, and that injecting glucocorticoid into the fetus hastened the beginning of labor.

While I was attending a meeting in Christchurch, New Zealand, in 1968, I heard Liggins report on his work. He noted that even though one lamb was born at 118 days (29 days short of the normal 147-day gestation), it still breathed normally. This seemed impossible to me. Two years earlier, George Brumley, one of my co-workers at Johns Hopkins University in Baltimore, had carefully documented the stages in the development of the lamb's lungs before birth. He found that lambs born before 125 days of gestation could not produce enough surfactant to breathe normally. How then could Liggins' lamb, born after only 118 days, survive?

While Liggins and I were chatting over tea after the meeting, the answer struck us both almost simultaneously. Somehow the glucocorticoid injected to hasten labor had also caused the lungs to mature more rapidly than normal.

When I returned to Baltimore, my colleagues and I immediately began comparative studies on twin lambs. We injected one fetus with glucocorticoid before birth and left the other untreated, as a control. We then sacrificed and examined pairs of twins at various periods ranging from 16 hours to 3 days after the injection. In every case, the lamb treated with glucocorticoid had more mature lungs than its twin.

In June, 1969, I moved to McGill University in Montreal, Canada. I was anxious to get on with my studies, but the problem with sheep is that they give birth only once a year—in the spring. I did not want to wait that long to go on with my research, so I switched my attention to rabbits, renowned for their rapid multiplication. Rabbits breed throughout the year and also have a much shorter gestation period than sheep—only 31 days. There was a more scientific reason, too. Repeating the experiments on rabbits would indicate whether the phenomenon occurred in species other than sheep.

In our first rabbit experiment, Robert V. Kotas injected glucocorticoid through the mother's uterus into one or more of several fetuses being carried by each of a number of rabbits. The injections were given on the 24th day of gestation. Two days later, we sacrificed the rabbits and found that lungs in the injected fetuses were about twice as

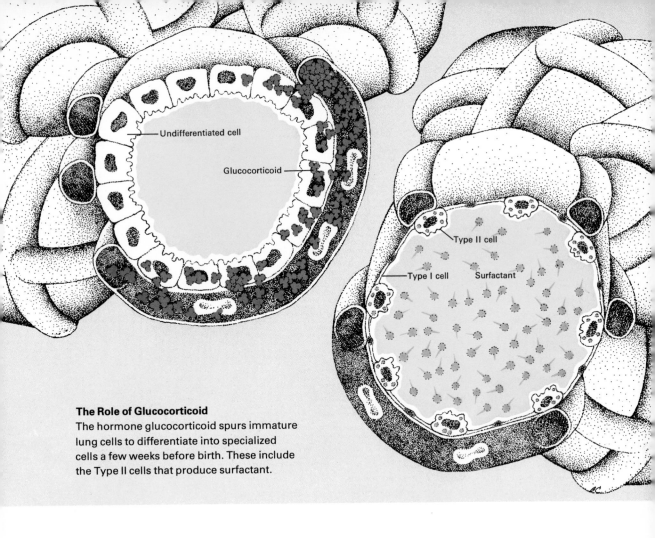

Undifferentiated cell

Glucocorticoid

Type II cell

Type I cell   Surfactant

The Role of Glucocorticoid
The hormone glucocorticoid spurs immature
lung cells to differentiate into specialized
cells a few weeks before birth. These include
the Type II cells that produce surfactant.

mature as the lungs of others carried by the same mother. Later studies by Kotas and H. William Taeusch, Jr., in my laboratory confirmed the benefits of injecting glucocorticoid before birth. They showed that treated rabbits were four times more likely to survive after premature birth than their untreated litter mates.

Meanwhile, in 1972, Liggins and his colleague Ross N. Howie conducted the first controlled clinical trial among 555 women who went into premature labor in Auckland, New Zealand. They injected the mothers with betamethasone, a synthetic compound that acts like glucocorticoid. They found that, if delivery could be delayed at least 24 hours, the betamethasone reduced the likelihood that a premature infant would have trouble breathing and also decreased the number of deaths from hyaline membrane disease.

Their results prompted a number of other investigators to look into how glucocorticoid influences lung cells to mature. Several studies since 1972 indicate that in the normal human pregnancy, the adrenal gland in the fetus increases its output of glucocorticoid at some time shortly before birth. The glucocorticoid then stimulates the lung cells to differentiate into different types of cells, including the kind that

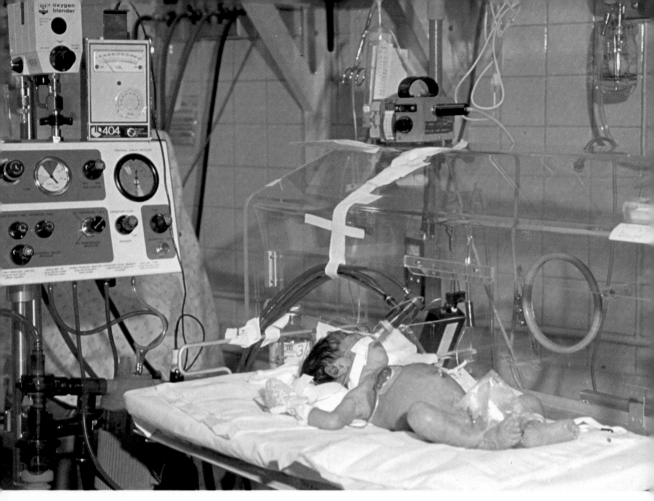

Special respirators keep a premature baby alive by pumping in oxygen under pressure to keep the infant lungs open. Temperature and carbon dioxide are also carefully monitored.

produce surfactant. A baby born before the rise in glucocorticoid is highly likely to have respiratory problems, and it may die. Giving the mother glucocorticoid at least 24 hours before delivery stimulates the lung cells to differentiate and produce surfactant before they would normally do so, thus preventing the disease.

It also seems that certain types of stress in the unborn infant are related to a lower-than-expected incidence of the disease. Such stressful conditions include low-grade infections, infant malnutrition, or labor preceding Caesarean section. Since stress is known to increase adrenal gland activity, it may be responsible for triggering the production of glucocorticoid needed to stimulate lung differentiation. In infants that recover spontaneously after birth, the stress of the disease itself may have stimulated the glucocorticoid production.

But what about babies who are unexpectedly born prematurely? Or those born in a hospital that is not equipped for prenatal testing? Or what about those few instances—about 1 in 6,000 births—in which the disease appears unexpectedly in infants born after enough time to normally take them out of danger?

Bringing new hope for such infants was the development in the 1960s of respirators adapted for tiny babies weighing as little as 1$\frac{1}{2}$

pounds, new methods to monitor the amounts of oxygen and carbon dioxide in the blood, special devices to maintain body temperature, and improved methods of nourishing premature infants.

The most critical problem is providing the infants with enough oxygen during the crucial first few days after birth, when they cannot get enough on their own. A major advance came in 1972 when anesthesiologist George A. Gregory at the University of California at San Francisco reported great success with a treatment that supplied continuous pressure inside the lungs during both inhaling and exhaling. The method cut the expected death rate in half among the 90 infants he reported on.

Gregory's treatment consists of keeping some air containing up to 90 per cent pure oxygen in the lungs under pressure. The air can be kept in the lungs by means of a tube inserted into the baby's throat or, preferably, by using an airtight hood over the infant's head. The objective is to keep some air in the infant's lungs at all times so that they will not collapse between breaths, and to increase the exposed surface area of the air sacs so that oxygen can be better absorbed into the blood. This treatment was used for Ryan Shaf, and it is now routine in most hospital nurseries to provide extra internal lung pressure in this or some other way until the infant can produce the surfactant needed to keep his lungs open on their own.

The treatment, however, is a delicate one. Too little pressure does no good, and too much will kill the infant. A similar danger exists in the high concentration of oxygen—too much oxygen can damage an infant's eyes and lungs, but too little oxygen will surely kill him. Thus, careful, continuous monitoring is needed to provide the baby with the appropriate amounts of pressure and oxygen until he recovers.

The incidence of hyaline membrane disease should continue to decline dramatically in the next decade as physicians make wider use of diagnosis and treatment before birth. Perhaps we will also find new ways to prevent the disease and to treat it after birth. But today, although great progress has been made in detecting and treating the disease, a number of critical questions are still unanswered.

We need to know more about how the lung cells function, what regulates the production and use of surfactant, and how to stimulate its production. And, what does administering glucocorticoid before birth do to the developing intestine, liver, or brain? Perhaps, in those infants who are close to birth, glucocorticoid helps in preparing other organs to adapt to life after birth, as it does with the lungs. But we do not know what the effect may be in very immature infants, those born two months or more prematurely—before they would normally produce extra amounts of glucocorticoid. Could glucocorticoid damage developing organs if administered too early? Until we can answer such questions, we must continue to search for other ways to stimulate the production of surfactant and to conserve it in those infants born with only limited supplies.

Agents of Slow Death

By Richard T. Johnson

The discovery that slow-acting viruses and viruslike agents cause some chronic human diseases may lead to control of such illnesses as multiple sclerosis

The solution to a medical mystery is beginning to unfold as researchers pick up clues to the causes of fatal diseases of the nervous system. Seemingly unrelated bits of information from all over the world are providing clues: studies of diseased sheep in Iceland and in England; reports of a strange illness among members of a primitive tribe in New Guinea; research on a brain disease in Siberia; and studies of a rare neurological disorder among children in all parts of the world who had measles at an early age.

Molecular biologists, neurologists, veterinarians, and virologists who have tracked down the clues have discovered that at least some of these diseases are caused by viruses or viruslike agents that can hide out in the bodies of victims for years before the first symptoms of illness appear. Then the disease grows progressively worse and finally ends in death. Since we began extensive research in the late 1960s, we have found that these slow-acting infectious agents cause four rare, chronic diseases of the human brain. We also suspect that slow-acting agents might be involved in at least a dozen other chronic human diseases,

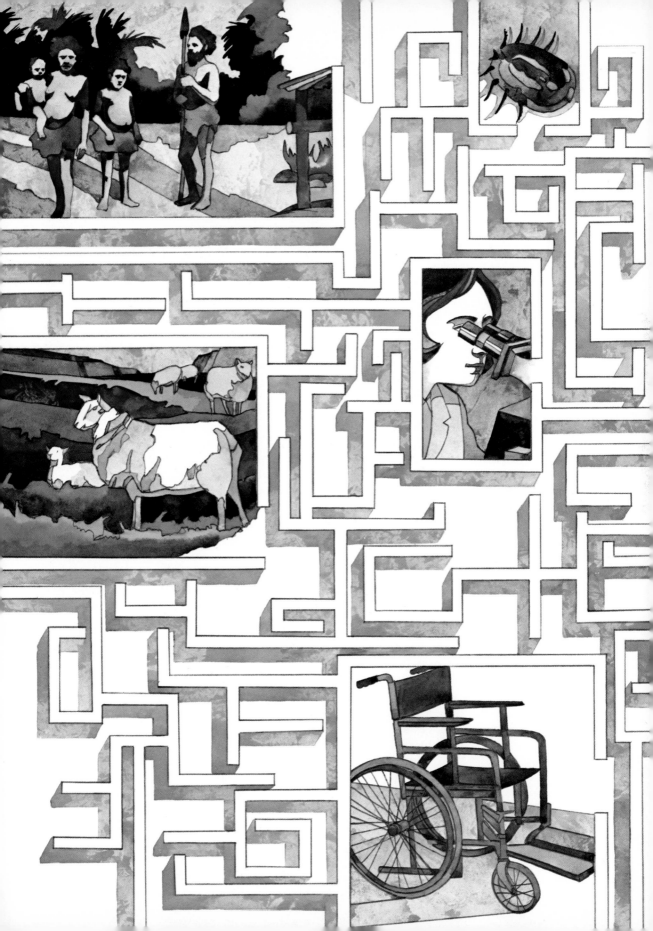

including multiple sclerosis (MS), Parkinson's disease, cerebral palsy, and some birth defects.

The term "slow infection" was originally coined in 1954 by Björn Sigurdsson, an Icelandic pathologist working on diseases of sheep at the Institute for Veterinary Pathology in Reykjavík. The process is somewhat like a slow-motion movie of acute viral infections, such as influenza or measles. When a victim is exposed to a virus that causes acute infections, the virus incubates in the body for only a few days. Then the patient becomes very ill for a short time, but recovers after a few days as antibodies attack the invading virus.

Curiously, the first slow virus infection of the brain discovered in humans was an extremely rare disease limited to one tribe living in an isolated mountain area in New Guinea. A U.S. anthropologist-pediatrician, D. Carlton Gajdusek, who was also trained in virology, went to New Guinea in 1957 to study children growing up in primitive cultures. There he met Vincent Zigas, a physician with the Papua New Guinea Health Service, who told him about reports of a strange disease among the isolated Fore tribe. Zigas thought the disease's symptoms sounded like those of Parkinson's disease, which causes a loss of muscular control and usually affects older persons.

Gajdusek and Zigas decided to investigate for themselves, and climbed the mountains of eastern New Guinea to examine the primitive people. They found that the Fore, who were living under almost Stone Age conditions, were being decimated by the mysterious malady. But it was not Parkinson's disease.

They were suffering from a previously unknown disease called *kuru*, the Fore word for trembling. Its victims would first have trouble walking steadily, then gradually lose control of all coordinated movements. They died about six months after the first symptoms appeared.

The disease was limited almost completely to the Fore, most of whom had never been in contact with the outside world. They lived in barricaded villages, waging war with neighboring villages and practicing magic and ritual cannibalism. The only members of neighboring tribes to contract kuru had married members of the Fore tribe.

Most of the victims were women. Children of both sexes were also affected, but it rarely struck men. This pattern was not typical of a hereditary disease. And initial laboratory studies of tissue from the victims failed to show any poisons, bacteria, viruses, or parasites. The victims did not have fevers or other clinical symptoms of infection. Furthermore, certain pathological changes in the kuru victims' brains resembled those of a degenerative rather than an infectious disease. There was damage to brain cells, but no inflammation.

The mystery of kuru was solved in a most unusual way. William J. Hadlow, an American veterinarian working in an English laboratory, had been studying the pathological changes in goats infected with scrapie, a chronic disease of sheep known throughout the world for more than 200 years. Scrapie was thought to be a hereditary disease

The author:
Richard T. Johnson is professor of neurology and microbiology at the Johns Hopkins University School of Medicine.

The Fore, a primitive tribe in New Guinea, were the main victims of kuru, the first known slow infection of man.

until the 1930s, when French researchers found that it could be transmitted by injecting tissue from an afflicted sheep's brain into a healthy sheep. The incubation period, however, ranged from two to four years. Then signs of severe disorder in the nervous system began to appear and progressively worsened until the sheep died three to six months later. Hadlow and other researchers had also transmitted this slow infection to goats.

The research showing that viruses could cause chronic disease was not limited to nonhuman animals. Since the 1930s, Russian scientists had been studying the possibility that a virus discovered in natives of Siberia might cause chronic brain and spinal cord diseases. This virus, carried by ticks, causes encephalitis, an acute inflammation of the brain. Many patients recovering from acute encephalitis developed chronic muscle weakness or recurring epileptic seizures. The Russian investigators suspected that the encephalitis virus continued to grow in the nervous system, causing these chronic symptoms, which progressively grew worse. In fact, they recovered the virus in several patients years after the patients had been ill with the acute infection.

The science of virology developed differently in Europe and the United States. Because early Russian virologists studied mainly tick-borne encephalitis virus, they were the first to suspect that viruses might cause chronic disease in humans. But early Western virologists concentrated on the acute infections—such as influenza and poliomyelitis—and never associated viruses with chronic diseases. However, the research on kuru and other chronic human diseases, together with the experiments on sheep diseases, eventually changed their minds.

In 1959, while he was working with scrapie in goats, Hadlow read medical journal reports on kuru by Gajdusek and Zigas and noticed remarkable similarities between scrapie in sheep and kuru in man. Both diseases occurred in groups—a flock of sheep or a tribe of New Guineans. Members of the group could develop the disease long after leaving their flock or tribe. And breeding infected sheep with sheep from healthy flocks, like marriage between Fores and members of neighboring tribes, seemed to spread the disease.

Tracking the Mysterious Agents

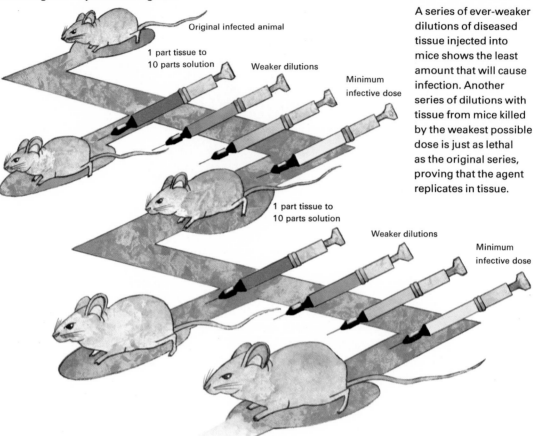

Original infected animal

1 part tissue to
10 parts solution

Weaker dilutions

Minimum
infective dose

1 part tissue to
10 parts solution

Weaker dilutions

Minimum
infective dose

A series of ever-weaker dilutions of diseased tissue injected into mice shows the least amount that will cause infection. Another series of dilutions with tissue from mice killed by the weakest possible dose is just as lethal as the original series, proving that the agent replicates in tissue.

Unlike normal monkey brain tissue (magnified 9,000 X), *below left,* tissue infected by unconventional agents (magnified 9,000 X), *below right,* develops large holes apparently filled with cell membrane.

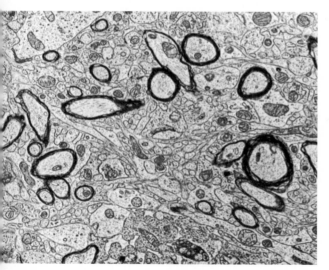

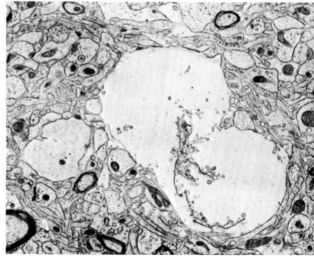

Hadlow also noted that the diseases had similar symptoms, beginning with unsteady walking and trembling and progressing to a loss of all coordination. Both diseases appeared to be infectious, but none of the victims developed fever. Within from three to six months, they invariably died. Also, the mysterious agents that caused scrapie and kuru left footprints in the form of pathological changes in the brain. In both diseases, the brain tissue degenerated into an unusual spongy form filled with holes, and the nerve cells were destroyed. However, researchers could detect no signs of inflammation in the tissue.

In 1959, Hadlow wrote a letter to an English scientific journal pointing out the similarities between scrapie and kuru. He proposed that kuru might be a slow infection of man similar to scrapie in sheep. To test this theory, Hadlow suggested that brain tissue from human kuru victims be injected into chimpanzees and other subhuman primates. These animals would have to be observed for years to see if symptoms of a neurological disease developed. If the signs appeared, scientists would know that kuru is a slow infection.

Members of the U.S. National Institutes of Health in Bethesda, Md., read Hadlow's letter and were excited by his suggestion. To pursue these experiments, the U.S. government established the Laboratory for Slow Virus Infections in 1962 under the direction of Gajdusek and virologist Clarence J. Gibbs, Jr. The scientists began injecting tissue from victims of kuru and other chronic brain diseases into chimpanzees and other primates.

The inoculated chimpanzees eventually began showing symptoms virtually identical to those of Fore people suffering from kuru. The researchers found that the incubation period in chimpanzees varied between 18 months and several years. They learned through later experiments that the disease could be transmitted from chimpanzee to chimpanzee and to a variety of other primates. So it was clear that kuru was a slow infection.

However, even before scientists learned that kuru could be transmitted, the number of kuru cases among the Fore people began to decline. Looking back, it is now clear that kuru was transmitted by the tribe's cannibalism. Cannibalism was practiced primarily by the women, who, as a ritual, ate members of the family who had died. Sometimes the children participated, but the men rarely did so. This explained the unusual pattern of the disease among tribal members. The Australian authorities, then governing New Guinea, suppressed cannibalism beginning in 1957, and by 1959, the practice had been almost completely stopped. The disease is gradually disappearing. The ages of the women afflicted with kuru, and who had practiced cannibalism before it was outlawed, suggest that kuru has an incubation period in humans of from 4 to 20 years.

Meanwhile, Gajdusek and Gibbs had inoculated chimpanzees with tissue from victims of another chronic human malady called Creutzfeldt-Jakob disease, a form of senile dementia that usually strikes per-

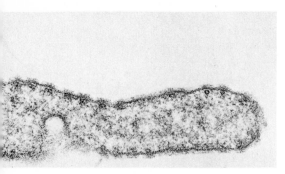

Fast and Slow Conventional Virus

Whole measles virus (magnified 50,000 X), *above,* with an outer protein coat enclosing nucleocapsids, causes the acute measles infection. Defective measles virus (magnified 110,000 X), *below,* consisting of nucleocapsid bundles without a protein coat, causes a slow infection, subacute sclerosing panencephalitis, which leads to nerve cell degeneration and death. It usually strikes children from 5 to 15 years of age who had measles before the age of 2, *right.* Maternal antibodies in the young child may create conditions under which the defective virus can develop, or an unknown factor may activate dormant measles virus. The defective virus stays inside the cells, safe from antibodies sent by the immune system.

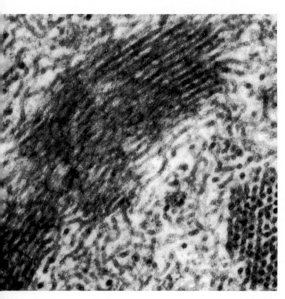

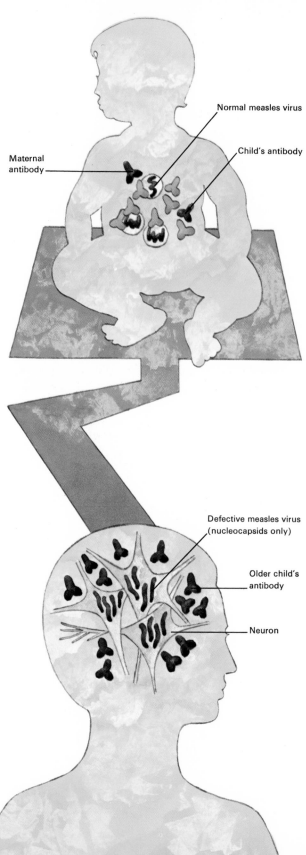

Normal measles virus

Child's antibody

Maternal antibody

Defective measles virus (nucleocapsids only)

Older child's antibody

Neuron

sons in middle age. The scientists had noticed that Creutzfeldt-Jakob disease causes changes in brain tissue similar to those caused by kuru. But unlike kuru, which was limited to one tribal group, this unusual form of senility is found throughout the world. Its victims rapidly lose their mental abilities and develop involuntary jerking movements. They usually die within a year after the first symptoms appear. Experiments by Gajdusek and Gibbs proved that tissue from afflicted humans can also transmit the disease to chimpanzees. Symptoms developed in the animals about one year after they were infected.

How long the infectious agent lives in human beings before causing symptoms or how people acquire this disease is still unknown, but occasionally it occurs within families. Some scientists speculate that the agent causing this disease may be a fairly common virus. Children may inherit genes making them more or less susceptible to Creutzfeldt-Jakob disease. We know that such a situation occurs in sheep, in which different breeds vary in their susceptibility to scrapie. On the other hand, some parents may transmit the dormant infectious agent to their children along with their genes.

The landmark studies of kuru and Creutzfeldt-Jakob disease inspired laboratory investigations throughout the world. Researchers immediately began looking for other chronic diseases that are slow infections and for the agents that cause and transmit them. We have now found two other chronic human neurological diseases that are slow infections—subacute sclerosing panencephalitis (SSPE), a disease that afflicts children, and progressive multifocal leukoencephalopathy

Fluorescent antibodies reveal that measles antigen is in the diseased brain tissue of a mouse, the only sign that the virus is present. The mouse had been injected with a deadly variation of defective measles virus that did not even form nucleocapsids.

(PML), which resembles MS. All three are characterized by a slow destruction of the nervous system.

The agents that are responsible for slow infections appear to fall into two general categories—conventional viruses, which cause SSPE and PML, and unconventional viruses or agents, which cause scrapie in sheep and kuru and Creutzfeldt-Jakob disease in humans. Conventional viruses all have certain characteristics by which we can identify them. They consist of a nucleic acid contained in a protein coat. Using an electron microscope, we can see the virus. Conventional viruses also have identifying molecules called antigens. The body's immune system recognizes these antigens as being foreign and sends out antibodies to destroy the virus. Conventional viruses can also be weakened or killed in the laboratory by exposing them to high temperatures, formaldehyde, or ultraviolet light.

The unconventional agents are a different story. We have never seen any trace of them while examining tissue with the electron microscope. Furthermore, their victims' immune systems apparently do not recognize them, because we also have never found antibodies that should have been sent out to fight them. And they are apparently unaffected by treatments that kill conventional viruses. Researchers have exposed infected tissue to heat, formaldehyde, and ultraviolet light, and found that the tissue can still transmit the disease.

If these agents have never revealed themselves in any way, how then do we know they even exist? First of all, we know that tissue from patients with diseases such as kuru is infectious. But we have also found that the unconventional agents replicate, or reproduce. Researchers discovered this by taking a specific amount of tissue from a diseased animal and mixing it with a salt solution. They then made a series of dilutions, down to only a few particles of the infectious tissue, and injected these into healthy animals. Not only did the animals come down with a slow infection, but their tissue contained as much infectious material as the original fluid.

Nerve tissue is not the only kind infected by these slow agents. Researchers have taken tissue from parts of diseased animals other than their brain and found that most tissue will transmit the disease to healthy animals. Therefore, we know that unconventional agents infect almost all tissue. But only brain tissue shows signs of disease. The disease may be so slow that most tissue, in following its normal cycle of replacing dead cells, repairs itself fast enough to keep symptoms from appearing. Nerve cells, however, cannot replace themselves. So the disease reveals itself in the nervous system. The infection also appears to be associated with cell membranes. Veterinarians and microbiologists in Compton, England, using the electron microscope, in 1971 saw curious coils that looked like membrane inside abnormal vacuoles, or holes, in the degenerating brain cells.

The characteristics of these unconventional agents have led scientists to debate whether they are viruses or some other form of infectious

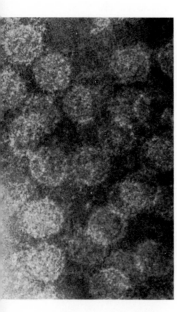

A papovavirus causes progressive multifocal leukoencephalopathy (PML), a chronic brain disease that strikes people whose immune response is deficient.

material. They are certainly different from the viruses we know about, and we may have to expand our definition of viruses or create a new category of infectious agents to encompass them.

In contrast, the agents causing the two other chronic human neurological diseases that are classified as slow infections—SSPE and PML—are clearly viruses. In these diseases, it is not the virus that is slow. An unusual interaction between the virus and its victim, or host, produces the slow infection. In typical virus infections, there are three virus-host interactions—the virus can kill the host fairly quickly; the immune response of the host can kill the virus and create immunity to further infection; or a symbiotic virus-host relationship can develop in which the virus continues to live in the host, but causes no disease. However, none of these apply when conventional viruses cause slow infections. This could be because either the virus is defective or the host's immune response to it is defective.

The common measles virus presents a fascinating example of the variety of possible virus-host interrelationships and the diseases this can cause. Normally, the virus makes itself known after an incubation period of about two weeks. The symptoms include coughing, a rash, fever, and sensitivity of the eyes. The disease lasts for only a few days, and after recovery the patient is permanently immune to measles. However, in about 1 of every 2,000 children the disease is more severe and affects the brain, causing acute encephalitis. This may lead to chronic deafness, convulsions, coma, or even death. Now we have found that measles virus can also cause SSPE. Although this disease is quite unlike measles, most of the children who develop this chronic measles infection of the brain have had typical uncomplicated measles several years before. Interestingly, most cases occur among boys from 5 to 15 years old living in large rural families. First, they begin to have trouble concentrating in school and they develop behavioral problems. Parents and teachers often mistake these symptoms as signs that the children are emotionally disturbed. But it soon becomes evident that their brains are actually degenerating. Other abnormal neurological signs develop, such as loss of coordination and involuntary jerking. Finally the children become paralyzed, and they eventually die. This may take a few months, or even 5 to 10 years.

In 1965, Belgian researchers discovered that the measles virus in the brains of these children is defective. With the electron microscope they saw that instead of forming fully developed virus particles, the measles virus forms only nucleocapsids, the internal structures of the virus. It has no outer protein coat. In 1967, British and U.S. researchers found measles antigen in the brains of SSPE victims, and in 1968 and 1969, U.S. researchers isolated measles virus from the tissue.

We do not know yet why the measles virus becomes defective or how it continues to grow in the body. Most of the children with SSPE came down with measles before they were 2 years old. At that age, they not only had their own measles antibodies, but also antibodies passed on

Brain

Neuron

Axons

Oligodendrocyte

Blueprint for a Virus Attack on the Brain
Brain cells called oligodendrocytes maintain the myelin sheaths that wrap around and protect the axons. Demyelinating diseases may be caused by viruses attacking these cells. In a direct attack, a virus simply kills off the cell. In an auto-immune attack, a virus emerging from a cell carries part of it away, sensitizing the immune system to both virus and cell, which are then both attacked by the antibodies (brown). In an immune attack, the virus leaves a piece of itself on the cell, and antibodies (blue) again kill both.

to them by their mothers. The presence of the maternal antibodies may have created a condition under which the defective SSPE virus could develop. Then again, the measles virus may have lain dormant in their bodies until activated by some unknown third factor. Or combinations of these and other circumstances may have caused the appearance of the defective virus and SSPE.

Researchers working with measles virus have taken the defective virus one step further. When they injected mice with certain strains of measles virus, the virus seemed to disappear. Not only did it lack a protein coat, but it also did not form nucleocapsids. The mice inoculated with this strain died, but there was nothing to indicate what caused their death. The cells simply stopped working without leaving a clue as to why. Researchers who examined the mouse tissue under the electron microscope found measles antigen, but nothing else to indicate the presence of measles virus.

Recently, measles virus has also been implicated in MS, another chronic and mysterious disease, which affects more than 100,000 persons in the United States. It is the major crippling neurological disease among young adults. Persons with MS suffer demyelination, a loss of the myelin membranes that surround and insulate axons. Axons project from neurons, or nerve cells, and transmit nerve impulses from one neuron to another.

Researchers have found that MS patients have unusually high levels of antibodies for measles virus. In 1974, Rockefeller University immunologist John B. Zabriskie found evidence that MS patients may have a defective immune response to measles virus. Many laboratories are now investigating the possible role of measles virus in this disease.

Scientists have studied demyelinating diseases similar to MS, such as PML. An extremely rare disease, PML usually occurs in persons whose immune response is already deficient, either because of some other disease, such as cancer, or because of immune-suppressant drugs administered after organ transplants. So PML appears to be a slow infection of the brain that develops because the host does not respond normally to eliminate a virus.

Using an electron microscope, neuropathologist Gabriele M. ZuRhein at the University of Wisconsin in 1964 detected viruslike particles in brain tissue taken from patients with PML. To everyone's surprise, these particles looked like papovaviruses, a class of viruses known to cause tumors in animals. Furthermore, the only human papovavirus known at that time caused warts, and it seemed improbable that warts virus could be related to this chronic brain disease.

Researchers were unable to recover these viruses in the laboratory until 1971, when virologist Billie L. Padgett and her colleagues at the University of Wisconsin succeeded in growing a new, small, human papovavirus from PML tissue. Subsequent studies have shown that most of us have antibodies against this virus. So it is probably a very common human virus that causes an infection without symptoms during childhood. However, if anything interferes with normal immunity later in life, this virus appears capable of selectively infecting the oligodendrocytes, the brain cells that form myelin. And the destruction of oligodendrocytes leads to demyelination. At the same time, neurologist-virologist Leslie P. Weiner and his colleagues at Johns Hopkins University in Baltimore, Md., grew another type of this virus from two PML patients. Tissue from patients with PML or MS does not cause these diseases in animals. Nevertheless, PML is the first human demyelinating disease in which researchers have consistently recovered viruses, thus increasing hopes of finding a solution to the MS mystery.

Even though the PML virus strains do not cause a demyelinating disease in animals, the researchers at Johns Hopkins and the University of Wisconsin found in 1973 that both cause tumors when injected into hamsters. And this again, as with measles, demonstrates the variety of virus-host interrelationships that are possible. These papovaviruses can cause infection with no symptoms in normal children, PML in persons with defective immune responses, and tumors in experimental animals. These findings demonstrate that we cannot categorize viruses as acute infectious agents or slow viruses or tumor-producing viruses, because the same virus may cause very different kinds of infections depending on the host and its immune response.

Now, researchers are trying to relate slow infections to other chronic human diseases. Molecular biologists are studying defective, incomplete, or mutant viruses and their ability to evade the immune system and survive in cells. Immunologists are studying the complex interaction of viruses and the immune responses. Sometimes a virus may alter the immune response by leaving a piece of itself on the host cell membrane or by taking a piece of the membrane with it when it leaves the cell. The immune system then reacts to antigens of both the virus and the cell, sending out antibodies to destroy both. So, the immune response might actually cause the disease in some cases.

Many virologists and pathologists are studying animal diseases, searching for new methods and insights they can use against chronic human diseases. Others are trying to confirm 1972 reports that viruses have been recovered from MS patients and patients with another strange progressive neurological disease of spinal nerve cells called amyotrophic lateral sclerosis, the disease that killed baseball star Lou Gehrig. In addition, scientists in 1974 found possible traces of an influenzalike virus in patients who developed Parkinson's disease. This raises the possibility that even the common flu virus might persist in the brain and cause a slowly evolving disease.

So far, scientists have convincingly linked only four chronic human diseases to slow viruses—kuru, Creutzfeldt-Jakob disease, SSPE, and PML. It would seem unlikely that only these diseases are caused by slow infections. There is still a great deal of mystery surrounding this area of investigation. But researchers continue to look for clues that will help us solve the enigma of these elusive agents of slow death.

New "Eyes" For the Blind

By Arthur J. Snider

**Sophisticated electronic devices promise
the blind a new sense of independence**

"I listen intently to automobiles and to buses as I try to direct myself along the curb, using my cane to keep a safe distance from mailboxes and lampposts. I concentrate on the traffic until I feel that I am approaching the end of the block. Then I have to listen to noises in front of me as well as noises at my side. I approach the curb very carefully. When I reach it, unless some person comes forward to assist me, I have to wait, hoping that someone with leather heels on his shoes will come by so I can follow him. People with rubber heels don't make enough noise to guide me."

This excerpt from the diary of a 28-year-old man who participated in a study of the daily travels of blind commuters in New York City describes how he makes his way along a block in Manhattan. It indicates how difficult just getting around is for the blind person, particularly the dangerous process of crossing a street. No matter where he turns, he faces barriers and dangers he cannot see. Homes and offices are filled with hazards such as misplaced chairs or half-open doors. Outdoors, telephone and light poles, awnings or pipes at head level, newsstands, fire hydrants, refuse boxes, stairways, and open manholes are all problems. Even children who stand in a blind person's path, fascinated and transfixed by his searching cane, become obstacles.

"Walking along 42nd Street, I try to keep myself a safe distance from the curb. When I feel that I am reaching the end of the block, I swing wide to the left so that I can avoid the stairs of the Third Avenue subway I know is there. I proceed to cross Third Avenue, remem-

69

bering that somewhere in the middle of the block and close to the curb, extending inward for a few feet, there is an obstacle that is 3 feet high. On occasions I have found myself hitting this with great force."

By most measures, sight is the most valuable of our senses. When it is lost, most of the nerve fibers that tell the brain about the outside world also vanish. Each eye has 1.2 million individual nerve fibers, many more than the combined total for hearing, touch, taste, and smell.

There is a common belief that blind persons develop a compensating "sixth sense," but authorities say this is nothing more than the sum of greater concentration on the four that remain. Along with this concentration, most sightless persons develop a greater ability to interpret input from the remaining senses.

"On getting out of the subway on Lexington Avenue, I must turn right to reach the corner before I cross. I continue walking toward the corner looking for a cue to tell me I have reached it. This cue is the hot-dog stand. It is easy to detect because of the odor of hot frankfurters. I take two more steps and then turn left to cross Lexington."

Rustling leaves, flapping overhead awnings, and the echo of traffic sounds tell a blind person much about the open spaces and buildings around him. The reverberation of his voice tips off the size, height, and shape of a room, and whether it is full or nearly empty. A snap of his fingers as he walks sets up echoes that indicate how close he is to a wall and whether doors are open or closed.

Until recently, the only navigational aids for the blind were canes and seeing eye dogs. Neither is wholly satisfactory. The long, thin "prescription" canes, used as antennalike probes to extend a blind person's reach, can sweep over only a limited area. In addition, the blind must be intensively trained to use them properly. Guide dogs require care, exercise, and affection. Only from 1 to 3 per cent of all blind people are qualified to use them, or wish to do so.

To better overcome the problem of navigation in a sightless world, engineers have designed electronic guidance devices to extend the capabilities of a blind person's remaining senses of hearing and touch. Some of these depend on reflections of light; others on reflected sound waves and electromagnetic radiation. Modern technology, primarily miniaturized circuits, have made these devices practicable. The most promising have been evaluated by the Central Rehabilitation Section for Visually Impaired and Blinded Veterans headed by John D. Malamazian at the Hines Veterans Administration Hospital in Maywood, Ill. All new instruments for the blind must be field-tested at Hines before they can be given U.S. government approval.

Perhaps the most successful new instrument is a sonarlike device created by Leslie Kay, dean of the faculty of electrical engineering at the University of Canterbury in Christchurch, New Zealand. A production model of Kay's device was introduced in late 1974.

The device, built into a pair of spectacles and powered by a rechargeable battery, transmits ultrasonic waves and receives their

The author:
Arthur J. Snider,
science editor of the
Chicago Daily News,
received the American
Medical Association's
1975 journalism award.

echoes. These echoes guide the user much as a night-flying bat is guided by the echoes from its high-pitched cries. Transmission of audible sound waves could conceivably aid the blind person in the same way, but these would also be heard by everyone else. Ultrasonic waves are inaudible to humans, because they have frequencies of 20,000 or more vibrations a second. Electronic circuits in the device convert them to a lower range of audible frequencies so that the user can hear them in small earphones set in the temples of the spectacles. The miniature ultrasound transmitter is mounted above the bridge of the nose, and two disk-type microphones above and to each side of the transmitter pick up the echoes. The spectacles' battery is carried in the user's pocket.

One microphone sends messages to the left ear and the other to the right, indicating to the blind person the direction from which the sound is reflected. The effect is similar to that of listening to a stereo recording with headphones. As long as an object is within its range – about 20 feet (6 meters) – the transmitter will produce a continuous echo, with its pitch increasing with distance. For example, an object 10 inches (25 centimeters) away will produce a pitch of about middle C on the piano. The frequency rises by one octave each time the distance from the object is doubled. The signal also becomes weaker as the distance increases, because the ultrasonic waves spread out and they are absorbed by the air.

Skilled users can get a great deal of information about nearby objects from the quality of the feedback. They describe the echo from a leafy bush as mushy, while a glass door signals a clear tone.

One of the first to use the Kay sonar device is Wayne Barker, a young New Zealand pig farmer blinded by an explosion in 1971. Today, Barker strides confidently around his farm, despite such obstructions as oil drums, fences, trees, and livestock. "I suppose I could still get around without the spectacles, but they make recognition of everything so much quicker and easier," he says. "I couldn't do nearly as much without them."

Marshall Pierce, 32, one of the first Americans to try the Kay device, says the same thing. He makes a regular 20-mile (32-kilometer) round trip by bus and subway between his suburban Brookfield home and a college in Chicago's crowded Loop in little more time than it takes a sighted person.

"I can pick my way through heavy pedestrian traffic by zigging and zagging," explains Pierce, who was blinded 10 years ago in an Army training accident in Germany. "I can pick out tree branches and dodge them. I still make six or seven 'contacts' on my trips, but that is about 80 per cent better than I can do with the cane alone.

"The signals are so distinctive, I can recognize the metal strips on a plate-glass window. I can distinguish a cyclone fence from a picket fence. I can walk toward an object and tell within several inches how far away it is. I can tell if a person is coming toward me or going away

The Kay sonar device, built into a pair of spectacles and powered by a pocket battery, uses ultrasound waves to help a blind person detect objects ahead.

Kay sonar spectacles, *above,* contain radarlike device that produces an echo of an object in front of the wearer, who can determine the distance and horizontal direction to the object. The Travel Pathsounder, *above right,* worn on the chest, also uses the echo technique, but it has limited range.

from me. I often use this ability to navigate, picking out a person ahead of me and following at the same pace."

Another device using ultrasonic waves is the Travel Pathsounder, developed by Lindsay Russell at the Sensory Aids Evaluation and Development Center of the Massachusetts Institute of Technology (M.I.T.) in Cambridge. It hangs from the neck by a strap that allows it to be worn at chest level. It picks up an echo through receivers on the neck strap. Unlike Kay's sonar spectacles, which locate objects up to 20 feet (6 meters) away, the Pathsounder only records obstacles in the user's immediate path. It makes a buzzing sound for all objects within 6 feet (2 meters). The traveler can distinguish between close and very close because the signals change abruptly to a higher-pitched beeping sound at 30 inches (76 centimeters).

"The user might walk for many minutes in an open area and hear no signal," Russell explains. "But if he approached something dead ahead—a telephone pole, for example—and kept walking forward, he would hear 'Buzz, buzz, buzz, buzz, beep, beep, beep.'"

The sonar spectacles and the Pathsounder do not reveal hazards at the blind person's feet, because the ultrasonic waves are transmitted straight ahead from the upper body rather than downward. The user must also carry a cane.

Another new device, a long cane with an electronic travel aid built into it, helps to overcome this problem. Called the laser cane, it provides a safe path straight ahead, at the feet and overhead, even though

Laser canes, *above left* and *above,* provide an optical echo at three vertical angles, warning the blind user of objects overhead and at his feet, as well as straight ahead. The laser cane communicates its information by both tones and touch.

it does not provide as much detail about the texture and structure of nearby objects as do the spectacles and Pathsounder. The laser cane sends out three pencil-thin beams of infrared laser light, which are invisible to the human eye. One beam is focused straight ahead. Barriers within a few feet actuate a stimulator that the user feels through the index finger. A tone may also switch on, if desired. A beam focused upward warns, by sound of a tone of a different pitch, of such head-high obstacles as tree branches, signs, and awnings. The cane's downward beam notifies the walker, by a third pitch, of any drop-off that is greater than 6 inches (15 centimeters) appearing two paces in front of him. These might include street curbs, flights of stairs, train-platform edges, and open manholes.

Users express great satisfaction with the laser cane. John Williams, a 24-year-old X-ray technician, says, "My cane-with-eyes has allowed me to travel at a much faster speed with more confidence." He makes a daily round trip of 26 miles (42 kilometers) by bus and elevated train to work in downtown Chicago. Blinded in April, 1970, by a booby-trap blast in Vietnam, he took a five-week training course to learn how to use the cane. "After the first two weeks, I was really good at it," he says with justifiable pride.

The second greatest concern of blind people is that of reading. Ever since Louis Braille, a blind French teacher, published his system of embossed writing in 1829, coded raised characters have given the blind direct access to literature. But only 1 out of 4 blind persons can read braille, according to the American Foundation for the Blind in New York City, and only 3 out of 100 read one or more braille books a year. Furthermore, only 1 out of 100 of the books published annually in the United States are translated into braille. Braille books are bulky. The most widely read braille book, the Bible, consists of 18 volumes, each of which is the thickness of an unabridged dictionary, while the average novel requires 3 braille volumes.

Scientists at the U.S. Argonne National Laboratory near Chicago have developed a "braille machine" that sharply reduces the cost and bulk, storing reading matter in coded symbols on magnetic tape. When this information is needed, the tape is placed in a machine that reproduces the symbols as a series of standard braille characters on a reusable plastic belt that feels to the reader much like embossed braille on paper pages. A 400-page book can be transcribed on a single reel of tape. The braille machine was developed by Arnold Grunwald, a researcher in nuclear reactor fuels at Argonne, whose son, Peter, has been blind since birth.

The M.I.T. Sensory Aids Evaluation and Development Center has developed another braille device, the Braillemboss, that uses a computer to convert coded electric signals into braille. The size of a teletypewriter, it is connected with computers either directly or through telephone lines. For example, if a teacher wants a braille version of an examination, he types the examination on the keyboard and receives it in braille almost instantly. The Braillemboss has become an essential tool for blind computer programmers. Paul Caputo, a blind television newscaster in Westfield, Mass., uses the Braillemboss to convert Teletype print into material he can broadcast on his news programs.

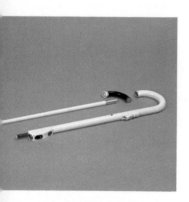

The laser cane probes with three pencil-thin laser beams from a point midway on the cane. The curved black battery that powers the device fits into the cane handle.

Long-playing records and tape recordings of someone reading books aloud are available to the blind. But the rate at which a person can hear and understand speech is usually much faster than the rate at which a reader can speak.

Speeding up a record or tape distorts the vocal pitch and quality, producing a "Donald Duck" effect. But technical advances can now overcome this. The harmonic speech compressor developed at the American Foundation for the Blind doubles the word rate of recorded speech while preserving vocal pitch and quality. It does this by electronically compressing the sound waves, removing part of each wave, and then restretching the waves to their original shape so that they will not sound distorted. At a rate of 275 words per minute, compressed speech enables blind people to hear books on records at rates comparable to the silent reading speeds of high school students.

Both the Braillemboss and the harmonic compressor are commercially available, but there are two reading machines, still in the experimental stage of development, that offer some exciting new possibilities. One is the Optacon (*Op*tical-to-*Ta*ctile *Con*verter), an 8-pound (3½-kilogram), electronic device that has a movable camera the size of a lipstick case and a tactile screen the size of a fingertip.

To use the Optacon, the reader moves the camera along the line of print with his right hand. It focuses on one letter at a time, and the image received is magnified and displayed on the tactile screen's mosaic of 144 pins. The reader places his left index fingertip on the screen, and feels a vibrating image of the letter from the printed page that the pins produce as they rapidly rise and fall. In this way, the blind reader feels the same shapes that a sighted person sees.

Beginning users who learned to read before they became blind can read up to 20 words a minute, and a few experienced users read up to 80 words a minute, about half the best braille reading speed. The device was developed by John G. Linvill, chairman of the Department of Electrical Engineering at Stanford University in Stanford, Calif., who has a blind daughter.

The Optacon has opened exciting new vistas for Robert Stilwell, professor emeritus of German and former chairman of the Department of Foreign Languages at West Virginia University in Morgantown. "Ninety-eight per cent of the material I have had to use in six or seven languages was not available in braille," Stilwell relates. "Almost nothing is available in braille above the high school level. I have had to have things copied by my wife or someone else. You cannot get large dictionaries in braille and, if you could, they would be incredibly large. My class textbooks always have been translated into braille, but acquisition of the Optacon enabled me to read textbooks of many different kinds in type for the first time."

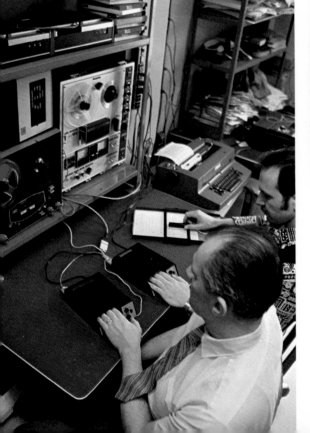

A blind person is instructed in how to use the Optacon, which converts printed letters into vibrations that can be "read" with the fingertips, much like braille. The lightweight, portable Optacon can be easily carried home for practice.

Blinded when a small child, Stilwell has never known print or the shape of letters, even though he has operated a typewriter most of his adult life. "Using the Optacon, I learned the shapes of letters for the first time and there were some surprises," he says. "I didn't know, for example, that an apostrophe was above the line, even though I had written it on the typewriter for 40 years. I had never thought about it. I never knew the first letter of a paragraph was sometimes larger. Yesterday, I picked up a book of short stories and discovered the four words in the title were placed one above the other. I had never realized it was done. I can picture the word as it is written, not as it sounds. You don't picture the shape of a word in braille."

By January, 1975, there were about 1,250 Optacons in use in 19 countries. Students from age 6 and up are the largest users of the instrument. Others range from computer specialists to housewives.

A reading device using sound rather than touch is being tested at Hines VA Hospital. The Stereotuner, developed by the Mauch Laboratories in Dayton, Ohio, uses a probe that is like a camera except that the film is replaced by a vertical column of 10 photocells that converts the image into electric currents. A straightedged tracking aid helps the beginner maintain the probe's position as it moves from left to right along a printed line. When part of the letter image darkens a photocell, electronic circuits in the control box produce a musical tone. A different tone is made for each of the 10 photocells. As the probe is moved across a capital letter V, for example, the black of the print actuates photocells that produce a high note at the top of the left side of the letter. A series of descending notes follows as the photocell probe moves to the right. After the bottom of the V is reached, the notes again rise in pitch until the probe reaches the top of the letter's right side. Then it is silent until activated by the next letter.

Harvey L. Lauer, electronic reading expert at the Hines hospital, and blind from birth, has found the Stereotuner useful for short reading tasks, such as memos, bills, notices, and labels, and for such purposes as checking whether clothes or other objects are light or dark. He can read 50 words a minute with the Stereotuner.

The cost of all these devices is a problem. The Kay sonar spectacles, for example, cost about $1,800; the Travel Pathsounder, about $1,000; and the laser cane, about $2,000. The reading devices, Optacon and Stereotuner, cost about $3,500 and $1,100, respectively. Most blind users could not afford this new equipment were it not for Veterans Administration subsidies and grants from private institutions. But if production costs go down, so will prices.

Of course, none of these instruments give a blind person anything resembling sight. The ultimate dream of doing this through by-passing the eye and providing artificial sight by patterned electric stimulation directly to the visual center of the brain has intrigued investigators in the United States and Europe for several years. Electric stimulation of a point on the cerebral cortex at the back of the skull causes an indi-

The stereotuner has an optical probe that converts printed letters into musical tones. The tones are triggered in a series according to the shape of the letters.

vidual to sense a point of light called a phosphene. Phosphenes are the "stars" one sees when hit on the head. They are believed by some scientists to be the building blocks of visual images. These scientists say that the visual world is made up of millions of phosphene light points, much as images on television are made of thousands of dots of light.

In theory, electrically stimulating a certain point in the visual cortex should produce a phosphene that will be "seen" by the brain just as it would see the phosphene if it were produced by vision. If a pattern of points were stimulated, multiple points of light would be sensed. Researchers believe that if a grillwork of stimulating electrodes were placed in the brain, phosphenes could be activated to produce outlines of objects—say, the letter F—so that the blind person would actually experience seeing the letter. Sequentially, the stimulating electrodes might spell out a message.

In July, 1967, Giles S. Brindley, physiologist at the University of London and Maudsley Hospital, implanted 80 electrodes in the visual center of the brain of a 53-year-old nurse who had recently lost her sight. Wires led from each electrode through a small hole in the skull to a package of 80 individual radio receivers placed under the patient's scalp but outside her skull. External radio transmitters carried signals to the implanted receivers, which then delivered short electric charges to the electrodes resting on the woman's brain.

Only 37 of the 80 implanted electrodes produced phosphenic light spots when stimulated. But data indicated that the nurse could generate and identify predictable simple patterns.

Biophysicist William H. Dobelle and his associates at the University of Utah Institute for Biomedical Engineering in Salt Lake City carried the Brindley work a step further in 1973 by demonstrating that the

Electronic Vision

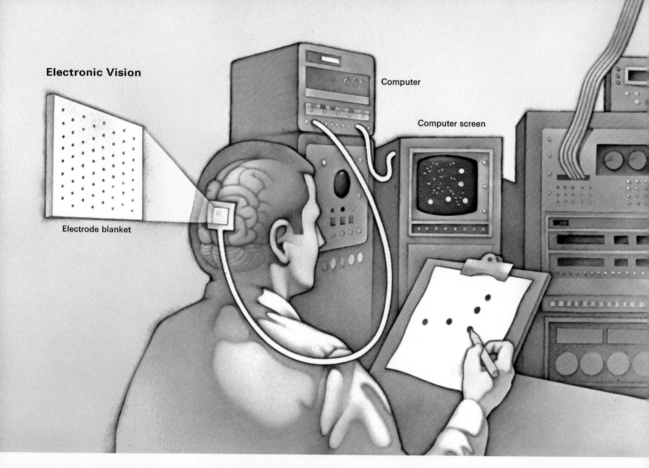

Computer

Computer screen

Electrode blanket

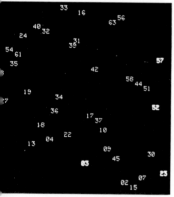

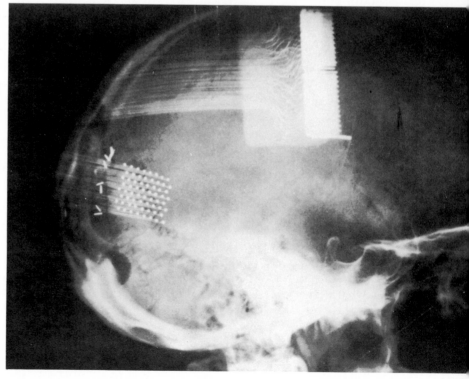

Electrodes implanted in a blind person's skull, *top* and *right,* are used to stimulate the visual centers of the brain and permit him to "see" phosphenes, or points of light. The pattern of the spots can be monitored on a TV screen, *above.*

electrodes would produce phosphenes even after the patient has been blind for many years.

One of their subjects was a 43-year-old electronics technician and piano tuner, born with congenital cataracts and blind for 28 years. The second was a 28-year-old graduate student who lost his sight at the age of 21 in an explosion.

Both arrays consisted of 64 platinum electrodes embedded in a thin strip of Teflon and placed against the visual center of the brain. The electrode wires were connected with a computer system that included a television screen and a stimulator to provide current to the implanted electrodes. The researchers triggered the electrodes with the stimulator and then asked each man where he "saw" the phosphenes. Then they altered the TV set so that it would flash a spot of light on its screen in the proper place for each electrode as the electrode was activated. This let the researchers see what the patients "saw."

The 43-year-old blind patient reported white flickering phosphenes "about the size of a coin at arm's length" from all responsive electrodes. The other patient saw two distinct types of phosphenes. The first were colorless and ranged in size from "a grain of rice to a coin at arm's length," with the larger ones appearing at the outer edges of the visual field. Some flickered, while others did not. The second group flickered in an orange color. The researchers could stimulate groups of phosphenes, creating simple squares, triangles, and letters of the alphabet, which the patients identified easily and the researchers monitored on the TV screen.

Admittedly, these are crude beginnings. Even the most optimistic observer would not expect anything more than a rough approximation of sight from this phosphene-stimulation system for many years, one somewhat like watching an extremely fuzzy television picture.

Instruments that allow the blind to get around and read almost as easily as sighted persons do are still a long way off, but scientific developments have already far surpassed the white cane and seeing eye dog. It is urgent that the progress continue, because according to the American Foundation for the Blind there are 400,000 totally blind persons in the United States alone.

"Even though the new electronic aids are quite useful," says Harvey Lauer, the blind reading specialist at Hines hospital, "we shouldn't forget that there are only a limited number of people that can now use them. Each one takes extensive training, and some people learn them faster than others do because many physical, mental, and emotional factors play a part in such a complex process as learning to read.

"The new aids are only supplements to more traditional aids. When canes, braille, and guide dogs were first introduced, it took nearly a generation before they were widely accepted and used. Let us hope that those who try the new instruments don't quickly become discouraged and reject them out of hand because they don't work miracles. After all, none of the aids yet invented can replace true sight."

Messengers
From the
Brain

By Wylie W. Vale

**Hormones discovered in the hypothalamus can be
synthesized and used to help fight a host of maladies**

In January, 1974, John E. Gerich and Peter H. Forsham of the University of California School of Medicine in San Francisco carefully introduced small quantities of a new substance into the veins of several people with diabetes. Diabetics are normally treated with insulin injections, which often only partially and erratically decrease the large quantities of sugar in their blood. Yet, within minutes, the blood sugar levels of these patients began to fall toward normal. This and other studies have led some specialists to speculate that the new substance, called somatostatin, offers great promise in the fight against diabetes.

Somatostatin is one of several hormones recently discovered in the hypothalamus, a region in the lower part of the brain. Through their action on the pituitary gland, which lies in a bony cavity at the base of the skull, these hormones help the body to regulate a variety of important biological processes.

The pituitary, often called the "master gland" because of its role in controlling several other glands, is made up of three parts. Some hypothalamic hormones act through one of these, the anterior pituitary, which is composed of several types of cells that release at least six hormones into the blood. These hormones are complex molecules that travel through the blood stream to act on specific organs. Some of the target organs are other glands—such as the thyroid, adrenal, and gonads, whose secretions are essential to many vital activities, including growth and reproduction.

The secretion of the target organ completes a feedback loop by controlling the pituitary when the secretion reaches a certain level in the blood. For example, Thyroid Stimulating Hormone (TSH) secreted by the pituitary makes the thyroid gland release thyroid hormones. When the blood contains a high level of thyroid hormones, less TSH is released. Conversely, when the thyroid hormone level drops too low, more TSH is released.

Until the 1940s, this type of feedback loop was the only known natural control of the anterior pituitary. Then anatomist Geoffrey Harris of the University of Oxford in England, neuroendocrinologist Charles H. Sawyer of Duke University in Durham, N.C., and several other scientists discovered that the brain also controls pituitary activity. Scientists focused more precisely on the source of this control when they showed that destroying the hypothalamus or transplanting the pituitary gland to a site farther from the hypothalamus drastically altered the gland's activity.

This fit remarkably well with what anatomists already knew. The hypothalamus and the pituitary are linked by a network of blood vessels called the portal vessels. The brain hormones could merely travel down the portal vessels and into the pituitary gland where the vessels disperse, branching throughout the gland to the specific cells that make pituitary hormones.

The first solid experimental evidence that brain hormones were involved in pituitary control came in 1955 from physiologist Roger C. L. Guillemin at Baylor College of Medicine in Houston and physiologist Murray R. Saffran and biochemist Andrew V. Schally of McGill University in Montreal, Que. Guillemin and his co-workers ground up animal hypothalamus, added the material to pituitary tissue cultures and whole pituitary glands that were carefully kept alive in their laboratories, and showed that the tissue cultures and glands produced more of one of the pituitary hormones. Teams led by Guillemin, Schally, physiologists Samuel M. McCann of the University of Texas Southwestern Medical School in Dallas and Joseph Meites of Michigan State University at East Lansing later reported that the ground hypothalamus could either stimulate or inhibit the secretion of each of the pituitary hormones.

The first hypothalamic brain hormone, TSH Releasing Factor (TRF), was not isolated and its chemical structure was not determined

The author:
Wylie W. Vale, a physiologist and biochemist, is an associate research professor in the Neuroendocrinology Laboratory at The Salk Institute for Biological Studies in San Diego.

until 1969. Again, Guillemin's group, which by then included biochemist Roger C. Burgus and myself, and Schally's group, which had by this time moved to the Veterans Administration Hospital in New Orleans, did the job. To get only a few milligrams of pure TRF, the two groups extracted material from the hypothalamus of about 1.5-million animals—sheep in our case, pigs in Schally's. Our laboratory reported that several synthetic substances, including pyroglutamyl-histidylprolineamide (pGlu-His-Pro-NH$_2$), were able to match the natural substance's activity in test animals. Then, both laboratories quickly identified TRF as pGlu-His-Pro-NH$_2$. This is a peptide, a substance composed of chemical units called amino acids.

As a peptide with only three amino acids, TRF is a good example of several of the advantages that the hypothalamic hormones have over the pituitary hormones as potential drugs. TRF can be synthesized easily and inexpensively, in sharp contrast to the pituitary's TSH, a molecule consisting of over 150 amino acids and with a portion not yet fully identified. TRF is also very potent. For example, as little as 0.000000001 gram (1 nanogram) of it injected into a mouse causes the animal's TSH level to jump more than 20 times. In addition, the hypothalamic hormones act far faster than the feedback mechanism. The rise in the mouse's TSH, for example, took about two minutes, whereas the feedback mechanism would take two hours.

When TRF increases the level of TSH, the TSH triggers the thyroid gland to produce thyroid hormones. These hormones play several important roles in regulating growth, development, and metabolism. For example, a severe thyroid hormone deficiency in infants causes cretinism, a condition that often includes mental retardation, slowed metabolism, and stunted growth. In addition, increased TRF stimulates the secretion of prolactin, a pituitary hormone that promotes milk production in mammals.

It is unlikely that TRF will ever be used to treat cretinism, because thyroid hormones that are inexpensive and effective against this malady are already in use. However, on the basis of its effect on prolactin, TRF has been successfully tested on women who are breast-feeding infants, but find they cannot produce enough milk. In addition, TRF has proved useful in determining the cause of thyroid-hormone deficiency. If treatment with TRF cannot trigger the production of more thyroid hormones, the difficulty probably lies in the pituitary gland or in the thyroid gland itself. However, if TRF does raise thyroid-hormone levels, the difficulty probably lies in the patient's brain. In several cases, brain tumors have been discovered after TRF has pointed to such a problem.

We understand only partially how TRF and the other hypothalamic hormones act on the pituitary gland. Receptors, most of them probably large protein molecules, in the outer membrane of pituitary cells will bind to specific hypothalamic hormones, the two molecules fitting together like a lock and key. When TRF interacts with its

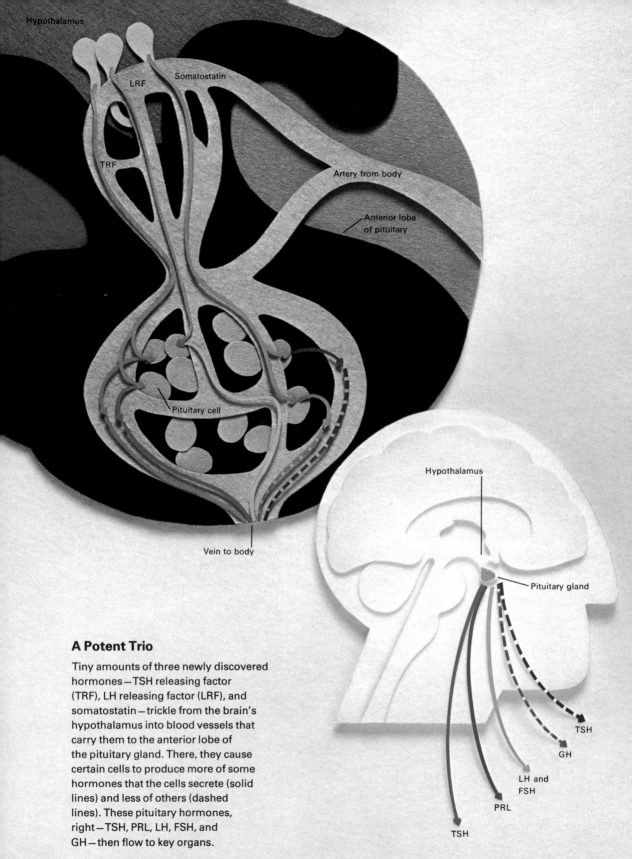

Hypothalamus

LRF

Somatostatin

TRF

Artery from body

Anterior lobe
of pituitary

Pituitary cell

Vein to body

Hypothalamus

Pituitary gland

TSH

GH

LH and
FSH

PRL

TSH

A Potent Trio

Tiny amounts of three newly discovered
hormones—TSH releasing factor
(TRF), LH releasing factor (LRF), and
somatostatin—trickle from the brain's
hypothalamus into blood vessels that
carry them to the anterior lobe of
the pituitary gland. There, they cause
certain cells to produce more of some
hormones that the cells secrete (solid
lines) and less of others (dashed
lines). These pituitary hormones,
right—TSH, PRL, LH, FSH, and
GH—then flow to key organs.

receptors, a rapid series of events begins within the cell, leading to the release of TSH that is packaged in granules concentrated near the cell's surface. Among these events is an increased concentration of calcium and cyclic AMP, a substance known to affect many cellular activities. Calcium is more concentrated in the blood and other fluids that bathe the cell than within the major portion of the cell itself. There are, however, tiny structures within the cell that contain large quantities of calcium. The calcium that builds within the stimulated cell could enter through the outer membrane from the fluid surrounding the cell or it could be liberated from the tiny structures. The cell forms cyclic AMP with the help of an enzyme, adenyl cyclase, which may be near or even a part of the receptor, and may be thrown into action by the arrival of the hypothalamic hormone.

The next hypothalamic hormone identified was Luteinizing Hormone Releasing Factor (LRF), which regulates reproduction. Schally and his colleagues determined its chemical structure in 1971, identifying each of its chain of 10 amino acids, from hormone isolated from extracts of pig hypothalamus. At the Salk Institute for Biological Studies in San Diego, where Guillemin had moved his group, we quickly found sheep LRF to have the same structure. LRF stimulates the pituitary gland to release two hormones, Luteinizing Hormone (LH) and Follicle Stimulating Hormone (FSH), which control the production of sperm and male sex hormones by the testes, and eggs and female sex hormones by the ovaries.

In clinical tests, LRF has helped otherwise infertile couples have babies. In these cases, the hormone raised the low levels of LH and FSH of one of the partners. LRF can also induce ovulation in females, which seemingly would make it an ideal drug for synchronizing ovulation with artificial or natural insemination in farm or zoo animals and even in human beings. But it has drawbacks as a drug, mainly because it acts for only a short time within the body. This makes multiple doses necessary, which are both expensive and inconvenient.

Hoping to overcome these drawbacks, our group and others are making LRF analogues, synthetic peptides that closely resemble LRF. Such substances are made by deleting or changing one of the amino acids in the hormone or by modifying its structure in some other way. We have already found analogues that are from 15 to 20 times more potent and act several times as long as natural LRF.

We have found other analogues with even more surprising properties, including the ability to inhibit natural LRF. These molecules, called antagonists, bind to the LRF receptors of pituitary cells, but do not trigger LRF secretion. When enough of these molecules are present to bind to all or nearly all the LRF receptors, they block the LRF molecules that would normally stimulate LH and FSH production. Researchers have used the antagonists to prevent ovulation in animals. Doctors hope that LRF antagonists will prove as effective and safe a contraceptive as the "pill," without its side effects. These ana-

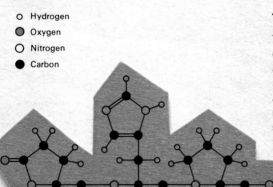

O Hydrogen
◐ Oxygen
○ Nitrogen
● Carbon

The Right Fit

TRF, left, the simplest of the hormones, is composed of only three amino acids. The hormone (1) triggers a pituitary cell by attaching to it at a specific site (2). The cell then increases its calcium content and its cyclic AMP, a substance that figures in many important cellular activities. Thyroid stimulating hormone (TSH) is released (3) until the TRF detaches itself from the cell's surface (4).

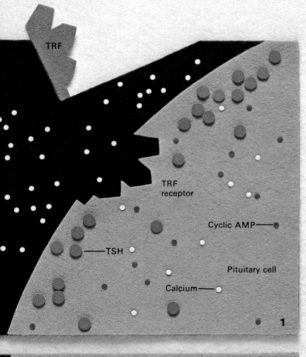

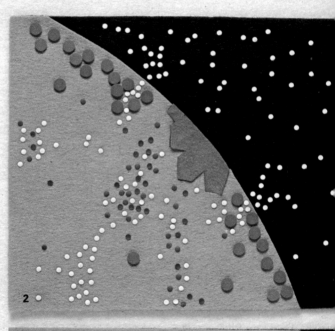

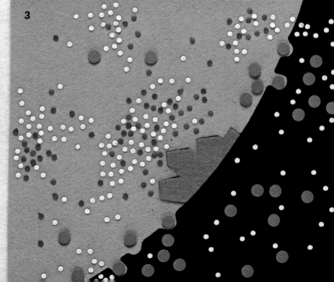

LRF

LRF analogue

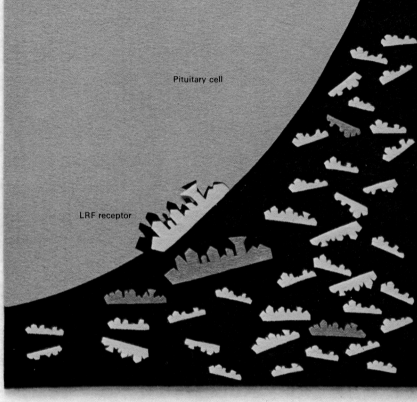

The Wrong Fit

One LRF analogue, or variation, is produced by removing its second amino acid, left. Experiments show that this analogue partially attaches to the LRF receptor site. The analogue does not fit well enough to trigger the cell, and blocks any LRF molecules that would.

Pituitary cell

LRF receptor

logues are being tested by Samuel S. C. Yen, chairman of the Department of Obstetrics and Gynecology in the medical college at the University of California at San Diego.

We began a new search in late 1971, this time for a hypothalamic hormone that triggers the release of Growth Hormone (GH) from the pituitary. Physicians might treat some types of dwarfism with such a substance, and farmers might use it to improve the rate and type of weight gain (protein as opposed to fat) in meat-producing animals. However, fate sent us in a completely opposite direction. When we added extracts from sheep hypothalamus to some pituitary cells and watched for an increase in GH secretion, we saw a decrease instead. We had discovered a GH-release inhibitor in our extracts.

About a year later, chemists Nicholas C. Ling and Jean E. F. Rivier, biologist Paul E. Brazeau, Burgus, and I isolated, determined the structure of, and synthesized the inhibitor. It is a peptide containing 14 amino acids. We called it somatostatin.

Somatostatin has inhibited the secretion of GH in all the many mammals tested, including human beings. A particularly promising fact is that somatostatin lowers GH levels in patients suffering from acromegaly, a condition in which the face, hands, and feet grow grotesquely large. Victims of this disease produce huge quantities of GH, which causes the physical deformities, high blood sugar, and kidney malfunctions. In addition, when we added somatostatin to rat pituitary cells that had been stimulated to produce all the known pituitary hormones in an experiment in 1972, we found that TSH production was depressed along with that of GH. Since then, somatostatin has been shown to depress TSH levels in man and many other animals.

In 1973, physiologist Charles Gale, physician John W. Ensinck, and several co-workers at the University of Washington in Seattle made an unexpected discovery. They found that somatostatin lowered the amount of sugar in the blood of baboons. There are two important sugar-regulating hormones, insulin and glucagon, both made in and secreted by the pancreas. Insulin lowers sugar in the blood. Thus, the somatostatin might have been causing the drop in blood sugar by increasing insulin levels. But, surprisingly, the scientists found that the somatostatin reduced insulin secretion in the baboons.

Glucagon raises the blood sugar level. The scientists found that the baboons' glucagon level was also down. This explained somatostatin's effect on the baboons' blood sugar.

This was the first time scientists could see what happens to animals in the absence or near absence of glucagon. The standard method of eliminating a hormone in experiments—eliminating the single gland that produces it—is not possible with glucagon, which is made not only in the pancreas, but also by cells scattered throughout parts of the intestines. So, somatostatin made possible the discovery that glucagon is at least as important as insulin in regulating blood sugar.

With this experiment in mind, Gerich and Forsham began the 1974 tests that showed that somatostatin added to standard insulin treatments controls blood sugar in diabetics better than does insulin alone. Then, in November, 1974, Gerich and Forsham went a step further—they gave somatostatin to diabetic volunteers and gave them no insulin. Blood sugar dropped to normal and held there.

If further trials confirm somatostatin's effectiveness against diabetes, it might be given to patients along with reduced amounts of insulin. However, there are some problems to overcome. Somatostatin acts for only a short time, and would require an expensive and inconvenient series of injections every day. In addition, it could not be used in young people, because it lowers the production of GH and would probably stunt their growth.

We are now synthesizing analogues of somatostatin, searching for one that is strong enough and acts long enough to reduce the cost and

Slowing Down Growth

When Q-shaped somatostatin, inset, attaches to its receptor on a pituitary cell, it slows production of cyclic AMP and growth hormone.

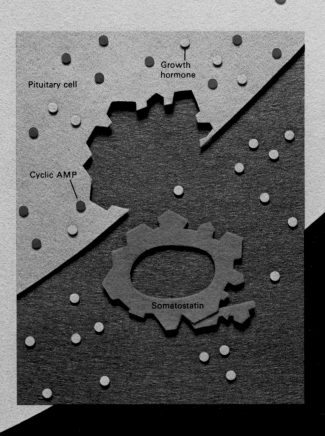

Pituitary cell

Growth hormone

Cyclic AMP

Somatostatin

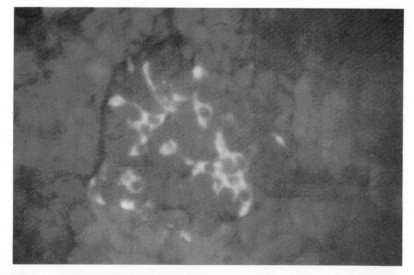

Green glow, *right,* marks somatostatin in sheep pancreas tissue magnified about 300 times. Bright areas, *below,* permit comparison, left to right, of somatostatin, glucagon, and insulin in pancreas tissue from a 64-year-old man. This is magnified 125 times.

 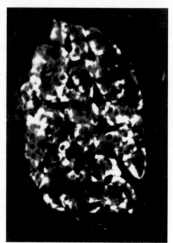 

inconvenience of using it and at the same time does not affect GH and TSH. We are encouraged by our success with LRF analogues, and the knowledge that insulin posed similar problems until the right analogue for it was found.

An extremely promising new dimension of the hypothalamic hormones, first hinted at by experimental data reported in 1972 by pharmacologist Nicholas Plotnikoff of Abbott Laboratories in North Chicago, Ill., was borne out in 1974. Plotnikoff's work strongly suggested that TRF could directly affect the brain. In March, 1974, psychiatrist Arthur Prange reported on experiments that seem to confirm this possibility. Prange dosed rats with barbiturates or alcohol, both of which decrease their activity and increase their sleeping time by acting on the brain. Then, when he injected the rats with TRF, he found that they were more active and slept less than rats that received only the barbiturates or alcohol. In November, 1974, physician Marvin R. Brown and I performed similar experiments at the Salk Institute. We found that TRF can protect rats against normally lethal doses of pentobarbital, a common barbiturate. This suggests that TRF might be useful in treating barbiturate overdose. In addition, there have been reports since 1972 suggesting that TRF may sometimes be effective against severe psychological depression.

In August, 1974, McCann reported experiments in which LRF seemed to stimulate mating behavior in rats by acting directly on the brain. This indicates that LRF may play some role in integrating mating behavior and reproduction. Somatostatin has also been shown to have direct effects on the brain. These effects generally oppose those of TRF. For example, psychiatrist Arnold Mandell of the University of California at San Diego showed late in 1973 that somatostatin inhibits the movement of rats. Then, Prange found that somatostatin increases rats' sleeping time. Our group found in 1974 that it lowers the amount of barbiturates that will kill a rat.

In view of these findings, it is not surprising that these new hormones, once thought to be confined to the hypothalamus, have now been found throughout the brain. What the future may hold is only hinted at by this discovery. It now seems possible that the hormones occur elsewhere in the body as well. For example, biologist Maurice P. Dubois of the Reproductive Physiology Station of the National Institute of Agricultural Research in Nouzilly, France, reported in August, 1974, that he had found somatostatin in cells within the pancreas of both animals and human beings. This strengthens the possibility that somatostatin normally helps regulate insulin and glucagon secretion just as it has been shown to do in animal experiments and tests on diabetic persons. So, synthesis of the new hormones and some of their analogues in large quantities may provide physicians with drugs to use against a wide variety of maladies.

Prospects for the more distant future are also encouraging. We have yet to track down and positively identify the brain hormones that stimulate the secretion of GH. The same holds for ACTH, the pituitary hormone that controls the secretion of the adrenal cortex hormones that regulate aspects of metabolism and the body's response to disease organisms and other invaders. There is probably also a brain hormone that inhibits prolactin secretion.

Even more exciting, there are probably brain hormones that regulate anterior pituitary hormones that have not even been discovered yet. For example, there is some evidence of new pituitary factors that can raise blood sugar, and this may further increase our understanding of diabetes and other ailments in which blood sugar levels are abnormal. There may also be pituitary factors that control cell proliferation. The brain hormones that control this activity might be useful against a disease that is characterized by the uncontrolled proliferation of cells—cancer.

The Nitrogen Fix

By Winston J. Brill

**In a world stalked by hunger, scientists may soon
be redesigning bacteria as minute fertilizer factories
to enrich soil, boost crop yields, and fight famine**

The world is in a food fix. In a constantly growing population, hundreds of millions of people do not have enough food. Their suffering ranges from malnutrition to outright starvation. There were once great hopes that new agricultural techniques, primarily the development of high-yield grains, would solve the problem. Unfortunately, the grains require large amounts of chemical fertilizer, and fertilizer factories have not been able to make enough. Worse, the costs of making it have soared.

Ironically, the soil contains billions and billions of tiny one-celled factories—bacteria—that make fertilizer for nothing. If scientists can find ways through which we can put these bacteria to work, the grave world food problem could be solved.

The key element is nitrogen. Along with sunlight and water, nitrogen is essential to plant growth. Although almost 80 per cent of the air we breathe is nitrogen gas, it cannot be used by plants in this form. It must be fixed—combined with other elements, generally with hydrogen to make ammonia, or with oxygen to make nitrates and nitrites—before plants can use it.

About 60 years ago, man learned how to make nitrogen fertilizer from natural gas by the Haber-Bosch process. This cumbersome industrial technique makes use of a metal catalyst to combine nitrogen and hydrogen at temperatures above 1020° F. (550° C) and at about

200 times atmospheric pressure to produce ammonia compounds that can be used as fertilizer. Because oil is a common energy source for making fertilizer, the cost of nitrogen fertilizer tripled in 1974 when the world's oil-exporting countries raised oil prices.

Certain bacteria in the soil fix atmospheric nitrogen as ammonia at room temperatures and normal pressure. These creatures play a critical role in the natural nitrogen cycle, because the ammonia they make can be used by plants when the bacteria die. The plants convert it to amino acids and combine these to make the protein they need to grow. Decaying plant matter and the wastes of plant-eating animals help to complete the cycle by returning nitrogen to the soil where it is reworked by other bacteria.

One group of bacteria called *Rhizobium* lives in close association with plants of the legume family, which includes such important crop plants as alfalfa, beans, peas, and soybeans. The plants produce nodules (tumorlike growths) on their roots. The nodules are packed with millions of *Rhizobium*, all converting nitrogen from the air into ammonia and then into amino acids. Thus, with a little help from their bacterial friends, legumes do not need nitrogen fertilizer.

Legumes are commonly used in crop rotation with nonlegumes to restore nitrogen to the soil. Corn, for example, leaves the soil deficient in nitrogen. A legume such as alfalfa may be planted in the field the following season. When the alfalfa is plowed under, soil bacteria break it down to ammonia or nitrate which can be used by corn replanted the third year. Farmers in China used this kind of crop rotation as far back as the 400s B.C. We do not know why *Rhizobium* prefer only legumes. However, recent research is helping us understand this important plant-bacteria relationship.

Each *Rhizobium* species lives with only one type of legume. For instance, the *Rhizobium* found in soybeans will not live with alfalfa or clover. Some experiments indicate that lectins, sugar-containing proteins that are commonly found in legumes, might play a role in this. In July, 1974, microbiologists Ben B. Bohlool and Edwin L. Schmidt of the University of Minnesota at Minneapolis showed that the soybean *Rhizobium's* outer surface has sites that stick to lectins produced by the soybean plant. An alfalfa *Rhizobium* does not stick to soybean lectins. The scientists suggest that the plant lectin fits onto a specific *Rhizobium*, like part of a jigsaw puzzle. Presumably, each type of legume produces a lectin on its root surface that is recognized by its *Rhizobium* species. Several laboratories are investigating this lectin binding in greater detail. When researchers discover the exact nature of the preference a given *Rhizobium* has for its host, they may be able to modify nonlegumes so that *Rhizobium* can also live with them.

Some plants other than legumes are associated with nitrogen-fixing bacteria. In 1972, Johanna Dobereiner, a microbiologist at the Agricultural Institute in Rio de Janeiro, Brazil, showed that several tropical grasses have large numbers of nitrogen-fixing bacteria on the root

The author:
Winston J. Brill is a bacteriologist at the University of Wisconsin, where he has been studying nitrogen-fixing bacteria since the 1960s.

Nitrogen's Circuitous Route

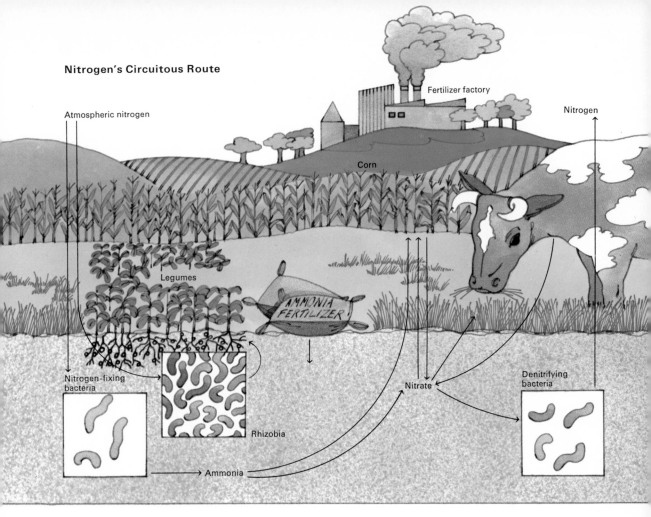

In a complex natural system, nitrogen is drawn from the air by bacteria and used by plants and animals, then returned to the soil in wastes. This system is supplemented by man-made fertilizers. Denitrifying bacteria return some nitrogen to the atmosphere.

surface. These bacteria are not *Rhizobium*, and no nodules are formed. Such grasses might be useful for fertilizing nitrogen-deficient soils in the tropics. In June, 1974, Dobereiner's laboratory found large numbers of a spiral-shaped, nitrogen-fixing bacterium living within the plant cells of a tropical grass called *Digitaria*. Dobereiner tested these bacteria to see if they would live in the roots of other grasses, especially corn. In July, 1975, she reported that they will grow in large numbers on corn-plant roots and that they appear to be fixing nitrogen for the plant. These important, though tentative, experimental results will be the basis for further tests.

Harold J. Evans, a plant physiologist at Oregon State University in Corvallis, is also looking for nitrogen-fixing bacteria that might grow naturally around the roots of some corn or oat plants. With such bacteria, these crops might not need fertilizer. Evans searches for the healthiest plants growing in nitrogen-deficient soil. If he should find that such plants are vigorous because they are attractive to nitrogen-fixing bacteria, he will isolate the bacteria, and recycle both the seed of the vigorous plants and the bacteria in other nitrogen-poor soil. Each cycle will further refine the selection of compatible plants and

95

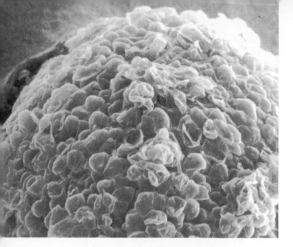

Scanning electron microscope reveals nodule cells, *top,* at 80X magnification. Opened cells, *middle,* at 2,100X and, *bottom,* at 5,000X, contain puffy bacteria. For a 3-D view of the bacteria in an opened cell, see page 134.

bacteria. Such experiments might yield a pure strain of nitrogen-fixing bacteria that grows well on or around specific varieties of oats or corn, thereby satisfying their nitrogen requirements. However, several more years of testing are needed to see if this approach is practical.

Nitrogen-fixing bacteria cooperate with some animals, too. In 1970, biologist Fraser J. Bergersen of the Commonwealth Scientific and Industrial Research Organization in Australia made the astonishing discovery that humans have nitrogen-fixing bacteria in their intestine. However, these bacteria fix only a little nitrogen. Making them increase their production could cut down the protein requirements of humans.

John A. Breznak, a postdoctoral fellow, and I, working at the University of Wisconsin in Madison, discovered nitrogen-fixing bacteria in the termite gut. Termites eat such nitrogen-poor food as wood and paper, and large numbers of microscopic, one-celled animals in the gut change the cellulose in the wood or paper to sugar. The nitrogen-fixing bacteria, which produce the components of protein required by the termite, feed on the sugar. We may someday be able to obtain cheap protein by eating termites that have, in turn, fed on nitrogen-poor industrial waste products.

Edward J. Carpenter and John L. Culliney of Harvard University reported in February, 1975, that wood-eating shipworms also harbor nitrogen-fixing bacteria in their intestines.

As a result of these findings, many laboratories are examining other species of animals that live on low-nitrogen diets to see if some have bacteria that fix nitrogen. However, most of the recent understanding of nitrogen fixation has come from studies with certain species of free-living bacteria commonly found in water and soil. Using high-pitched sound and other techniques, the scientists break the bacteria open and test thousands of different compounds that are released. Researchers have isolated those compounds specifically involved with nitrogen fixation.

The bacteria require two proteins to form nitrogenase, the enzyme that fixes nitrogen. Nitrogenases from a variety of different kinds of nitrogen-fixing bacteria (including *Rhizobium* in nodules) are very similar in size and composition, and all contain iron and molybdenum as well as amino acids.

Bacterial Visitors Lend a Hand

When rhizobia bacteria recognize sugar-containing protein lectins on the roots of soybean plants, they bind themselves to the roots, left. The host plant builds a nodule of cells, each housing millions of the rapidly reproducing bacteria. In exchange for shelter, the guest bacteria use two enzymes, nitrogenase and glutamine synthetase, to fix nitrogen gas as ammonia and then make amino acids from the ammonia. The plant uses these acids to make protein. Oxygen, which stops nitrogen fixation, is blocked by another enzyme, leghemoglobin.

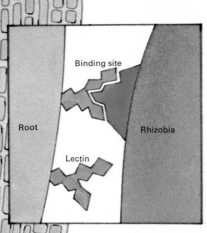

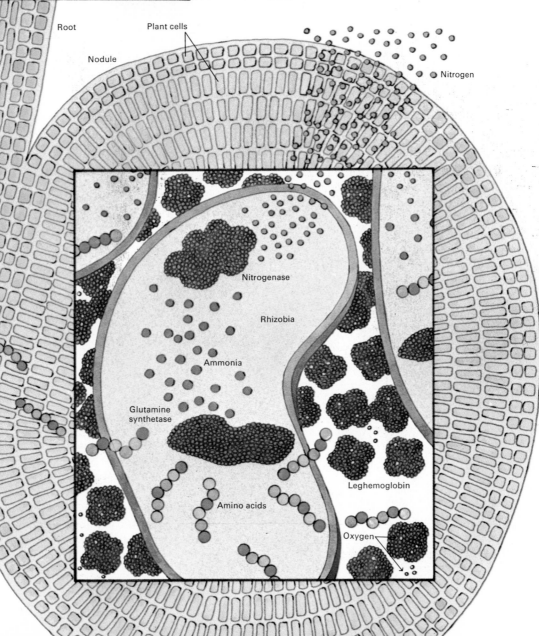

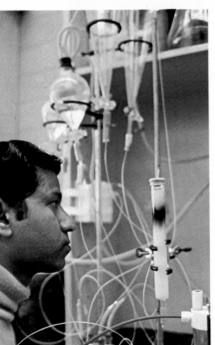

Laboratory-grown soybean plants, *left,* provide bacteria in their root nodules from which nitrogenase is extracted. Vinod Shah purifies the dark brown nitrogenase, *below left,* in an airtight purifying column, while the author, *below,* studies the enzyme's chemical action with a gas chromatograph.

Once we understand how nitrogenase efficiently converts nitrogen gas to ammonia at room temperatures and pressures, we may be able to use the enzyme as a model for more efficient processes for making commercial fertilizers. In early 1975, William H. Orme-Johnson and Vinod K. Shah, biochemists at the University of Wisconsin Center for Studies of Nitrogen Fixation, made a major contribution toward understanding how nitrogenase works. They placed samples of purified nitrogenase in magnetic fields at temperatures near absolute zero. Being metals, the iron and molybdenum in the nitrogenase responded to the magnetic fields. However, an actively fixing nitrogenase sample responded differently from an inactive sample. This indicated that the iron and molybdenum were playing an important role in the nitrogen-fixation process. Scientists are now trying to determine how iron and molybdenum are attached to the enzyme and what their exact roles are in the reaction. Interestingly, these metals are also used as catalysts in the Haber-Bosch process.

Each such advance in our understanding of nitrogen fixation has stimulated chemists to synthesize new catalysts, and in January, 1975, Joseph Chatt of the University of Sussex in England reported making a catalyst containing molybdenum that will convert nitrogen gas to ammonia at ordinary temperatures and pressures. The synthesis of such catalysts should eventually make it possible to produce commercial fertilizer much more cheaply.

But fixing nitrogen gas at normal pressures and temperatures is not the only advantage bacterial nitrogen fixation has over commercial fertilizer. Nitrogen-fixing bacteria convert nitrogen gas to ammonia only when the soil around them contains no fixed nitrogen. Consequently, nitrogen-fixing bacteria do not produce excess ammonia and do not contribute to the pollution of rivers and lakes. In contrast, a farmer must apply enough fertilizer to last for many months in a field of nonleguminous plants. Rain run-off carries some of this fertilizer to nearby bodies of water, encouraging plant and algae growth which reduces oxygen levels in the water. Water pollution from fertilizer use in the United States has increased in the last 10 years, according to a July, 1974, report by the Environmental Protection Agency.

Because bacteria make just enough ammonia for their own use, they must have a mechanism that turns nitrogenase synthesis on and off. Scientists are just beginning to understand what regulates nitrogenase synthesis. In October, 1974, Roy S. Tubb, who was doing graduate work at the University of Sussex, and Stanley L. Streicher of Massachusetts Institute of Technology in Cambridge, Mass., independently found that some mutant strains of *Klebsiella pneumoniae*, a free-living, nitrogen-fixing bacteria, could not synthesize nitrogenase and failed to grow in some nitrogen-rich mediums. They traced this failure to a defective enzyme in the mutants. The enzyme, glutamine synthetase, was already known to play a key role in converting ammonia into amino acids. Tubb and Streicher now believe it has another important function: glutamine synthetase turns on the synthesis of nitrogenase when no ammonia is available. When ammonia is available, glutamine synthetase does not do this.

Joyce K. Gordon, a graduate student in my laboratory, was studying how bacteria convert ammonia into amino acids when she wondered whether the enzyme methionine sulfoximine, which is known to block that conversion, might not also alter glutamine synthetase so that it continues to trigger the production of nitrogenase, even after enough ammonia had been made. She set up two experiments in June, 1974. In the first, she grew cultures of the nitrogen-fixing *Azotobacter vinelandii* and *Klebsiella pneumoniae* in vessels containing ammonia. She added methionine sulfoximine to the culture, and found that the bac-

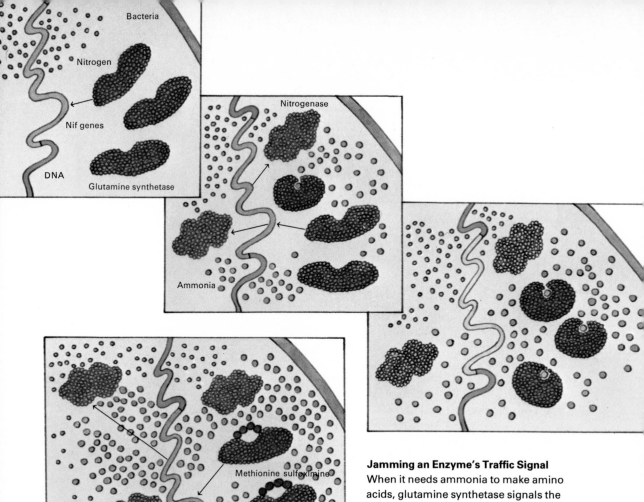

Jamming an Enzyme's Traffic Signal
When it needs ammonia to make amino
acids, glutamine synthetase signals the
nif genes, top left, to make nitrogenase,
which then fixes nitrogen as ammonia,
center. When enough ammonia is made,
ammonia molecules block further
glutamine-synthetase activity, above.
Methionine sulfoximine can jam the
blocking site, left, and the enzyme
still triggers ammonia production.

teria began fixing nitrogen, even though sufficient ammonia was available. In the second experiment, she grew the same bacteria in a nitrogen-poor medium, in which the bacteria fixed enough atmospheric
nitrogen for their own use. When she added methionine sulfoximine,
the bacteria were unable to stop fixing nitrogen and synthesized more
ammonia than they needed, excreting the excess.

This discovery excited us and we wondered if methionine sulfoxi-
mine-treated cells could be used to produce ammonia commercially
from nitrogen gas. Unfortunately, methionine sulfoximine is too expensive to be practical for ammonia production. However, Gordon's
work challenged us to consider other approaches that could by-pass
use of the expensive compound. We isolated mutant bacteria contain-

ing glutamine synthetase that resembled the glutamine synthetase which had been altered by treatment with methionine sulfoximine. These mutants also continue to synthesize nitrogenase, and therefore fix nitrogen gas, even when ammonia is present. The mutants are currently being modified even further by a number of genetic techniques to increase the amount of ammonia they form. Whether such mutants will have an economically important effect on agriculture is not yet clear. Perhaps they can be grown in huge fermentation tanks to convert nitrogen gas to ammonia, or even spread directly on farm fields to fertilize the soil.

Workers in biochemist Raymond C. Valentine's laboratory at the University of California at San Diego and in my laboratory have now

To test bacteria survival, Joyce Gordon sterilizes an inoculating loop before she adds mutant strains of bacteria to a sterile growing medium.

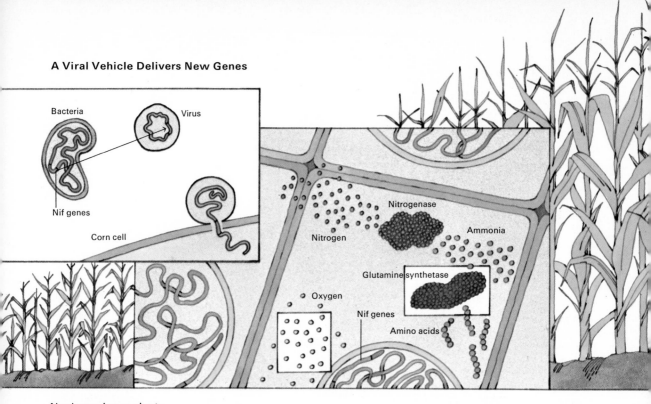

A Viral Vehicle Delivers New Genes

Bacteria

Virus

Nif genes

Corn cell

Nitrogenase

Nitrogen

Ammonia

Glutamine synthetase

Oxygen

Nif genes

Amino acids

Nonleguminous plants, such as corn, may be made to fix their own nitrogen. For example, bacterial nif genes could be transferred to a virus, which would infect corn plant cells. These could then be grown into plants and reproduced, if corn's glutamine synthetase will turn on nif genes and something is found to keep oxygen in the cells from destroying the plant's nitrogenase.

also isolated mutants of *Klebsiella pneumoniae* and have showed that the genes responsible for producing nitrogenase are located close together on the single chromosome of this organism. This cluster of genes, called nitrogen-fixing (nif) genes, comprises less than 1 per cent of the total cell chromosome.

At the University of Sussex, geneticist Ray A. Dixon successfully transferred the nif genes of *Klebsiella pneumoniae* to other bacteria in 1972. Dixon used bacterial plasmids, small circular pieces of the genetic material deoxyribonucleic acid (DNA) that are found in many bacteria. Plasmids have the ability to snip off small pieces of the bacteria's chromosomal DNA and incorporate it into themselves. They can also transfer relatively easily from one bacterium to another, even from one species to another, thus transferring genetic material. Dixon took advantage of these abilities, plus the fact that nif genes are located in a small region of *Klebsiella pneumoniae*'s chromosome. After extensive test runs in which he mixed large numbers of bacteria and plasmids, he found a strain of *Klebsiella pneumoniae* with plasmids that had taken on just the nif genes. He mixed this strain with a culture of *Escherichia coli*, which does not normally fix nitrogen gas. The plasmids transferred the nif genes and *Escherichia coli* began to fix nitrogen. Our own intestines harbor millions of bacteria of various species. It might be possible to put nif genes in some of them, perhaps satisfying some of our own protein requirements. However, such genetic manipulation might cause excessive production and result in ammonia poisoning. The transfer of nif genes from *Klebsiella pneumoniae* to *Escherichia coli* seems to have been aided by the fact that the two bacteria are very closely related. Transferring these genes to unrelated bacteria may prove more difficult.

Scientists are also studying the possibility of nif gene transfer into the cells of nonleguminous crop plants. Researchers in several laboratories have shown that they can transfer a bacterial gene to a plant cell using a certain kind of bacterial virus as a vehicle. No one has yet attached nif genes to that type of virus, but several geneticists are working on it. Should we be able to produce plant cells that fix nitrogen, they can be developed into mature plants in which all cells have the nif genes. Such experiments could produce nitrogen-fixing corn or wheat plants, but many problems must be overcome first.

Even if nif genes are transferred to a nonleguminous plant, making these genes work requires a specific form of glutamine synthetase. One way of getting around that problem is to first create nif genes containing a mutation that would allow nitrogenase to be produced without glutamine synthetase.

Another serious problem is that nitrogenase is destroyed by oxygen. Bacteria that fix nitrogen gas either do so in environments that contain no oxygen, such as a termite gut or a stagnant pond, or they have some means of keeping oxygen away from their nitrogenase. For example, leghemoglobin, a red protein found only in root nodules, surrounds the *Rhizobium* in legumes. Leghemoglobin binds oxygen and prevents it from inactivating the nitrogenase in the *Rhizobium* cells. If we can develop a way to make corn or wheat fix nitrogen, we will also have to find a way to protect the nitrogenase.

It will probably be many years before we make significant headway in such complex research. For the near future, Tom Wacek, a postdoctoral fellow in my laboratory, is testing relative rates of nitrogen fixation in soybean varieties collected from around the world. He wants to determine whether some varieties fix nitrogen gas more rapidly than others. Strains that are most effective in nitrogen fixation will be bred with commercial strains to produce better plants. Other researchers are trying to isolate and genetically modify *Rhizobium* strains to allow legume plants to fix nitrogen gas more rapidly.

Still other possibilities for improving nitrogen fixation in legumes are being studied in many laboratories. A team of researchers at the Du Pont Experimental Station in Wilmington, Del., recently showed that high levels of carbon dioxide will greatly increase nitrogen fixation and protein content in the soybean plant. Adding carbon dioxide to an open field is not practical, but the research may lead to other more useful applications.

There are no easy, simple solutions to the complex problems of food, population, climate, and energy. But governments are becoming more aware of the crucial role nitrogen plays in producing food, and they are sponsoring increased research. With imagination, hard work, and international cooperation, scientists may persuade the bacteria to help us out of the fix we are in.

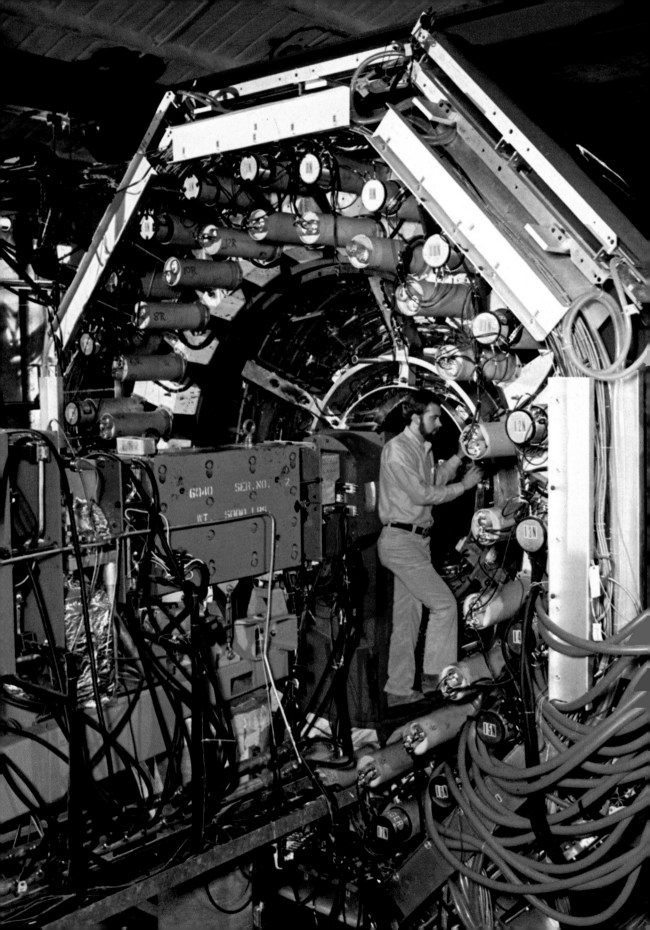

A Subatomic Surprise

By Robert H. March

**A new and unpredicted elementary particle has
striking properties that have physicists baffled**

It is a common fantasy among scientists. You are working in the
laboratory on a Sunday morning (such dedication), rechecking inconsistent data, when the instruments suddenly go wild. You can scarcely
believe your eyes, but there, plain as day, is something totally unexpected, yet so obvious that no one can doubt it is real. A chill runs
through you as you realize that your discovery will turn your field of
research topsy-turvy.

Of course, scientific discovery rarely happens this way. Especially if
you belong to a huge research team in elementary particle physics,
where the significance of a day's work emerges only after months of
laborious computations. Yet, the fantasy became reality on Sunday,
Nov. 10, 1974, for Adam Boyarski, Carl Friedberg, and Petros Rapidis
–three physicists on a 35-member team of researchers from the Stanford Linear Accelerator (SLAC) in Palo Alto, Calif., and the University of California's Lawrence Berkeley Laboratory (LBL). By the luck
of the draw, they were serving as the day crew at a new SLAC experimental facility, the Stanford Positron-Electron Accelerating Ring
(SPEAR), when it revealed a subatomic particle whose unpredicted
existence and stability startled theorists and experimenters alike.

That same Sunday morning, Professor Samuel C. C. Ting of the Massachusetts Institute of Technology (M.I.T.) boarded a jet at Boston's Logan International Airport on his way to California to attend a routine SLAC committee meeting—or so he believed. But he was linked unknowingly to the drama then unfolding at SPEAR by a graph that only he and a handful of his co-workers had seen. Ting's graph showed a prominent peak, which meant that his group had also discovered a subatomic particle. Their discovery might have been nothing to get very excited about, because more than 200 new particles had been found in this way since 1960. But Sam Ting suspected he might have found an extremely narrow peak. To particle physicists, a narrow peak indicates a particle that lives a long time. An unpredicted narrow peak means that physicists must invent a new theory to explain why the particle exists and why it lives so long. And if the explanation results in new rules and formulas for understanding the subatomic world, someone involved might win a Nobel prize.

Ting had been puzzling over his graph since August, when results from his experiment at the Brookhaven National Laboratory (BNL), in Upton, N.Y., first showed the peak. Was it as unusual as it seemed? More important, was it real, or just the result of some machinery malfunction? Leon M. Lederman, a Columbia University physicist, had performed a similar experiment five years earlier and found nothing spectacular. This would have been enough to make any scientist pause and ponder before claiming a new discovery—especially Sam Ting, who has a reputation for jealously guarding his data until he feels the moment is ripe to publish it. In this case, he even bet $10 with a visiting SLAC physicist that no peak would be found. He made the bet long after he knew a peak was there, and later admitted that he would have gone as high as $50 to squelch the rumors.

But Ting kept his data under his hat too long. He was somewhere above the Great Plains when the SPEAR team also discovered a narrow peak. Though found with a different apparatus, SPEAR's particle was the same as Ting's.

On that fateful Sunday morning, SPEAR was operating as usual, steering electrons into head-on collisions with their positive counterparts in the antimatter world, positrons. This microcosmic demolition derby takes place on a race-track course, slightly larger than the standard ⅛-mile (⅕-kilometer) running track. The core of the track is a tube pumped to a very high vacuum. The curved ends are enclosed in magnets to hold the particles on course.

Every few hours during an experiment, physicists replenish the particles in SPEAR. About 10 trillion electrons enter from the 2-mile (3-kilometer) linear accelerator that gives SLAC its name. Trillions of positrons are also added. They are produced by smashing SLAC's electron beam into a special target located a third of the way along its length. Once in SPEAR, the electrons and positrons circulate in opposite directions at nearly the speed of light, in bunches about the size of

The author:
Robert H. March is a professor of physics at the University of Wisconsin. He wrote "The Quandary Over Quarks" for the 1975 edition of *Science Year.*

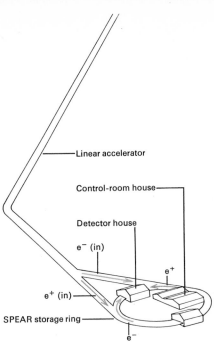

Linear accelerator

Control-room house

Detector house

e^- (in)

e^+

e^+ (in)

SPEAR storage ring

e^-

The Birthplace of Psi
An aerial view of
the Stanford Linear
Accelerator shows the
SPEAR storage ring
in which electrons
(e-) and positrons (e+)
circle in opposite
directions at nearly
the speed of light.
They collide head-on
inside the detector
house, producing other
particles that leave
tracks in the detector.

a pencil stub. They meet head-on on the straightaways, and a few particles collide each time. Violent collisions—in which an electron and positron annihilate each other in a flash of pure energy hotter than anything else in the universe—occur about once every minute.

Because energy and matter are always conserved, even in such violent reactions, the annihilation energy must go somewhere. Usually, it rematerializes in the form of particles from the hadron family. Hadron is a term applied to the 300 or so particles that interact via the strong force, that which holds protons and neutrons together in atomic nuclei. Electrons and positrons belong to the lepton family, a group of four particles that do not interact strongly.

The SPEAR experimenters were simply checking on how often hadrons form following the annihilation of positrons and electrons. They were using an elaborate array of particle detectors that surrounds one of the straight sections of SPEAR. The detectors track every charged particle that emerges from the collision region at an angle of more than 45 degrees to the colliding beams. Position reports go to a minicomputer, which records them on tape for later analysis and immediately reconstructs a picture of the particles' flight-paths on a radar-type display in the SPEAR control room. With a little practice, a physicist watching the display can spot the telltale pattern of hadrons—several tracks emerging from a common center.

Simply counting the number of annihilations that yield hadrons each minute seems almost trivial. But it is one of the most important things physicists do at SPEAR. In January, 1974, the joint SLAC-

LBL team led by Burton Richter reported that the counting rate drops slowly as the speed (and therefore the energy) of the colliding particles is raised. This was a complete surprise; existing theories predicted a more rapid fall-off. The decrease was smooth and steady except for one inconsistent data point near 3 billion electron volts (GeV). Either the apparatus was acting up, as seemed most likely, or there was some sort of peak near that energy. A recheck in June proved inconclusive. If there was a peak, it would be ridiculously narrow.

Roy F. Schwitters of SLAC and Gerson Goldhaber and Gerald S. Abrams of LBL were particularly insistent that another recheck be made. They could find no indications of experimental error in the June data tapes. Of course, it was still an odds-on bet that the apparatus had caused the unusual reading, but the SLAC-LBL team decided to check its data near 3 GeV one more time.

The run began on Saturday, November 9. Like any complex machine that has to be retuned, SPEAR was slow to settle down. Still, enough data had trickled in by Saturday evening to make it clear that the peak on a graph of the data was caused by a new particle. Just how unusual it was remained to be seen. At 9:30 P.M., physicist Vera Lüth left the lab to buy a magnum of champagne, so there would be something on hand for a victory toast if the big moment came on her owl shift. But machine trouble developed after midnight, and discovery passed that shift by.

The day crew took over and began fine-tuning SPEAR to make small changes in the energy of its colliding beams. The system for doing this was new and Boyarski and control-room engineer Ken Underwood struggled to make it work. Rapidis, a graduate student working on his first experiment, plotted the data. Friedberg frantically recorded the events in the laboratory logbook to make sure that they could later reconstruct exactly what had happened.

As the morning wore on, the team's senior members began arriving to check on the progress. The news quickly spread by phone, and soon other senior physicists drifted in. To Underwood, this was the tipoff that they were onto something big. He had never seen so many gray heads in one place—especially on a Sunday. Richter was one of the first to arrive. For him, more than for any other member of the team, a momentous discovery would be sweet. Richter is a "hardware man," one of the gamblers in particle physics who stakes his reputation on expensive, high-powered machinery. SPEAR was his brainchild, the culmination of more than 10 years of work on colliding-beam devices.

By 9:30 A.M., the peak had begun to emerge on a graph of the data. It was seven times higher than the smooth curve from which it rose.

The Spectacular Spike
A sharp, narrow peak (red) rises about 3.5 feet (1.1 meters) above the top of the page, dwarfing the rest of the SPEAR graph. It represents psi, a new particle that lives much longer than a typical particle (blue).

There could be no doubt that they had found a new particle. Goldhaber had never seen anything so spectacular. A peak only twice as high as the smooth curve would have been a sensational discovery. He went off to a back office to scrawl a rough draft of an article describing the find on the back of some used computer printout paper. When he returned at 11 A.M., he found his colleagues staring at the screen in disbelief. Hadrons were forming from electron-positron annihilations so fast that the minicomputer could not show them all on the display screen. Instead of one every few minutes, they were coming nearly every second. The summit of the peak had been found at 3.105 GeV, and it towered 100 times above the rest of the graph. It was as narrow as it could possibly be – 1 part in 3,000 – the limit to which the energy of the colliding electrons and positrons could be controlled. The new particle was christened psi.

Somebody remembered the champagne in the refrigerator, and the happy scientists began celebrating, even as they continued to take data. By then, everyone on the team who could be reached by telephone had been rounded up. Underwood remembers SLAC's director Wolfgang Panofsky standing in front of the display screen muttering over and over, "Oh my God!"

Historic Day at SPEAR
In the control room when the psi particle was discovered on Nov. 10, 1974, an engineer monitored the controls, *top left,* while physicists plotted the data, *top right,* and group leader Burton Richter checked the computer printout, *above.* Later, Richter and others recalled the eventful discovery day, *above left.*

SPEAR's Smashing Success

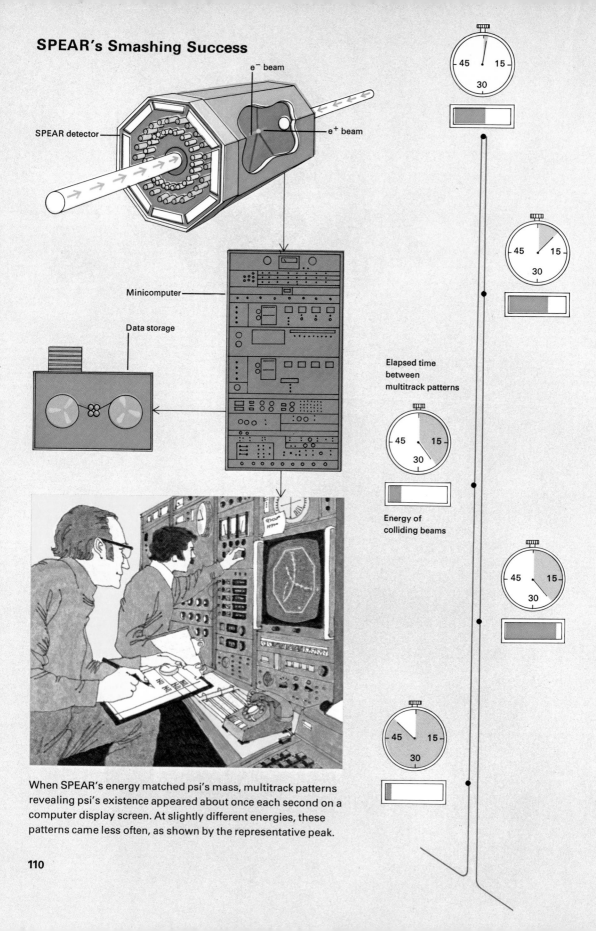

When SPEAR's energy matched psi's mass, multitrack patterns revealing psi's existence appeared about once each second on a computer display screen. At slightly different energies, these patterns came less often, as shown by the representative peak.

Why does a tall, narrow peak draw such an emotional response and cause such jubilation among physicists? Part of the answer lies in Albert Einstein's famous equation $E=mc^2$ (energy equals mass times the speed of light squared). It tells physicists that energy (E) and mass (m) may be converted into one another. To particle physicists, the c^2 is just a conversion factor between energy units and mass units; in fact, they normally use energy units for mass and write the formula $E=m$.

According to this mass-energy equivalence, the mass of the psi is 3.105 GeV. When the combined energy of an electron and positron exactly match that value, they can annihilate to produce one psi, with no energy left over. Producing one particle at a time is easier than producing two or more hadrons. So, electrons and positrons of exactly the psi's energy are more likely to annihilate than those at other energies. The psi particles formed in this way spontaneously break up into three or more hadrons that leave tracks in the SPEAR detector array, and the counting rate rises dramatically. The higher it rises, the more often is the psi forming from the annihilation energy.

Narrow peaks are particularly exciting because of the uncertainty principle, a basic rule of the quantum theory, which states that a particle must live a long time in order to have a definite mass. A particle that spontaneously breaks up too quickly has a mass, or energy, that varies each time it is measured. Consequently, its peak is spread out over a wide range of energies. Such was the case for most of the particles discovered since 1964.

Of course, the phrase "live a long time" must be interpreted on the bizarre time scale of particle physics. The long life of the psi particle is 0.00000000000000000001 second. However, the strong force normally acts in a time period a thousand times shorter than this. Because the psi particle is so stable, it must have some characteristic that blocks the fast breakup, perhaps some new internal property.

Sam Ting and his team of 1 BNL and 13 M.I.T. researchers knew all about the significance of tall, narrow peaks. But they found their peak with equipment that could not be tuned as finely as SPEAR. They steered an intense beam of protons from BNL's Alternating Gradient Synchrotron into a target of beryllium. About once every billion times that a proton from the beam collided with a proton in a beryllium nucleus, a pair of electrons and positrons emerged from the debris. As the pair sped away from the collision zone at an angle of about 15 degrees, each traveled down an arm of a mass spectrometer that was 70 feet (21 meters) long. Three magnets on each arm bent the particles' paths, allowing physicists to determine the energy of each by measuring its trajectory through multiple detectors. Special counters positively identified the electron-positron pair.

Ting had been designing and building his complex system of magnets, detectors, and counters since 1971. When it began operating at BNL in April, 1974, his mass spectrometer was buried under 10,000 short tons (9,072 metric tons) of concrete, 25 short tons (23 metric tons) of lead, 5 short tons (4.5 metric tons) of uranium, and 5 short tons of borax soap to shield the researchers from intense radiation. Six physicists manned the detector while the experiment was in progress.

Ting's BNL experiment was, in a sense, the reverse of the SPEAR experiment. If an intermediate particle that could subsequently break up into an electron-positron pair were to form from part of the proton-proton collision energy, then the counting rate would jump when the magnets were tuned for the energy that matched the particle's mass. Ting saw such a peak in August, then again in October, 1974. He named it the J—a letter that is coincidentally nearly identical to the Chinese character for Ting. Then he carefully rechecked his experiment to erase all doubt about his J particle.

But at SPEAR, on November 10, there was no room for doubt. The peak there was about 40 times narrower than Ting's. By afternoon, word of the discovery was beginning to spread over long-distance telephone lines. Ting, resting in a Palo Alto motel a stone's throw from SLAC, got the word that evening via a phone call from a colleague at M.I.T. The following morning, he and Richter compared notes and concluded that the psi and the J must be the same.

The news aroused particular interest in Frascati, a laboratory that is situated among picturesque vineyards near Rome. This laboratory has a machine named ADONE that is very much like SPEAR. But ADONE had been designed to go to a maximum energy of exactly 3 GeV. Had the psi been just a touch lower in mass, it would probably have been found there first. Upon hearing the news, Italian physicists squeezed 3 per cent more energy out of their machine, and by the end of the week had observed the psi.

The news reached my home base, the University of Wisconsin at Madison, on Monday morning. At a hastily convened seminar the following afternoon, rumors were exchanged and compared while sev-

SPEAR's Demolition Derby

An e− and an e+ can annihilate each other in a flash of energy — hotter than anything else in the universe — from which a psi (J) particle can form. In a hundredth of a billionth of a billionth of a second, the psi in turn becomes three or more hadrons that leave telltale multitrack patterns in the detector.

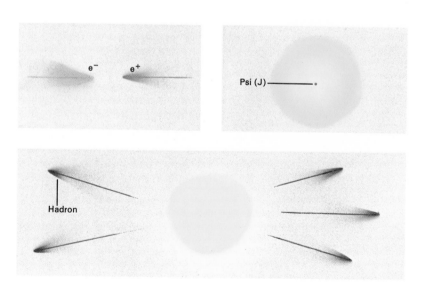

eral theorists thought aloud about the possible significance of the discovery. The atmosphere in the crowded seminar room was so electric that three new graduate students, who had been wavering in their choice of research, opted on the spot for particle physics. By this time, there was hardly a particle physicist anywhere on the planet who did not know what had happened.

The SLAC-LBL team, however, was not content to rest on its laurels. On November 21, they found another sharp peak at 3.695 GeV. Because the newer peak seemed similar to the psi and there are so many particles to remember, they also named it psi. The particles are now designated psi (3105) and psi (3695). The team found a third peak at 4.1 GeV in January, 1975. Though far shorter and broader than the first two peaks, it may be related to them.

Throughout the rest of the winter, a visitor at any accelerator laboratory could almost feel the earth rumble as experimenters reshuffled their cumbersome detectors, hoping to be the next for whom lightning would strike. From past experience, they knew that new particles usually come in flocks.

While experimenters continue their searches, theorists try to explain what the discoveries mean. As usual in this peculiar field, where there is one theorist for each experimenter, there are plenty of suggestions. Many of these schemes can account for the psi particles, but one immediately became more popular than the others because it makes rather specific predictions as to what else experimenters may find.

It is a standard trick in particle physics to explain real particles by inventing imaginary ones. For a decade, the most successful particle theory has accounted for all the known hadrons by supposing them to be made up of three basic building blocks called quarks. The quarks are designated u, d, and s, for up, down, and strange. The strong force can recombine quarks to produce new hadrons, but it cannot change

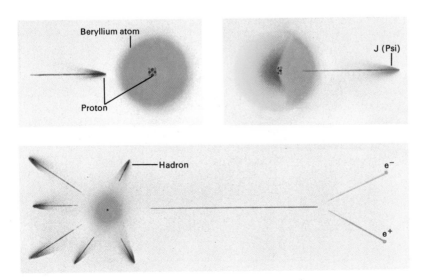

BNL's Proton Billiards
In a reverse reaction, a speeding proton smashes into another proton in the nucleus of a beryllium atom, producing hadrons and a J(psi) particle. As the J particle speeds away, it turns into a moving e−-e+pair.

one kind of quark into another. That takes a weaker and slower nuclear force, the so-called weak force. The terms *strange* and *quark* were first used by Nobel prizewinner Murray Gell-Mann of the California Institute of Technology, whose sense of humor has done much to enliven scientific terminology.

The quark model gave a natural explanation for an earlier dilemma in particle physics—the existence of eight particles that were discovered between 1947 and 1955. Like the psi particles, they seemed to live too long. It turned out that these particles were simply the lightest ones that contained one or more s-quarks. No simple recombination of quarks could produce lighter hadrons, so in order to break up, the particles had to live long enough for the weak force to convert s-quarks to u-quarks at its leisurely pace.

The three-quark model accounts for hundreds of particles that have since been discovered. However, despite 10 years of intensive searching, no experimenter has yet detected a free quark, one unattached to others. And no theorist has ever satisfactorily explained why.

A number of theorists, especially Sheldon L. Glashow of Harvard University in Cambridge, Mass., expressed dissatisfaction with the three-quark scheme as early as 1967. They proposed a fourth quark for a variety of reasons. First of all, this would make the number of quarks equal the number of leptons, suggesting that leptons may themselves be alternate forms of quarks, thus uniting the hadrons and leptons. Second, the u- and d-quarks are near brothers, while the s-quark is a

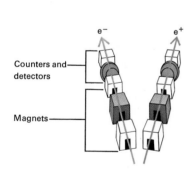

Emerging from one out of about one billion proton-proton collisions, an e⁻ and an e⁺ travel down arms of a mass spectrometer at Brookhaven National Laboratory (BNL). The system of magnets, counters, and detectors enabled researchers to discover the J particle and measure its surprising properties.

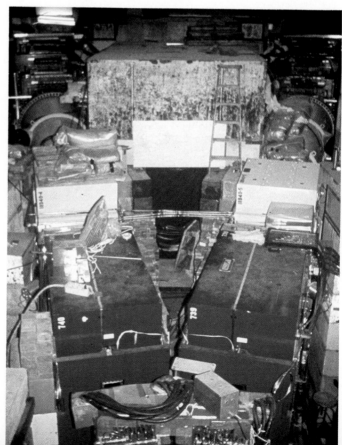

Surrounded by readouts that link him to his mass spectrometer at BNL, Samuel Ting ponders a graph of the data that revealed the unexpected, and still unexplained, new subatomic particle.

loner. These theorists felt it should have a companion. Finally, a fourth quark could clear up a few loose ends in the weak force theory.

Not to be outdone by Gell-Mann, Glashow named his new quark c for charmed. Charmed quarks added whole new families of hadrons. The theory has progressed little, however, because experimenters have found no hadrons containing a charmed quark. The charmed-quark theorists resorted to a typical escape hatch to explain this. Charmed hadrons must be too heavy to have been produced by the energies available at earlier accelerators, they suggested. This is the usual way of justifying the failure to find predicted particles.

The psi (3105) fits the charmed-quark scheme like a hand in a glove. It would be the simplest member of the charmed hadron family–the combination of a c-quark with its own antiparticle. The psi (3695) would be an excited state of the psi (3105). Because the mass of the psi is known, the theory can be used to predict the masses of other hadrons containing charmed quarks. These should live even longer than the psi, and be reasonably easy to find with existing equipment.

Whether this scheme will stand the test of experiments remains to be seen. In a sense, it will be more exciting if it does not. That might mean that particle physics needs a new rulebook, not just a new rule.

In any event, the psi has already proved its worth and given a group of hard-working young physicists a thrill they will never forget. Which leaves us with a thought about Petros Rapidis, the graduate-student co-discoverer of the psi (3105) at SPEAR. This was his first taste of experimental work. What will he do for an encore?

Energy on
The Shelf

By Robert K. Johnson

**New energy banks will help balance electric power
supply and demand, let sunshine light our homes at
night, and pave the way for practical electric cars**

A man-made reservoir near Ludington, Mich., pumped full at night, provides hydroelectric power as water flows back to Lake Michigan.

At 8 A.M. on a typical weekday in early 1975, Mrs. Gilman P. Cyr switched off the water heater in her all-electric ranch house in Rutland, Vt. For the next three hours, she sewed, dusted, and did other manual household chores. At 11 A.M., she turned the water heater back on and vacuumed until the water was hot enough to do a wash. In the early afternoon, she cooked dinner on her electric range, then, at 5 P.M., moved the hot food to the wood-burning Franklin stove her husband bought in 1974 to help beat the energy crisis. Once again, she switched off the water heater.

Mrs. Cyr was saving money. Her family and a few dozen others were participating in a test program offered by their electric utility. The Central Vermont Public Service Corporation charged less for electricity used between 11 A.M. and 5 P.M. and between 9 P.M. and 8 A.M. – the periods when the utility had excess generating capacity. By rearranging their daily routine to reduce the use of electricity at other times, the Cyrs expected to save about $200 during 1975.

Central Vermont's reduced-rate program is one possible solution to a serious problem facing United States utilities—matching the supply of electricity to the demand for it. Customer demand varies widely by the hour, day, week, and season, and is different in each region of the United States. For a Midwestern utility serving an industrialized urban area, the peak demand for electric power on a hot summer afternoon can be twice as heavy as it is during the night, and it can be 50 per cent greater than the peak daytime levels typical during spring or fall days when few air conditioners or heaters are operating.

To meet such wide variations in demand, the utilities rely partly on large new fossil- or nuclear-fueled power plants that ideally operate continuously except during maintenance or other unavoidable shut-down periods. They supply about 70 per cent of a typical electric utility's yearly energy production, but they represent only about 45 per cent of the total power-generating capacity that the utility needs in order to meet peak demands. (Power is the rate at which energy is generated.) Older fossil-fueled plants add another 40 per cent of the generating capacity. They are normally started up in the morning and turned off again at night.

To meet peak daytime demand and provide reserve power, most utilities use small, easy-to-install, peaking generators. The most common peaking generator is the gas turbine, which is much like a jet engine. In 1975, gas turbines supplied about 6 per cent of the electric power-generating capacity in the United States. But as fuel costs climb and reserves of oil and natural gas dwindle, the inefficient gas turbines are falling out of favor as the best solution to meeting peak demands.

Many utility planners believe that reduced-rate programs, which shift some demand to otherwise slack periods, are not the best solution either. Consequently, they are looking for a technological solution to their peak-power-generating problem that will not require their customers to change their habits drastically. By storing excess electricity generated by the large, efficient nuclear- or coal-fueled power plants and using fewer of the less efficient peaking generators, the utility planners believe that they can meet peak demands and thus prevent brownouts while drastically reducing oil imports. "By the 1990s," says Fritz R. Kalhammer of the Electric Power Research Institute (EPRI) in Palo Alto, Calif., "replacing peaking generators with energy-storage systems could cut annual oil imports by more than 100 million barrels, thus saving more than a billion dollars each year. Development of efficient, long-lasting, and inexpensive energy-storage systems could become a key to making the United States self-sufficient with energy from domestic supplies."

Those who propose using windmills or solar cells to generate electricity also need to develop efficient energy-storage systems, without which their power-generating systems work only when strong winds

The author:
Robert K. Johnson is a senior editor for *Science Year* and *The World Book Year Book.*

blow or when the sun shines brightly. And if a lightweight energy-storage device is developed, electric vehicles could replace many of today's petroleum-powered automobiles and trucks, saving additional millions of barrels of oil.

But in 1975, less than 2 per cent of all the electric power that was used in the United States came from storage. This power was drawn from pumped hydroelectric storage (hydrostorage) plants. These are turbine-driven hydroelectric generators that become electric pumps when run in reverse. They pump water uphill at night to a storage reservoir, using electricity from nearby nuclear- or fossil-fueled plants that would otherwise be virtually idle. The next day, the water flows back downhill, turning the turbine-generators to meet the increased demand for electricity.

Hydrostorage plants are about 67 per cent efficient; that is, for every three units of electric energy used to pump water uphill, only two units are returned to the power network. But, even with such large conversion losses and the expense of building hydrostorage plants, they produce cheaper electricity than do gas turbines.

One of the largest of these plants was built on the eastern bluffs of Lake Michigan near Ludington, Mich., by Consumers Power Company and Detroit Edison Company. The upper storage reservoir is a man-made lake that covers 1.3 square miles (3.4 square kilometers). When filled, the lake's water surface is 364 feet (111 meters) above Lake Michigan, which serves as the lower reservoir. The $350-million system took nearly 15 years to plan and build. It generates nearly 2-

The demand for electric power varies widely, by hour, day, and season. Utilities run their most efficient generators around the clock and turn on their less efficient units only to fill peak demands.

The Uneven Call for Electric Power

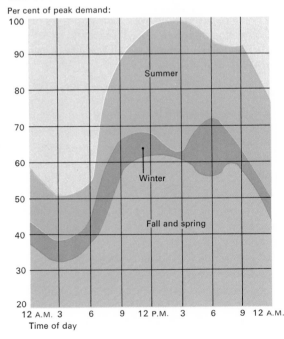

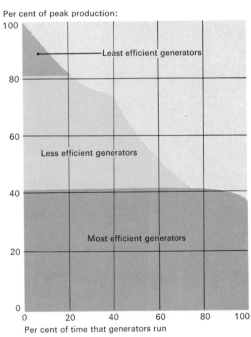

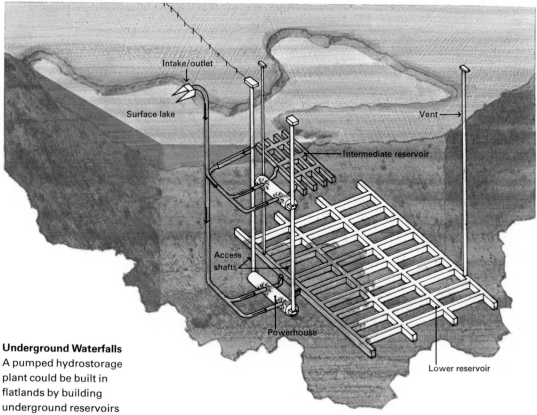

Underground Waterfalls
A pumped hydrostorage plant could be built in flatlands by building underground reservoirs and powerhouses. Water from a small surface lake would drop in two or more stages, turning turbines to generate hydroelectric power.

million kilowatts (kw) of electric power for nearly 9 hours as 34 million gallons (128 million liters) of water per minute rush back into Lake Michigan. This is enough generating capacity to replace all the peaking generators for a city about the size of Chicago.

United States utilities plan to triple their hydrostorage capacity during the next 10 years. However, these systems are not without problems. In addition to their great cost, they must be built near water in hilly terrain. Because some electricity is wasted in transmission, they must be relatively close to the power plants that replenish them each night. And their large reservoirs sometimes raise environmental problems. A New York utility's plan to build a hydrostorage plant on the Hudson River has been blocked in the courts since 1965 by naturalists who fear the plant might spoil the river's beauty, and by commercial coastal fishermen who fear it might kill striped bass, which spawn almost exclusively in the Hudson.

One way to reduce the environmental impact of new hydrostorage plants is to go underground, according to Frank M. Scott of the Harza Engineering Company in Chicago. He and others have suggested exploiting the fact that a small volume of water falling a great distance to turn turbines can generate as much hydroelectric energy as a large amount of water dropping only a short distance. Their plans call for a small surface lake, perhaps one-tenth as large as Ludington's upper reservoir, that would empty into a complex of tunnels thousands of feet, or hundreds of meters, underground. The water could actually drop in two equal stages, first filling a small subterranean reservoir.

Although none have yet been built, Scott believes that underground hydrostorage plants do not present unsolvable engineering problems. Many modern hydroelectric generating plants have their turbine-generators partly or completely underground. Water drops vertically through giant pipes, turns the turbines, then rushes horizontally into the lower reservoir. The world's largest underground hydroelectric power plant is the 5-million-kw Churchill Falls project in Newfoundland, which began operating in 1975. In all, some 2.5 million cubic yards (1.9 million cubic meters) of rock was excavated from a depth of about 1,000 feet (300 meters). By comparison, five times as much rock would have to be lifted from a depth of over 3,000 feet (915 meters) in excavating the lower tunnel complex of the 2-million-kw hydrostorage plant that Scott proposes.

Underground hydrostorage is not the only intriguing scheme the utilities are considering in their search for better energy-storage systems. High on a plateau overlooking the Rio Grande in New Mexico, government researchers at the Los Alamos Scientific Laboratory are perfecting another system. Rather than store energy in water that has been pumped uphill, they plan to store it in the magnetic field of an electromagnet. Current flowing through the coiled wires of an electromagnet creates a field of recoverable electromagnetic energy surrounding the wire loops.

To eliminate large energy losses due to electric resistance in the wire, the Los Alamos researchers plan to wind their electromagnet with wire made of titanium-niobium strands embedded in copper. The strands become superconductors, losing all resistance to the flow of current once they are cooled to temperatures near absolute zero. "With no energy losses in the magnet wires, the current could flow endlessly in circles until needed," says William V. Hassenzahl, a Los Alamos project leader. "Because some energy is used in refrigerating the magnet and inserting and withdrawing the stored energy, a superconducting magnet system would be about 90 per cent efficient."

Hassenzahl's research team and another group led by Roger W. Boom at the University of Wisconsin in Madison found that the energy field of a large superconducting magnet would exert virtually uncontainable forces on the wire and its housing. "Storing energy in a magnetic field is like storing energy in compressed air in a barrel," Hassenzahl explains. "Forcing more air into the barrel not only increases its ability to spin a turbine when released, it also increases the pressure on the container. Eventually, there is too much pressure, and the barrel bursts. Likewise, forcing more current through the magnet windings creates more pressure on them."

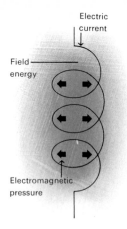

Electric
current

Field
energy

Electromagnetic
pressure

Electromagnetic Storage
Current flowing through
a wire coil creates a
field of recoverable
electromagnetic energy.
The field's pressure
pushes the loops out.
This creates a problem
of containment in an
energy storage system.

A cylindrical magnet large enough to store as much energy as the Ludington system would be about 130 feet (40 meters) high and about 330 feet (100 meters) in diameter. To make the magnet, about 750 miles (1,200 kilometers) of superconducting wire would be wound around the inner circumference of a giant cylindrical container. With 50,000 amperes of current flowing through the wire, the magnetic field at the coil's center would be nearly 100,000 times stronger than the earth's magnetic field. More than 624 million pounds (283 million kilograms) of reinforcing material would be needed to withstand the magnetic pressure. A stainless steel cylindrical reinforcement container strong enough to do this, for example, would cost over $1 billion, more than twice the cost of the complete Ludington system.

To avoid such astronomical costs, the designers propose burying the magnet in a cylindrical tunnel dug in bedrock, perhaps several hundred feet or meters underground. The superconducting magnet's immense stresses would be transmitted via refrigerated, fiberglass-reinforced, plastic bumpers from the insulating thermos container surrounding the wire to the rock surface of the outer tunnel wall. This rock would have to withstand a pressure of about 100 times normal atmospheric pressure when the magnet was fully energized. Surveys

Physicists test a coil
made of a continuous
superconductive wire.
When cooled in liquid
helium, the wire has no
resistance to the flow
of electric current.
Once started, current
flows undiminished
until the energy in
its field is tapped by
withdrawing the current.

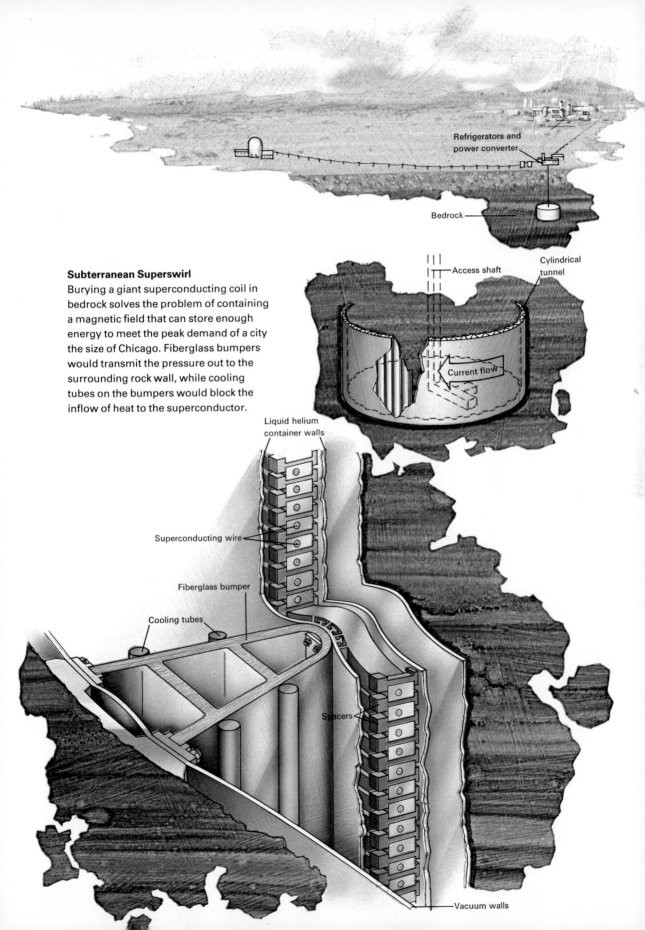

Refrigerators and
power converter

Bedrock

Subterranean Superswirl
Burying a giant superconducting coil in
bedrock solves the problem of containing
a magnetic field that can store enough
energy to meet the peak demand of a city
the size of Chicago. Fiberglass bumpers
would transmit the pressure out to the
surrounding rock wall, while cooling
tubes on the bumpers would block the
inflow of heat to the superconductor.

Access shaft

Cylindrical
tunnel

Current flow

Liquid helium
container walls

Superconducting wire

Fiberglass bumper

Cooling tubes

Spacers

Vacuum walls

Spinning Brushes

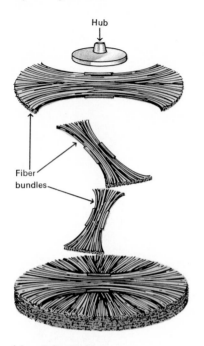

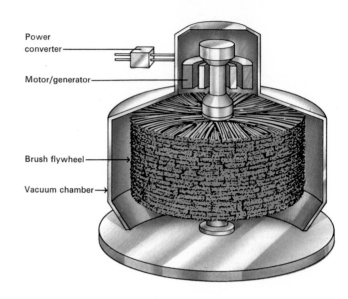

A brush flywheel, made by clamping and stacking bundles of strong fibers, could be revved up to high speed in a vacuum chamber by an electric motor when power demand is low. When demand rises, the motor can be used as a generator, providing electric power while the flywheel spins down.

show that bedrock formations dry and strong enough to house giant superconducting magnets exist within 100 miles (161 kilometers) of nearly every urban area except those in the Gulf Coast region.

The Los Alamos team plans to build a $3-million test system above-ground from 1977 to 1979 to prove the worth of their concepts. "We can then test the refrigeration and the electronic circuits that insert and withdraw energy—every concept except the underground bedrock containment," Hassenzahl explains. The Los Alamos test magnet will be about 5 feet (1.5 meters) high and 16 feet (5 meters) in diameter—large enough to provide a few minutes of power. A superconducting system that could store as much energy as the Ludington system in one-hundredth the space might be possible by the mid-1990s.

Both superconducting magnet and hydrostorage systems operate most economically when built in giant sizes. Spinning flywheels, however, may provide storage on both a large and small scale. A flywheel energy-storage system contains a motor-generator that is connected to a flywheel. The motor sets the flywheel spinning in a partly evacuated container that reduces the drag of air friction. When energy is needed later, the motor is used as a generator and the flywheel spins down. Flywheel systems, like superconducting magnet systems, are expected to be about 90 per cent efficient.

Flywheel systems that use steel flywheels are not new. For example, many physics laboratories use them to provide pulses of power for fusion energy experiments. A research group headed by engineer William F. Weldon at the University of Texas in Austin built and tested a system during 1974 and 1975 in which the electric motor's rotor is itself the flywheel. The system delivers a 1,000-kw burst of electric power for 3 seconds as its 1,600-pound (725-kilogram) steel rotor slows to a stop from 5,600 revolutions per minute.

Delivering 2 million kw of electric power for nine hours like the Ludington system, however, may well be beyond the reach of single steel flywheels. A flywheel's energy-storage potential is limited by its weight and the everpresent danger that it could tear itself apart if it spins too fast, ejecting large chunks of flying shrapnel. "For practical reasons, steel flywheels in use today usually operate at less than half their calculated storage capacity," explains David W. Rabenhorst, an aeronautical engineer at Johns Hopkins University's Applied Physics Laboratory (APL) in Silver Spring, Md. "A steel flywheel large enough to store energy from a power-generating plant would operate at only about one-fiftieth its theoretical capacity."

Rabenhorst believes that the future of flywheel energy storage lies in the effective use of lightweight materials that, weight for weight, are much stronger in at least one direction than steel. "High strength-to-weight ratios are typical of such filament materials as fiberglass," Rabenhorst says, "and many dried woods and bamboo are excellent flywheel materials where low cost is important."

The trick to using these materials, however, is to align the filaments with the directions of greatest stress in the spinning flywheel–either radially outward from the center or in concentric rings. Rabenhorst proposes a brush flywheel, a design in which strands of high-strength fiber pass unbroken through a hub and radiate like the spokes of a rimless bicycle wheel. Layers of such strands are stacked and clamped together to form each section of the flywheel. "The brush flywheel would not be used where size is of paramount importance–for example, to replace batteries in electric cars of the future," explains Rabenhorst, "but a flywheel unit that a utility could use would still be only one-thousandth the size of a hydrostorage reservoir of equal capacity." Because flywheel systems are efficient whether large or small, a utility might build 10 smaller flywheel units instead of one giant unit and place them at substations to minimize distribution costs. Even smaller flywheel systems may be used in the home.

Rabenhorst says the brush flywheel is safe: "When it spins too fast, it crumbles into very small pieces." Studies at APL show that most of the stored energy changes into heat when a brush flywheel disintegrates. The studies suggest that a container strong enough to withstand the reduced air pressure inside will be able to absorb the impact force of any particles that break away.

The filamentary flywheel storage concept sounds too good to be true. Because the concept has yet to be proven, the keepers of the research purse strings are cautious. "To be acceptable, filamentary flywheel system costs must match their advocates' most optimistic estimates," EPRI's Kalhammer says. "At present, we do not have

A Featherweight Contender

A technician assembles a lightweight battery, *right,* that uses lithium and sulfur as its active materials, *below,* then attaches leads to test its performance, *below right.* Such batteries may power electric cars.

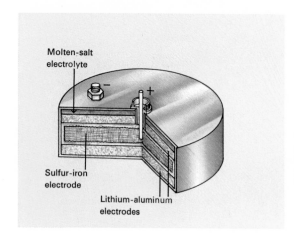

Molten-salt
electrolyte

Sulfur-iron
electrode

Lithium-aluminum
electrodes

enough practical experience with them to know whether these goals can be met." Consequently, EPRI is funding a research program to build and test a flywheel that forms fibers into rings rather than rods. This design can store five times more energy than a brush flywheel in the same space, but it is more difficult to construct.

EPRI is also taking a long look at batteries, particularly lead-acid storage batteries, which were once used in direct-current power systems. Like the familiar automobile batteries, lead-acid storage batteries consist of individual cells in which reversible chemical reactions convert electric energy to chemical energy, and vice versa. Each cell is made up of two electrodes, lead plates in this case. The positive plate is the cell's anode, the negative is its cathode. An electrolyte, a solution of sulfuric acid, provides the charged particles that move from one electrode to the other.

For a utility storage system, lead-acid batteries would be wired together into banks like those used by telephone companies to provide emergency power. Even though such systems would be compact and about 70 per cent efficient, they may be too costly. The problem is that relatively inexpensive lead-acid automobile batteries are too short lived, while such long-lasting batteries as those used in electric golf carts cost too much for utility energy storage.

While battery manufacturers try to combine long life and low cost in lead-acid batteries, EPRI is funding the development of more advanced batteries. "We hope that several of these will cost less and last longer than lead-acid batteries," Kalhammer says, "but we won't know for another five years how good the chances really are."

Many researchers throughout the world are already testing such new types of batteries that use relatively low-cost, lightweight materials as the active cell components. For example, these batteries use aluminum, lithium, sodium, or zinc for the anode, and sulfur or chlorine for the cathode.

The lithium-sulfur battery under development at a number of research centers, including the Argonne National Laboratory (ANL) near Chicago, is a typical example of the promise and problems of these new batteries. Studies by Paul A. Nelson and his team of 50 scientists and technicians suggest that, if mass-produced, its materials would cost only about 60 per cent as much as those used in a lead-acid battery of comparable capacity. Because of the lighter-weight components, it could store six times as much energy as a lead-acid battery of equal weight, making it practical for future electric vehicles.

But the exotic lightweight metals have certain drawbacks. When ANL researchers began their tests in 1971, they found that the use of pure lithium caused short circuits and severe corrosion. So they substituted a less reactive, porous anode made from an alloy of lithium and aluminum. They also switched from pure sulfur to an iron-sulfur compound for the cathode. These changes solved their problems, but the new cells are heavier.

The cell's electrolyte is a mixture of two salts, lithium chloride and potassium chloride. It must be heated to at least 665°F. (352°C) in order to melt and ooze into the pores of the electrodes. The cell itself is heated to above 750°F. (400°C). Both a blessing and a curse, the high operating temperature makes it easier to remove heat generated by the chemical reactions but it also adds an insulation problem. Despite this, ANL's new test cells appear to meet all of a utility's energy-storage needs except long life, which has become the focus of research.

With so many competing battery systems being developed, ANL, EPRI, and others are designing a facility for testing them in an actual utility network. The facility is to be built by 1978 at a site not yet determined. It will first test lead-acid batteries with a total power capacity of about 1,000 kw. Nelson estimates that lithium-sulfur batteries will be ready for testing by 1980. Other advanced-battery systems could be installed and tested at about the same time.

Energy-storage technology today is largely a mix of these and other imaginative but unproven schemes. But someday, when you switch on an air conditioner or an electric heater, a giant utility company flywheel may slow down imperceptibly in response or water may spill into a deep underground cavern as energy generated and stored the night before goes to work for you.

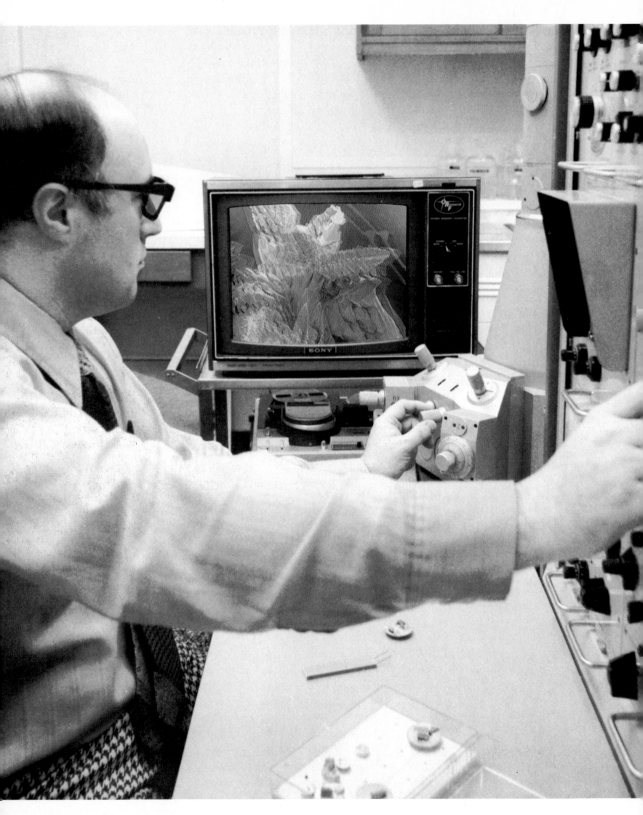

A New Dimension in The Microscope

By Eric J. Chatfield

A modified scanning electron microscope gives us an instant 3-D view of the world of the supersmall

A jagged ledge and precipitous drop into a valley below interrupted the otherwise uniform topography of ridges and fissures. The precipice, although only 10 thousandths of an inch deep, was a very significant find, because it was in the "landscape" of a shattered turbine blade taken from a turbojet engine that had failed. With any conventional microscope, the flaw would almost certainly have gone unnoticed. But I could see in three dimensions (3-D) with the aid of a new device attached to a scanning electron microscope (SEM).

As Head of the Electron Optical Laboratory at the Ontario Research Foundation, in Mississauga, Ont., I began the research in 1973 that produced the new device. Several techniques already existed that could produce a 3-D picture from a microscope. But none allowed real-time viewing in 3-D–that is, they all required photographs to be taken and developed to produce the 3-D effect. And the microscopist, seeing the specimen in only two dimensions as he scanned it, had to guess which features were worth photographing for 3-D viewing. The results were much wasted time, film, and money. Worse yet, important features were often missed completely.

The device I had in mind would display in 3-D, on a TV screen, exactly what the microscopist saw, at the moment he saw it. I was

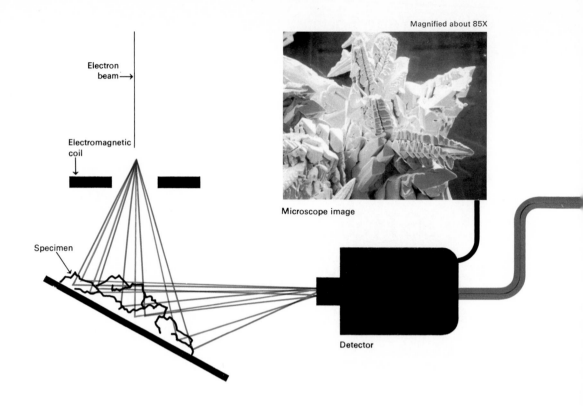

Magnified about 85X

Electron beam→

Electromagnetic coil

Specimen

Microscope image

Detector

joined in the effort to create such a device by Verner H. Nielsen, an electronics engineer, and Jonathan More, who was completing his first year in engineering physics at the University of Toronto.

We were, of course, not the first to seek a better view of the microscopic world. Man has long explored methods of magnifying objects in order to learn more about them. He did not progress very far until he discovered some of the properties of glass lenses in the late 16th century. This led ultimately to modern optical, or light, microscopes, most of which are limited by certain properties of light to a magnification of about 2,000 times.

These limitations, along with several discoveries about the properties of electrons and how to manipulate them, set the stage for the development of the electron microscope in the 1930s. The first of these devices, the transmission electron microscope, passes a beam of electrons through the specimen. This scatters the electrons, some of which are then focused into a clear image on a fluorescent screen by a series of magnetic lenses. Modern electron microscopes produce useful magnifications of several million times. But, because the electrons must pass through it, the specimen must be very thin, thinner than some specimens can be cut for useful viewing. This was one of the factors that led scientists to develop the SEM, which uses its electron beam quite differently from the transmission electron microscope.

The SEM's electron beam quickly scans repeatedly across the specimen's surface, causing it to emit secondary electrons. The secondary

The author:
Eric J. Chatfield, who heads the Electron Optical Laboratory at the Ontario Research Foundation, led the team that developed the new device that creates 3-D images with the scanning electron microscope (SEM).

Uncorrected Corrected

Uncorrected Corrected

160X

3-D TV image

The angle of the SEM electron beam is repeatedly switched by an electromagnetic coil, as the beam scans the specimen. This creates two slightly different views that become a green image and a red image. Corrected individually and merged, these become a single 3-D TV image, which gives more detailed spatial information than does a conventional, two-dimensional, image.

electrons are detected and translated into a signal that is amplified and creates a picture of the specimen on a TV screen.

Although it can magnify only up to 20,000 times, the SEM allows a scientist to view a far thicker specimen than can be viewed with a transmission electron microscope. It also provides a well-focused picture of details over a great range of depth. This depth of focus turns out to be a mixed blessing. With all, or nearly all, features in the image in focus regardless of their relative depths, it is difficult to judge or even detect depth variations. Gouged or raised areas usually cannot be deduced by looking at the two-dimensional image that the SEM provides. A scientist wanting a 3-D view usually takes two photographs of the image, tilting the specimen slightly between the exposures. He then views the two photographs simultaneously, using a device that presents the view from the first angle to one eye and the view from the second angle to the other eye. This technique simulates binocular vision and produces a 3-D picture.

We knew in 1973 that it was impractical to try to produce a real-time, 3-D image by rapidly tilting the specimen back and forth as it was examined. Such energetic rocking would present problems for both the specimen and the SEM. So we tried to find a way to tilt the electron beam instead. We finally did it by adding an electromagnetic coil that deflects the electron beam after it passes through the final lens of the microscope. As the device now functions, the beam is deflected in one direction as it scans the specimen, producing a red image on the

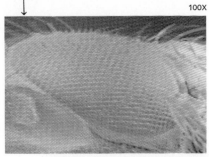

100X

Zooming in on a fruit fly's eye to get the best 3-D view of the tiny hairs that project between its lenses takes only a few seconds. The old method might have required several time-consuming steps. The function of the hairs is still being studied.

130X

2,000X

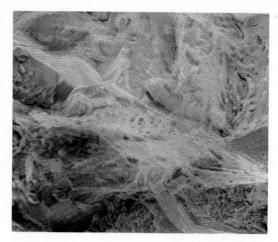

350X

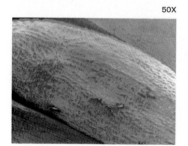

50X

screen with the aid of the color-TV set's "red gun." Deflected in a second direction, it produces a green image with the aid of the "green gun." Each image is focused and corrected for distortion independently. Then the two are alternated on the screen every $\frac{1}{60}$ of a second. Persistence of vision, the same phenomenon that allows us to see separate frames of a motion picture as a continuous scene, enables us to see both images at the same time. The 3-D view is achieved by viewing the screen with glasses that have a green lens, which filters out the red image for the right eye, and a red lens, which filters out the green image, for the left.

The 3-D device is particularly useful in examining metal surfaces for structural defects. But its promise may be greatest in biology, where it offers the opportunity of performing very precise microdissection and manipulation of specimens. With both specimen and tool seen in 3-D, a scientist can more accurately judge the tool's distance from the specimen he is working on.

The 3-D TV image can be recorded for later reference or display in several ways. A videotape can be made and played back through a color-TV set. Also, photographs taken directly from the TV screen can be viewed in 3-D and even projected for 3-D viewing in an auditorium or theater. And, of course, they can be printed on a page.

The standard SEM view of the sheared surface of a broken stainless steel screw, *top left,* gives little information on depth and distance. The same view in 3-D, *top right,* gives so much information that most people must concentrate for a long time in order to see it clearly with the special glasses. A continuous 3-D view of a seed, *above,* as it is squeezed, cracked, and broken can give structural details that are difficult to obtain by other methods.

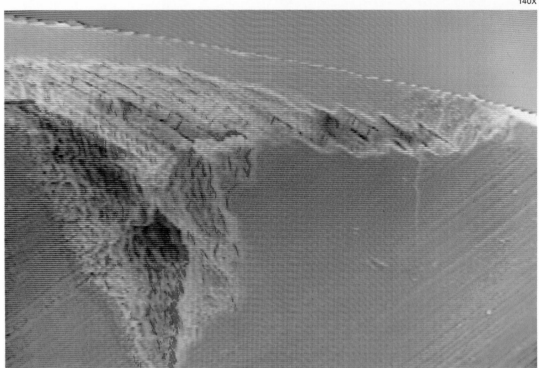

A detailed cross section of a tooth with a cavity that began on the right, then meandered left to where it really dug in, is an excellent example of how useful the 3-D device can be in medical research. A conventional SEM view of the same tooth appears on page 297 in the "Medicine (Dentistry)" article.

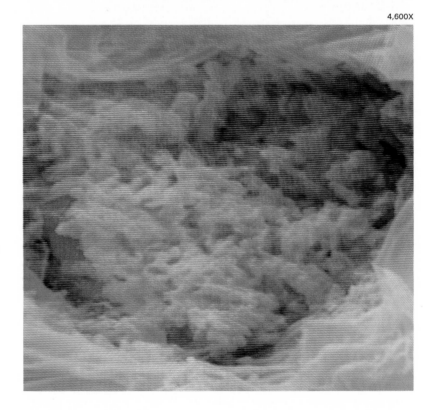

Nitrogen-fixing bacteria glut a soybean root cell, part of the nodule that such bacteria form. A conventional SEM view of the same cell appears on page 98 in "The Nitrogen Fix."

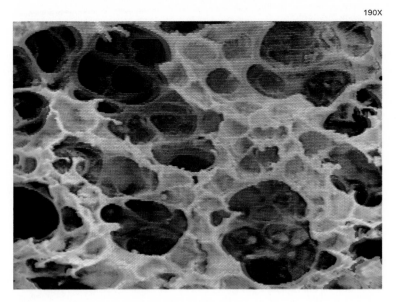

A healthy infant's lung tissue, *above,* has a lacy, filigreed texture created by myriads of tiny air passages. The dense lung tissue of an infant with hyaline membrane disease, *below,* resembles a mouse-gnawed loaf of bread, because only a few air passages are open. Conventional SEM views of the tissues appear on page 45 in "The Breath of Life."

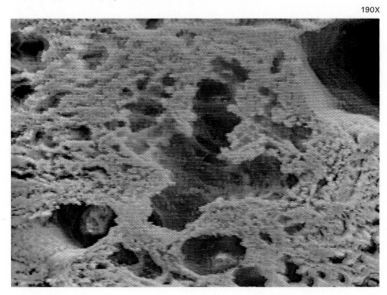

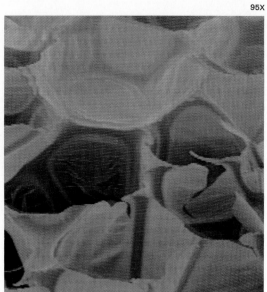

A cross section of cork, *above left,* compared in 3-D with cross sections of rigid plastic foam, *above right,* and soft plastic foam, *below,* reveals the typical way in which man's synthetic materials mimic those of nature. Such detailed views of materials can aid manufacturers in determining how products will perform under certain conditions or why they have failed in some cases.

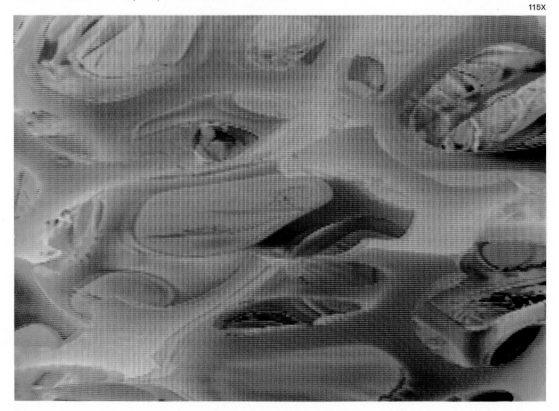

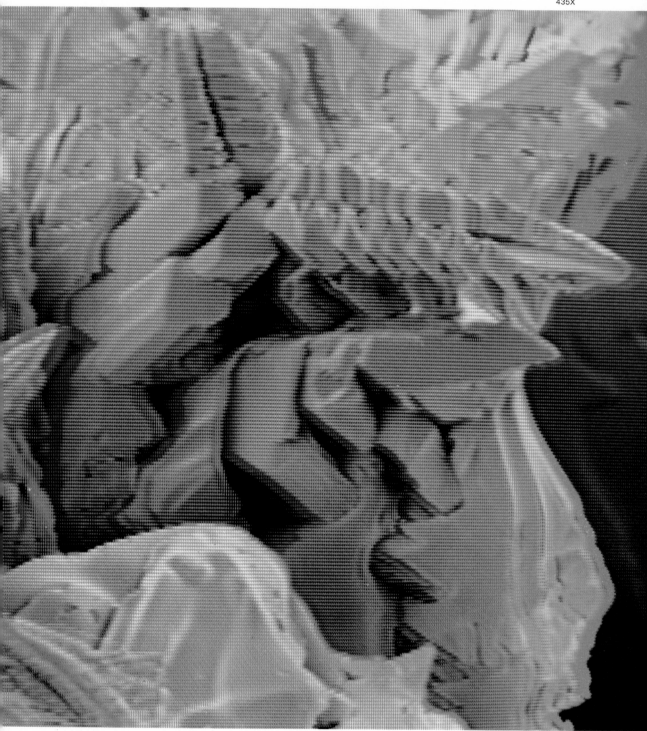

A bizarre but strangely beautiful landscape is revealed on the surface
of a burned tungsten filament. The discovery of unexpected beauty is
a frequent bonus for a researcher who works with the new instrument.

Journey to The Birthplace Of Continents

By Alexander Dorozynski

Scientists descending to the Atlantic Ocean floor saw for the first time the unhealed scar through which the earth's crust is constantly renewed

On a warm July day in 1974, in the middle of the Atlantic Ocean, an hourlong countdown approaches zero. The hatch of the bright-yellow underwater craft, *Cyana*, is tightly bolted down, sealing three men inside. As the saucer-shaped capsule gently rolls with the ocean swells, two frogmen pop to the surface nearby, having freed it from the buoys that keep it afloat. Weighted with ballast of iron pellets, the submersible tips down at a sharp angle and starts toward the bottom of the ocean, spinning slowly, like an autumn leaf in motionless air.

The crew on the surface-support ship watches the vessel sink silently into darkness. From now on, they can follow its journey only by watching a squiggly line drawn on paper by a computer-guided pen.

For the pilot, navigator, and scientist in *Cyana*, a voyage to a new world has begun. They are enclosed in a steel sphere only 2 meters (6.5 feet) in diameter that is designed to withstand pressure 300 times that on the surface and protect them from the cold underwater world. Turning on the sub's powerful floodlights, the men peer through two portholes. They can see a kind of snowfall in reverse, as pale fish and other marine life appear to tumble upward.

For an hour and a half, the *Cyana*, designed by French oceanographer Jacques-Yves Cousteau, spins down at the rate of about two turns a minute. However, the aquanauts have no sense of motion at all unless they look through a porthole.

The pilot watches a sonar device that bounces sound waves off the ocean floor about 3,000 meters (9,800 feet) below the surface. As the *Cyana* nears the bottom, he releases some ballast, and the vessel levels off and begins to maneuver under the power of its own engines. At this depth, the sub's floodlight beams can penetrate from 15 to 20 meters (50 to 65 feet) into the darkness. And as the *Cyana* slowly moves ahead, strange, never-before-seen shapes appear.

The crew spots an object that looks like a huge broken egg, the outpouring of its yolk frozen in time and motion. Thousands of years ago, a "pillow" of lava probably pushed up from deep inside the earth, and when it contacted the cold water, a shell quickly solidified around its molten interior. Then the shell broke, spilling its contents. The crew also sees deep fissures in the sea floor, slabs of glassy crust, rocky mushroom caps pushed up on stems of lava, and large lava formations that look like tubes, sheets, ropes, cones, bundles of yarn, and squeezed-out toothpaste.

This undersea journey is no mere visit to a circus of strange shapes. It and more than 40 other dives made during the summer of 1974 were the climax of a three-year oceanographic project–the French-American Mid-Ocean Undersea Study (FAMOUS). The dives were made by three research submersibles, one American and two French, in an area of the Atlantic Ocean about 124 kilometers (200 miles) southwest of the Azores. The U.S. submersible *Alvin* was tended by the catamaran *Lulu* and accompanied by the *Knorr*, a laboratory ship of the Woods Hole (Mass.) Oceanographic Institution. The *Cyana* was supported by the surface ship *Le Noroit*; the French bathyscaph *Archimède*, by the *Marcel le Bihan*.

The Woods Hole Oceanographic Institution organized U.S. participation under the direction of oceanographer James R. Heirtzler. In France, the project was supported by the French National Center for the Exploitation of the Oceans (CNEXO). Director of CNEXO's Man in the Sea program, Claude Riffaud, and oceanographer Xavier Le Pichon supervised the French team.

The goals of FAMOUS were to examine the place where the earth's crust is formed and to verify the theory of continental drift–the belief that the American continents are drifting farther apart from Europe and Africa. According to a theory called plate tectonics, the continents drift apart because of the motion of about 20 gigantic plates that form the earth's crust. These plates float on a layer of partially molten rock. Where two plates meet, one of four things happens: They may slide past each other; one may slide under the other and be pushed down to become molten rock; they may collide, pushing up rock to form mountains; or they may separate and drift apart. Where two plates separate, lava wells up, forming new crust and volcanic mountains.

The North American and African plates separate at a point along the Mid-Atlantic Ridge, a narrow underwater mountain range. A valley, or rift, runs down the center of this mountain range, and on

The author:
Alexander Dorozynski
is a free-lance science
writer based in France.

FAMOUS dive area

Azores

Rift valley

Mid-Atlantic Ridge

Where Continents Part
A portion of the rift valley that cuts down the center of the Mid-Atlantic Ridge was the site of the FAMOUS exploration.

either side, cutting through the valley at right angles, are deep canyons. The center of the rift valley has little or no sediment deposits, an indication that the ocean floor in this area is made of newly formed rock. And so the Project FAMOUS scientists chose to explore the small section of the ridge where the two plates meet.

Few expeditions have been prepared with such care and thoroughness. Beginning in 1971, oceanographers made detailed relief maps of the target zone by bouncing sound beams off the ocean floor. The results showed that the bottom of the rift valley is less than 3 kilometers (2 miles) wide and about 2,800 meters (9,000 feet) deep.

In the summer of 1972, the French and U.S. teams began making test dives, checking each submersible's fail-safe security system. The *Archimède* carried five tons of ballast that could be instantly released in case of trouble, sending the craft quickly to the surface. In the *Cyana*, the pilot could jettison the submersible's battery, weighing more than a ton, and other heavy equipment. The *Alvin* could shed part of its hull. According to Le Pichon, the only risk might be in getting caught under an overhanging cliff. Even then, the submersibles could help each other. Each of the French submersibles simulated a breakdown at the ocean bottom. Then the other hooked a steel cable to the "crippled" vessel, so one of the supporting ships on the surface could pull it free of the simulated cliff.

In August, 1973, the *Archimède* began preliminary exploration of the Mid-Atlantic Ridge, including tests of equipment to be sure that scientists in submersibles could make geological maps and take rock and water samples at such great depths. On one preliminary dive, the *Archimède*'s fail-safe system was put to a real test. After more than three hours on the ocean bottom, a short circuit caused an electric failure. The magnetic doors of the safety-ballast compartment automatically opened, dumping out the metal pellets. The bathyscaph soared to the surface, while smoke from burned-out wires filled the cabin. The crew had to use emergency breathing equipment, but they were safe.

On its preliminary dives, the *Archimède* explored Mount Venus, a hill rising 250 meters (820 feet) above the very center of the rift valley floor. Mount Venus was partly blanketed by a thin layer of sediment, like a hill covered with light snow. Cylinders of lava were scattered over its slopes, along with fresh, unbroken glassy crust. Analysis indicated that rock samples collected on Mount Venus were less than 10,000 years old. This means that it is the youngest part of the rift valley floor. So it seems that new crust was formed at the very center of the rift valley by recent upwellings of lava.

Surface support ships *Le Noroit,* with the *Cyana* hoisted aboard, and *Marcel le Bihan* dock at Ponta Delgada on São Miguel Island in the Azores, about 200 miles from the dive site.

The American submersible *Alvin* returns to its support ship, the catamaran *Lulu, above,* after a dive to the rift valley. *Le Noroit* crewmen lower the French diving saucer *Cyana* into the ocean, *below.* Frogmen in rubber rafts help prepare the French submersible, *Archimède, right,* for one of its dives to the bottom of the Atlantic.

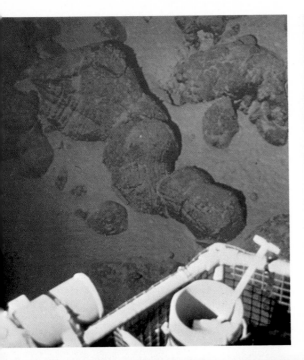

Scientists inside the *Alvin, opposite page,* keep a close watch on the vessel's instruments as it slowly descends to the Atlantic Ocean floor. Through portholes, the scientists observe strange lava formations that resemble squeezed-out toothpaste, *top left,* and fat pillows, *above.* The *Alvin*'s mechanical arm, *top right,* grasps a rock sample. The *Cyana*'s arm samples a hydrothermal mineral deposit, *above right.*

In June, 1974, with all preliminary work completed, French and U.S. submersibles and support ships sailed into Ponta Delgada on São Miguel Island in the Azores. More than 200 scientists and crewmen took part in the operation. The French and U.S. scientific teams each consisted of about a dozen geologists, geochemists, geophysicists, rock specialists, and volcanologists.

Each submersible was assigned a particular area to explore. The *Archimède*, capable of diving 11,000 meters (36,000 feet)—as deep as the ocean floor anywhere in the world—explored the bottom of the rift valley and mapped the deepest part of an intersection between it and a canyon perpendicular to the valley wall. *Cyana* and *Alvin*, smaller and more maneuverable than the *Archimède*, did some underwater mountain climbing. The *Cyana* explored the canyon; the *Alvin* inspected the floor and walls of the rift valley.

Each submersible was equipped with a mechanical arm, drills, and hammers that the aquanauts manipulated from inside to chip off and pick up rock samples. Occasionally, the submersible crews guided the surface ships to good areas for rock collecting. Dredging equipment aboard the surface ships was then lowered into the area to bring up great quantities of rock. The rocks were all carefully labeled so that scientists could study them later.

Cyana successfully completed 14 dives; *Alvin*, 17; and *Archimède*, 12. FAMOUS collected tons of rock samples and took thousands of photographs and motion pictures. Throughout the joint operation, French and U.S. scientists exchanged data, either immediately after the dives by radiotelephone, or during informal meetings aboard one of the ships. Scientists of both teams often used the laboratory facilities aboard the *Knorr*.

Although the FAMOUS scientists made no attempt to collect living creatures from the bottom, they were struck with the wealth of marine life, even in the deepest parts of the rift valley. They observed fish of all sizes, some more than 1 meter (3.3 feet) long. Apparently blind white fish live on the bottom, their vestigial eyes unperturbed by the powerful floodlights. They also saw tiny crabs, octopuses, starfish, sea stars, shrimps, sea anemones, sponges, and coral. There were even some brightly colored fish, all dressed up but with no one to see them in the darkness of the deep.

On August 6, the *Alvin* completed its part of the mission and was hoisted aboard the *Lulu* for the return trip to the United States. The *Cyana*, nested aboard *Le Noroit*, left only a few days later for its home port, Toulon, France. The *Archimède* stayed on awhile to complete its assignment; it almost did not.

On one of its last dives, *Archimède*'s pilot, French Navy Lieutenant Gilbert Harismendy, opened the lock that lets water into the hull, and the *Archimède* started to descend. Little more than an hour later, it touched bottom. The landing was a bit rough, and a cloud of sediment arose, obstructing the aquanauts' view.

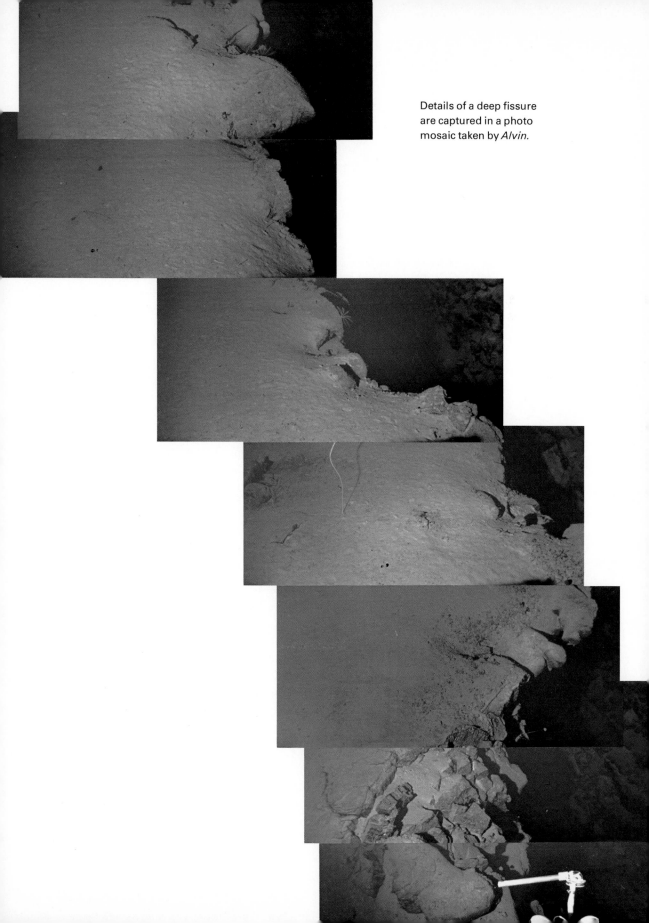

Details of a deep fissure
are captured in a photo
mosaic taken by *Alvin*.

Finally, some of the sediment cleared and geologist Jean Francheteau examined the ocean bottom, illuminated by twelve 750-watt floodlights. "There's not a single rock," he commented. "It's the Sahara." The front, rear, and left side of the *Archimède* were buried in sediment. The landing had caused a small landslide that partly covered the submersible.

Lieutenant Harismendy switched the main 20-horsepower engine to full speed ahead, but the bathyscaph did not budge. Engineer Jean-Louis Michel tried to move the submersible with the remote-control steel arm used to collect samples, but it simply sank into the sediment. The men were not overly worried, however. They knew that by releasing ballast they could probably free the submersible and send it floating up to the surface. But this would mean aborting the dive.

Lieutenant Harismendy changed his tactics. He began running two 5-horsepower engines that operated a horizontal and a vertical propeller, trying to rock the submersible loose—full speed right, full speed left, full speed up. Nearly two hours later, the *Archimède* rocked itself free, continued its mission, and then returned to the surface 13 hours after submerging.

Two days later, the *Archimède* was stuck again—between two rocks in the wall of the rift valley. The engines did not have enough power to move it. But this time, the crew was able to brace the steel collecting arm against solid rock and push the submersible free.

The *Archimède*'s last dive on August 31 ended the FAMOUS expedition. But the research work had just begun. Scientists are now studying the tons of rock samples, hundreds of reels of movie film, and thousands of photographs.

"My impression of the sea floor," says Heirtzler, who was on two dives, "is that a giant has been walking around crushing things. It's obviously a horrendous place to be, extremely busy with earthquakes and volcanic eruptions."

Heirtzler believes one of the most significant finds of Project FAMOUS was evidence that new material from deep inside the earth is injected by volcanic action along the center of the rift valley. "It isn't injected everywhere at the same rate along that line," he says. "There are hills, such as Mount Venus and another, Mount Jupiter, that seem to be formed of this new material. Maybe a few thousand years from now, the material will be injected between the hills. But it is almost always injected exactly along the center line, which is a line no wider than 100 meters [328 feet]."

The scientists also found fissures cutting across the valley floor, slicing sharply through small rocks and widening near the valley walls. And they are deep. "We could see as much as 24 meters [80 feet] down into one fissure," recalls Woods Hole geologist Robert D. Ballard.

Heirtzler believes the depth of these fissures and the way they open out strongly suggest that there is a moving plate beneath, pulling and cracking a thin crust as it moves, ripping the sea floor apart. "This

U.S. project head James R. Heirtzler, aboard the *Lulu, above,* checks film taken on an *Alvin* dive. Crewmen on the research ship *Knorr* lower a wide-angle camera to photograph the ocean bottom, *below,* while others man a drill, *right,* to take cores from the ocean floor.

crust," says Heirtzler, "probably isn't more than a couple of hundred feet thick. So we felt we were really very close to the moving plate that is separating the continents."

"Imagine you are pulling at both sides of a soft toffee," says French geologist Pierre Choukroune. "At the place where it will break, it begins to get thin, creating a central depression. This is roughly what happens in the rift valley." The lava below finds weak spots and fissures in the crust to seep through.

Another significant finding was made by the French scientists. They located hydrothermal deposits of metals boiled out of newly formed rock several kilometers below the ocean floor by water that seeped down through fissures. The metals are brought up by hot geyserlike spouts and sprinkled around the ocean floor. Choukroune discovered the first deposit in a canyon and brought back a sample. Scientists aboard the *Knorr* reported that it contained large concentrations of manganese and iron oxides. On other dives, the FAMOUS scientists found more sites believed to contain metals.

Geologists think that many of the world's mineral deposits may have been created by these metal-laden hot springs. Until now, such deposits had been found only in the Red Sea. Geologists also believe that the vast copper, silver, and tin deposits in Chile, Bolivia, and Peru may have been produced by a similar system of underwater geysers along the Pacific Ridge. The deposits slowly migrated inland with the movement of the area's plates.

But perhaps the most immediately significant result of the FAMOUS expedition is the development of a new technology. "People will be talking in terms of 'before FAMOUS' and 'after FAMOUS,' " says Le Pichon. "We did on the ocean bottom essentially what geologists and geophysicists have been doing on land."

Le Pichon does not expect a quick payoff in economic terms—only in knowledge. And FAMOUS was a significant step toward such knowledge. Continuing analysis of the rocks may yield more information. And scientists are already talking about another expedition, perhaps in the Pacific Ocean, west of South America, where a different tectonic plate relationship exists. There, in deep underwater trenches, the earth's crust is not created, but melted as one giant plate pushes another back into the earth.

According to Le Pichon, further exploration of these areas is necessary. "Land formations on the continents are caused by something happening in the earth," he says. "If nothing were happening, in 10 to 15 million years the earth would be completely flat and probably entirely covered with water. There would be no life anymore, except perhaps sea life. So I think we must find out about the earth, find out why it works the way it does."

Scientists on the *Lulu* carefully examine, label, and weigh rock samples gathered by the Project FAMOUS submersibles and dredges. Analysis of these rocks may add to our knowledge of how the continents formed.

Earthquake Early Warning

By William J. Cromie

Using monitors that range from seismographs to nervous animals, geoscientists may soon be predicting the storms within the earth's crust

The informal gathering of the Pick and Hammer Club on the evening of Nov. 27, 1974, turned out to be more than just a social event. Meeting in Palo Alto, Calif., these earth scientists, all from the United States Geological Survey's National Center for Earthquake Research at Menlo Park, Calif., listened with growing interest as Malcolm Johnson told them of recent magnetometer and tiltmeter readings near Hollister in west-central California. According to Johnson, the instruments had provided the sort of data one would expect to see before a quake. And he suggested that the quake would occur in "the very near future." No one at the gathering tried to predict exactly when the earthquake might take place. But John H. Healy suggested, mostly in jest, "Maybe tomorrow."

The next afternoon—Thanksgiving Day—the quake occurred, about 10 miles (16 kilometers) north of Hollister. It registered a magnitude of 5.2 on the Richter scale, a moderate tremor that only broke a few

Large earthquakes cause extensive damage to the earth's surface as did the 1971 quake that damaged the Golden State Freeway near Los Angeles.

windows and cracked some dishes. Yet, the quake was important because it was the latest in a series of tremors anticipated by earth scientists, who are beginning to believe that earthquakes can be predicted at least as accurately as the weather.

Earthquakes do not occur randomly. Almost all of them are within well-defined zones of active volcanoes, island chains with adjacent deep-ocean trenches, and submerged ocean ridges. These zones lie at the edges of the 20 slowly moving plates that form the outer crust of the earth. The tremendous forces generated as the plates slowly bump, push, and slide against and over each other produce earthquakes.

Los Angeles lies on the eastern edge of a plate that includes most of the Pacific Ocean floor. San Francisco lies on the western edge of another plate that includes all of North America and part of the western half of the Atlantic Ocean floor. These two plates come together along a system of cracks, or faults, that run from the Gulf of California to the northern part of California. Most California earthquakes occur along these faults.

The largest of the faults in California is the San Andreas Fault, which is 600 miles (970 kilometers) long. Plate movement along this fault in 1906 produced the quake that devastated San Francisco. Scientists fear that the forces built up by the plates have once again stressed the rocks along this fault system enough to cause another catastrophic earthquake in California before the end of the century.

The data that the Mènlo Park scientists analyzed before the Hollister quake included records gathered for about a year from seven sensitive magnetometers in the region. The earth's magnetic field between two of these instruments showed marked changes a few weeks before the earthquake. In one day, it changed as much as it had previously changed in six months, and then gradually returned to its original value in a week. The records of tiltmeters, which detect slight elevations and tilting of the earth's surface, also indicated that subtle changes in topography had begun at about the same time.

Following the Thanksgiving Day earthquake, the scientists examined a third indicator—the record of the velocity of seismic, or shock, waves created by motions deep within the earth produced by smaller nearby quakes and larger distant ones. These, too, showed variations of the type that have preceded earthquakes elsewhere. It was the first time in the United States that data from three different sources were associated with an earthquake prediction.

There are two types of seismic waves—P, or pressure, waves, and S, or shear, waves. P waves, acting like sound waves, compress and expand rock particles as they move away from the quake source. The effect is somewhat like that of a pulse traveling down a soft spring, such as a slinky toy. S waves vibrate the rock particles at right angles to the direction of the pulse, causing the ground to shake from side to side. In a strong earthquake, seismic waves crack buildings and other structures. Since P waves travel about 1.75 times faster than S waves,

The author:
William J. Cromie is Executive Secretary of the Council for the Advancement of Science Writing. He was the author of "Research in Orbit" in *Science Year,* 1975.

Tectonic Plates and Major Earthquakes of 1974

• Quakes stronger than 6 on the Richter scale

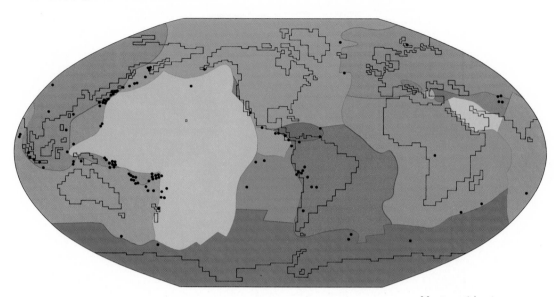

Most earthquakes occur near the edges of the 20 plates that form the crust of the earth.

scientists can pinpoint the origin of an earthquake by comparing the arrival times of waves recorded on three or more seismographs.

Seismologists once believed that P and S waves traveled through each type of rock at a constant speed, as unchanging as the speed of light through air. In the late 1960s, however, Russian scientists observed that the waves slowed down, then returned to normal shortly before earthquakes occurred.

With this in mind, U.S. seismologists began to re-examine their records of P- and S-wave velocities, and they set up experiments to look for such wave-speed variations. Geologists from Columbia University's Lamont-Doherty Geological Observatory in New York City measured the speed of shock waves that originated elsewhere, then passed through the Blue Mountain Lake area in the Adirondack Mountains in New York. The shock waves came from microquakes too weak to be felt without sensitive instruments. They discovered that shock waves did slow down and then return to normal speeds in an area where an earthquake was soon to occur. Using this observation, the Columbia group accurately predicted a small quake in the Adirondacks in September, 1973.

Scientists at the California Institute of Technology (Caltech) in Pasadena studied the records of seismic waves that preceded the San Fernando quake that killed 62 persons in February, 1971. The records were from microquakes, distant larger quakes, and quarry blasts, which generated shock waves picked up by instruments in a Caltech network of seismographs. These revealed that P waves passing through the San Fernando area slowed down significantly in 1967. "Normally, P waves travel through granite at 6.3 kilometers per second [14,000

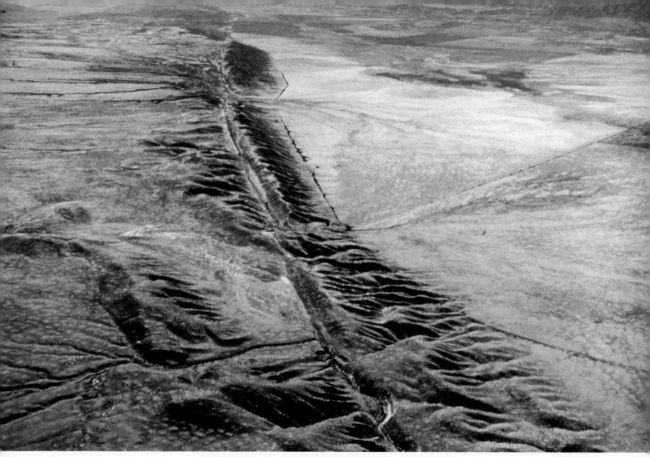

Earthquakes along
the San Andreas Fault,
with surface evidence
visible, *top,* can open
the earth, *above.* This
can devastate areas such
as Selmar, Calif., where
a 1971 quake toppled
a hospital, *right.*

miles per hour], but we discovered a sudden decrease to about 5.8 kilometers per second [12,600 miles per hour]," says Don L. Anderson, Caltech professor of geophysics. "This was followed by a slow return to the normal speed just before the quake hit and did $500-million damage." Had seismologists known then what they know now, they might have predicted the quake, and precautionary measures might have saved lives and property.

James Whitcomb, also a Caltech professor, then started studying the history of P-wave velocities in the area around Riverside, Calif. "We saw that P waves started slowing down in 1972 and stayed that way through most of 1973," he explained. "In November, 1973, the velocity began to increase again and so, in December, we predicted a quake would occur within the next three months. At the end of January, it happened, right in the middle of the area we had outlined." But the prediction was not a complete success. Whitcomb forecast a quake much larger than the one that actually occurred.

How do scientists explain these precursory changes? The dilatancy model, or theory, has offered the best explanation to date. Dilatancy simply means the ability of a substance to expand in volume when its shape or structure is altered under stress. It was first discovered in the late 1800s by Osborne Reynolds, an Irish-British engineer and physicist, who was studying the qualities of granular materials. William F. Brace and his associates at the Massachusetts Institute of Technology (M.I.T.) in Cambridge showed in laboratory studies in the mid-1960s that tiny cracks open in a rock and it swells when forces squeeze it or pull it apart. Brace also showed that dilatancy slows down the seismic waves passing through rocks, just as it presumably does in the earth's crust. Dilatancy also changes the electric resistance of rocks, another effect noted to occur just before an earthquake.

There are two versions of what happens next. Most U.S. experts believe that water seeps into the cracks. This would change electric conductivity and account for the waves' return to normal velocity before a quake. According to this theory, the water also weakens the rocks so that the number of tremors multiplies. Such increases in the number of tremors have been detected and are considered another warning signal.

On the other hand, water plays no part in the theory developed by Russian researchers. They envision an avalanche effect as more and more of the tiny cracks develop at increasing speed. This causes seismic wave speed to drop, electric resistance increases, and magnetism changes. As stress and cracks continue to deform the rocks, more cracks open in the immediate fault area. However, other cracks in the surrounding region partly close, producing the effect of seismic waves returning to normal velocity. In both cases, the rocks finally break, causing the quake.

Not all earth scientists accept the dilatancy theory. Jerry Eaton, a U.S. Geological Survey geophysicist, points out a major problem with it. "Dilatancy should produce a swelling that would show up as a bulge at the earth's surface. Tiltmeter readings that we took before more than 25 earthquakes don't show this bulge," Eaton says. "What may happen instead is that a large zone along a fault becomes strained until the rocks tear. They then slide or 'creep' past each other, without producing quakes. Such displacement would change the velocity of the seismic waves until friction stops the creeping. Then strain builds up against this resistance. Eventually, the rocks pull apart with a sudden release of energy that characterizes a quake."

Although seismologists disagree about what produces these effects, they have used them to predict earthquakes. In China, at least 10 quakes have been publicly predicted. Government representatives used radio broadcasts, posters, and public meetings to urge people to remain outdoors or in tents until after the quakes. American seismologists who visited China in October, 1974, were surprised at the large-scale effort and the progress that Chinese scientists have made in earthquake prediction. Frank Press, chairman of earth and planetary sciences at M.I.T., called the Chinese effort "an enormous, well-supported, high-priority program."

The American group visited the Hsing Feng Kiang Hydroelectric Dam, 150 miles (241 kilometers) northeast of Canton. This dam, 300 feet (91.5 meters) high, has become a focal point for many seismic studies. As soon as the reservoir was filled in 1960, small quakes were felt by farmers in the surrounding countryside. Chinese scientists concluded that a major earthquake was likely, and the dam was reinforced so that it would not collapse. A major quake occurred soon afterward. The dam cracked, but held. Press says it might well have collapsed had it not been reinforced. It has now been even more heavily strengthened.

China's earthquake-observation system now has 10,000 full-time workers and 5,000 observation points. In addition, many thousands of students and farmers and their families check the behavior of animals and the level and cloudiness of well water. Chinese scientists believe that unusual changes in these factors indicate approaching quakes. "Every report we heard of earthquake prediction in China included mention of the unusual behavior of animals," C. Barry Raleigh of the U.S. Geological Survey, who was part of the American group visiting China, noted. "They told of rats leaving houses and barns, snakes coming out of their holes, and dogs barking all day long."

The Chinese also use scintillation counters to monitor the radon content in wells. Radon is a radioactive gas given off by many rocks. The gas seeps into underground water and well water when the strain cracks rocks deep in the earth before an earthquake. In addition, Chinese scientists monitor the more conventional indicators, such as the changes in sound and shock waves as they move through rocks, in electric conductivity, in magnetism, and in the tilt of the ground. They give prediction efforts a high priority because of the tremendous

A Theory About Dilatancy

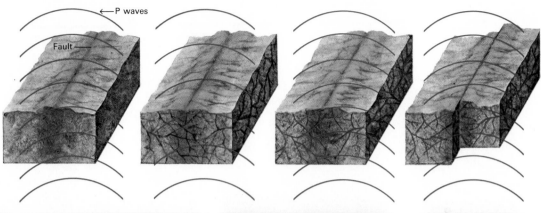

According to one version of the dilatancy theory, *top,* tiny cracks formed by forces that squeeze and stretch rocks slow down seismic P waves until water fills the cracks, allowing the P waves to return to normal speed, a change that occurs just before a quake. The effect has been tested on rocks, using mechanical stress, *left* and *far left.* The cracks in the magnified sample of granite, *above,* came from such tests.

loss of life caused by earthquakes in China. Major shocks in Kansu Province, for example, took about 180,000 lives in 1920 and about 70,000 in 1932. As recently as 1966, a quake at Tsing Tai in Hopeh Province killed thousands of people.

Dense concentrations of people still make such losses a constant possibility. For this reason, the government in heavily populated Japan began financing a major prediction program in 1965. Japanese scientists conduct many extensive surveys to monitor subtle long-term changes in sea level and ground tilting. Prior to a large 1964 quake in Niigata, their tiltmeters measured a rise in the ground of 0.04 inch (1 millimeter) a year. About six years before the quake occurred, this rate began to increase markedly. This led Japanese scientists to believe that future quakes might be predicted by monitoring such changes. Their belief seems to be well founded. For example, in 1966, when Taka Hagiwara of the Earthquake Research Institute at the University of

Tools used by earthquake predictors include magnetometer, *top,* and tiltmeter, *right,* both part of the remote sensing network of instruments for quake detection in California, as well as seismograph, *above,* at a recording station near Hollister.

A tiltmeter of such sensitivity that it can record the shift of a few thousandths of an inch at the top of a building in Las Vegas, caused by two men rhythmically shifting their weight from leg to leg, is used to test for hidden building damage following each big earthquake.

Tokyo noted sudden increases in ground tilting in central Japan, he accurately predicted damaging quakes there.

Russian seismologists have been operating a network of earthquake-monitoring stations since 1949, when landslides triggered by tremors killed 12,000 persons in Khait, Tadzhikistan, near the border with Afghanistan and China. Earthquakes in this area are caused by the plate on which India is located moving against the Asian plate. When the two plates crashed together about 25 million years ago, they created the Himalaya and other mountains to the west. Pressure from this collision has caused the Tien Shan Mountains on China's border and the Pamir Mountains along the border of Afghanistan to move together, with resulting quakes and landslides.

Robert Wesson, a U.S. Geological Survey geophysicist who worked in Tadzhikistan for six months in 1974, described the seismic stations responsible for collecting data on P and S waves there. "The stations are individual houses with instruments nearby. A technician and his or her family live in the house and tend the instruments. Once a week or so, a truck brings supplies and collects records." Analyses of these records led to the Russians observing that seismic waves slow down in the area of an impending tremor.

The Russian, Japanese, and Chinese programs for earthquake prediction are larger and more advanced in many ways than the U.S. program. The major reason for this is that those countries have many more earthquakes than the United States, and they have therefore been willing to pay the high costs required to develop a good earthquake-monitoring program.

In the United States after the great Anchorage, Alaska, quake, which killed 131 persons in March, 1964, a presidential panel urged a 10-year, $220-million, U.S. program of earthquake research with emphasis on prediction. Scientists wanted to place from 1,000 to 1,500 seismographs, tiltmeters, and magnetometers along all of the active faults in California and an equal number along faults in Alaska. These instruments would send readings to a computer center where they could be analyzed. However, this ambitious program was not carried out, largely because of its cost. There are now instruments in the active fault area of California, but far fewer than earthquake scientists believe are needed. At the beginning of 1975, Caltech was operating about 170 instrument stations in southern California, and the U.S. Geological Survey was operating an equal number at various places in central California.

With the present level of U.S. support, several potentially important methods for detecting earthquake warning signals–changes in land level, water-well levels, and radon gas content in wells–cannot be applied on a large enough scale to be of much value. Warning signals such as animal behavior before a quake, considered so important by the Chinese, are not given much credence by most American seismologists. However, Raleigh interviewed several farmers living near Hollister after the Thanksgiving Day earthquake to see if their animals behaved abnormally in any way before that quake. One person reported unusual animal behavior.

"This lady was working with some horses that day," Raleigh says. "She told me that an unusually well-behaved hunter [horse] would not eat the morning of the quake. A colt became so hard to handle that it had to be put back in the pasture. Other horses ran around erratically, and one slipped and fell."

The only way that scientists can explain such behavior, if it ever is linked to upcoming quakes, is to assume that animals perceive something human senses cannot detect. The most likely phenomena are foreshocks, ground tremors too slight to be noticed by humans. Animals may also hear high-frequency sounds produced by rocks cracking before they break. Some animals may also be able to perceive and interpret changes in the earth's electric or magnetic field.

Most seismologists agree that a major shock will probably occur somewhere in California sometime before the end of the century. The President's Office of Science and Technology issued a report in 1968 that predicted such a disaster would probably kill hundreds, perhaps thousands, of people and cause many billions of dollars worth of prop-

Sensitive seismometer, *top,* near the crest of volcanic Mount Asama, is checked for a record of earth tremors. Such data enables Japanese scientists to predict not only earthquakes but also the eruption of volcanoes. Water-tube tiltmeter is held in place by concrete blocks in a long tunnel at the Matsushiro Laboratory near Tokyo, *below.* It is also part of Japan's quake-warning system.

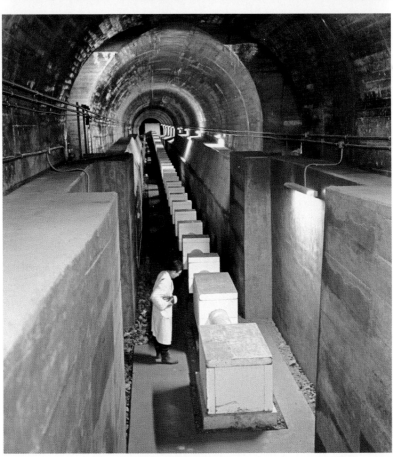

Two Russian earthquake researchers, *above,* test the seismic resistance of a silicate block to determine its usefulness as a building material in quake-prone areas. Chinese seismologists in Yunnan, *right,* measure the radon content of well water, which may foretell an earthquake.

erty damage. Geologist William T. Pecora, who headed the panel that prepared the report, told a Senate subcommittee that a great California earthquake would "certainly occur in the next 30 years and probably within the next decade."

According to a formula calculated by Caltech's Whitcomb, Anderson, and Jan D. Garmany, the earlier abnormal instrument readings occur and the wider the area in which they register, the larger the quake will be. The moderate tremor that struck near Hollister on Thanksgiving Day exhibited premonitory effects three months in advance. Going back over records, scientists have discovered warning changes that occurred three years before the 1971 San Fernando earthquake, which was more than 10 times as powerful as the Hollister tremor. According to the formula, shocks the size of the one that devastated San Francisco in 1906, about 100 times more powerful than the San Fernando tremor, should begin to produce warnings 10 or more years in advance.

Ten-year warnings are not precise enough to be very meaningful to the public. "Time is the parameter we have the poorest grasp on," Anderson admits. However, he and others believe the problem can be solved. "Once you isolate an area where anomalies occur, you can start moving in with more instruments and narrowing down the time." But Anderson adds that more funds are needed. During 1975, the federal government spent $12 million on reducing earthquake hazards, of which $3 million was directed specifically to prediction research. "We need $3 for every $1 now being spent, or we set back the day when prediction becomes a reality." Press and Clarence R. Allen, Caltech professor of geophysics who heads a National Academy of Sciences subcommittee studying the status of earthquake prediction, believe such a reliable capability "is at least 10 years away."

Experts are now wrestling with the problem of what to do in the meantime if warning signs appear but scientists are not sure a quake will occur. "On the basis of instrument readings, it is possible to prepare for a quake without alarming the public, or risking indifference generated by too many false alarms," Anderson says. "We can be sure fire, police, and other emergency services are readied, schools warned, hospitals and ambulances prepared."

"Seeing premonitory signals obligates us to say something in public whether we are right or wrong," says Healy. "But an education program is needed so that people will understand that we are not sure of all these premonitory effects." False alarms in such situations could have adverse economic effects and condition residents to disregard subsequent warnings.

"The problem of how one communicates an earthquake prediction and the consequences that flow from such warnings [and possible false alarms] are being studied," Press points out. "Hopefully, progress will be made on the social aspects of prediction as we progress toward a physical solution."

Dealing With Disaster

By Robert W. Kates

Powerful and unpredictable, nature has always been risky, but, in an increasingly crowded world, man must come to terms with her hazards, and his own

"At midnight we heard a great roar growing louder from the south-east. I looked out. It was pitch black, but in the distance I could see a glow. The glow got nearer and bigger and then I realized it was the crest of a great wave," said Kamaluddin Chodhury, who managed to survive the great wave on Nov. 13, 1970. During that night, high tides and a tropical cyclone with winds up to 150 miles per hour (240 kilometers per hour) combined to slam a crushing wall of water as high as a two-story building onto the shores of East Pakistan (now Bangladesh), the most densely populated nation in the world. On the morning after, at least 200,000 people were dead, most of them women, children, and the aged. Unofficial estimates ran as high as 600,000. Crops worth $63 million were destroyed and about 280,000 cattle were killed. It was the worst natural disaster of the 20th century.

Few places were hit harder than the chars, a farming area of sand bars and reclaimed land in the Ganges-Brahmaputra Delta. In the

Mississippi Delta, such land is considered fit only for waterfowl. But crops raised on the chars support a population of up to 1,300 people per square mile (520 per square kilometer).

On a warm, dry, winter day two months after the disaster, I sat in a small hut that served as a tea shop on Char Jabbar, talking with 10 men who survived the cyclone. They had lost a total of 130 relatives. My colleagues, Bangalee geographer M. Aminul Islam of the University of Dacca, and U.S. economist Howard C. Kunreuther of the University of Pennsylvania, and I were piecing together the unnatural history of this natural disaster. We were seeking answers to some pressing questions: Why do people live in areas they know to be hazardous? How do they measure hazard? How do they cope before, during, and after a disastrous event?

Our work on Char Jabbar was part of a 40-year-old quest that began when two geographers, the late Harlan H. Barrows of the University of Chicago and Gilbert F. White, now at the University of Colorado at Boulder, asked these questions about the people who lived in the American Dust Bowl and the flood plains of the Ohio and Mississippi rivers during the 1930s. Today, over 100 economists, engineers, geographers, psychologists, and other scientists in 20 countries, linked through the International Geographical Union, continue the study. We share certain working definitions: There are natural events in the complex cycles of weather and geological change. These events become hazards in the presence of people, and they become disasters when they greatly affect people, causing death, injury, and loss of property. To live with nature, as we all must, is to risk hazard. But the extent of the hazard or disaster is determined only in part by the forces of nature. The actions people take—or do not take—in the face of these natural events are important as well.

Since 1969, we have been making field studies to determine how people adjust to hazards. We have spoken to people threatened by drought in Australia, Brazil, Kenya, Mexico, Nigeria, and Tanzania; floods in Great Britain, India, Malawi, Sri Lanka, and the United States; tropical cyclones in Bangladesh, Puerto Rico, the United States, and the Virgin Islands; volcanic eruptions in Costa Rica and Hawaii; earthquakes in Nicaragua; and erosion, frost, high winds, and heavy snow in the United States.

The frequency of damaging events varied widely from place to place. Residents of Boulder, Colo., reported windstorms exceeding 50 miles per hour (80 kilometers per hour) on an average of three times per year; farmers on the Ganges flood plain in India noted serious floods that caused crop damage and loss of property on the average of once in five years. By contrast, San Francisco, which was almost destroyed by a 1906 earthquake and the fire that it caused, has not had a serious tremor since.

Just as doctors distinguish between chronic and acute illness, we classify those natural events that are potentially hazardous. We class

The author:
Robert W. Kates, professor of geography at Clark University, has been studying how people perceive and respond to natural hazards for many years.

as *intensive* those sudden, unpredictable events that have great but brief
impact in a relatively small area. Intensive hazards include earthquakes, tornadoes, landslides, hailstorms, volcanic eruptions, and ava-
lanches. *Pervasive* hazards are those in which the impact is spread over
a large area and a longer period of time. They develop more gradually
and can be predicted more accurately. Drought, fog, heat waves, and
excessive rainfall or snowfall are pervasive hazards. Floods, high
winds, big sea waves, sandstorms, dust storms, and tropical cyclones
are *intermediate*, with both pervasive and intensive features, depending
on local geographic features as well as on such factors as the location
or stage of the flood or storm.

For each study, we prepare a general description of the land and its
use, the vegetation, and the people and their work. We also examine
the historical record of the hazards in the area and assess their impact
on the economy and social organization. In addition, we conduct and
record detailed interviews with inhabitants of the area. Most of our
talks are with heads of families—businessmen, craftsmen, farmers, fishermen, government workers, laborers, manufacturers, and teachers—
at sites as varied as the lush citrus-growing areas of Florida and the

Tragedy Comes to the Chars

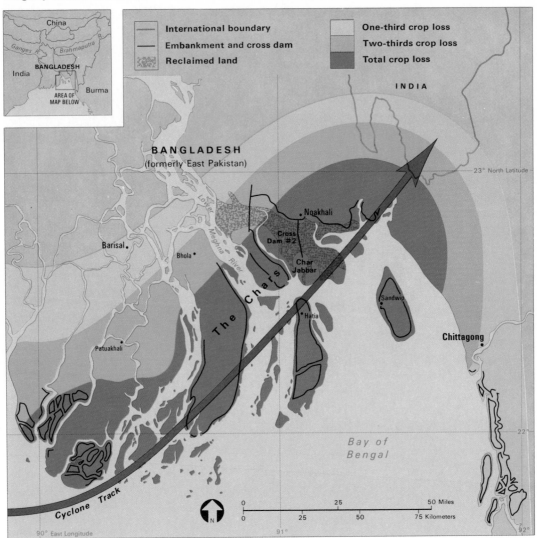

As it moved northeast through the heart of the Chars, the cyclone threw a 25-foot wall of water from the Bay of Bengal onto the reclaimed land.

arid scrublands of South Australia. Educational backgrounds range from almost total illiteracy among herdsmen in northern Nigeria to high school and college levels in the United States. Average income ranges from perhaps $2,000 per person per year in Shrewsbury, England, to less than $200 per year in Sri Lanka. Despite the many differences among the people we interview, we try to standardize the interviews as much as possible so that we can make comparisons across national and cultural lines. Each of the almost 5,000 interviews we have completed so far has added to our knowledge of how people feel about natural hazards and how they cope with them.

Islam, Kunreuther, and I went to the ravaged Char Jabbar because Islam had made a field study there in October, 1970, just before the cyclone struck. A month later, perhaps one-third of the villagers he interviewed were dead. His work remains the only such study of the perceptions, attitudes, and adjustments of people just before a major disaster occurred. We returned to link his findings with information gathered from the survivors.

Our interviews reinforced many of the conclusions we have formed over the past 20 years. Perhaps the most important of these is that human needs and choices have as much to do with causing a major disaster as do natural forces.

The men of Char Jabbar are not innocents. Like others we have talked to all over the world, they understand the hazards of the place they live in, and have done what they can to cope with them. They accept the dangers of cyclones–there had been seven other severe storms in the area in the previous 10 years–because of the benefits of living on the chars. The population of this land of hunger has quadrupled during this century. These men recognize the advantages of the larger farmholdings available on the reclaimed land, and of supplementary fishing in the delta. "It is true many have died," one farmer told us, "but where else than by the sea can we both fish and farm?" The shores of sea, lake, or river have always been good places for human settlement. That is why so many of the world's people live in such places. Shores are boundaries between ecosystems, and provide fish from the sea, rich plant and animal life in the tidal zone and marshlands, and, in the deltas of great rivers, enriched and renewed farming land. People live on the chars of the Ganges-Brahmaputra Delta not out of ignorance and foolhardiness but primarily because they can fashion a better life there.

In every pervasive hazard site we studied, we found that at least 80 per cent of those interviewed consider the hazard a significant one. Even where major disasters have not occurred for many years, 40 per cent of the people rate the potential hazard as significant and few are totally unaware of it. Yet, no more than a few of the people suggest moving permanently or changing their livelihood as a way of dealing with the hazard. Within limits, they adapt and adjust, employing practices, devices, and plans which might all be lumped together as "folk strategies," or "folk wisdom." With these strategies, they survive and even prosper in areas of high and recurrent danger. We found this store of folk wisdom at all our sites, though it was much more prominent in less developed countries. But as these preindustrial societies become more developed, much that is done in the name of improvement may be harmful if it leads to technological or social change without compensating adjustments. Tragedy came to Char Jabbar when reclamation of the land was unaccompanied by plans for emergency evacuation of the people living on it.

The farmlands of the chars are protected from tidal water and its salt and shifting river channels by some 2,000 miles (3,200 kilometers) of earthen dikes and cross dams, standing 15 to 30 feet (4.5 to 9 meters) high. This earthworks system, an impressive technological feat, was planned by Pakistani government officials, designed by Dutch engineers, and financed by international bankers. It created new farmlands and helped greatly to increase the population of the area and thus the potential for tragedy. But the system was not designed to protect against fierce cyclones, and the local and national governments had no organizations or programs that could have moved the people out of the threatened area.

A weather satellite, an even more complex technological triumph, provided early warning of the storm, but the radio station that relayed the satellite's message was using a new storm-warning system with terminology and definitions devised by officials but never explained to the people. The engineering works that made habitation possible could have provided an excellent escape route. The 11-mile (17.6-kilometer) -long Cross Dam 2 led inland to higher ground. But no warning reached Char Jabbar. There are no telephones or telegraphs there. A policeman would have had to bring the warning by bicycle from the town of Noakhali, about 15 miles (24 kilometers) away. Even if a warning had been sent, the unorganized farmers might not have responded to it, afraid of leaving their homes for distant places on a dark, uncertain night. Many thousands died on the chars because new technology was not followed by social planning.

Two years later, I sat with other colleagues by the railroad station in Masaya, Nicaragua, talking with the survivors of the 1972 Managua earthquake, another major disaster. A series of quakes on Dec. 23, 1972, destroyed most of the city and killed an estimated 10,000 persons in this capital city of 400,000. Seven of every 10 residents were left homeless. Managuans have always lived with danger. The city was built on a fault in the earth where many tremors occur and it has been struck by disastrous earthquakes four times in the last century.

Managua was chosen as the capital of Nicaragua in 1855 as a political compromise; it stands between the less hazardous cities of Granada and León, whose citizens had been feuding. The Indians of the area probably informed the new settlers of the site's instability, but the men of Granada and León overlooked the geophysical implications of their political compromise. In 1931, an earthquake leveled 4 square miles (10 square kilometers) of the city, which had more than 50,000 people. Nicaraguans spurned the opportunity to rebuild elsewhere or to build better; the city was rebuilt on the same site in substantially the same form. Over the next 40 years, Managua's population increased almost 8 times and so did its potential for disaster. Despite its seismic history, there was practically no planning for disaster. Instead of decentralizing its fire-fighting equipment, for example, the authorities kept the entire fleet of seven fire engines in a single central fire station—which was flattened in the 1972 earthquake.

Nevertheless, the interviews made in Managua reinforce those in Bangladesh. The advantages of living in hazardous areas often out-

The Paradox of Fewer Disasters and More Deaths

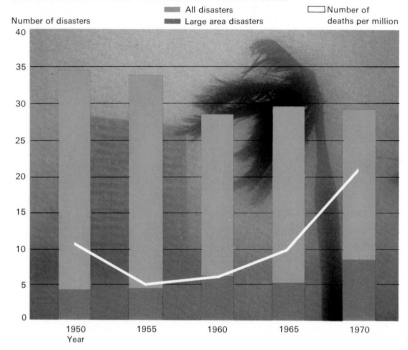

While the annual total of disasters decreased from 1950 to 1970, the number of large-area disasters and disaster deaths increased. This was caused in part by increased population density and also by the increasing use of risky, marginal land.

weigh the high risks. But a specific location may eventually burden a society to a degree far beyond the initial advantage or gain. Or, the benefits of living in a certain place and exploiting its resources may change with time. Although a city can be moved, once it is built it takes on an independent existence of its own. The investment in building, the particular activity of its people and their attachment to the place make moving difficult. In 1972, at a cost 10 times greater than in 1931, Managua again had a choice—to move, to build better, or to repeat its mistakes. The Nicaraguans have chosen to remain, but, by establishing a series of parks and open areas along the fault lines, and constructing stronger public and commercial buildings they hope to avoid another disaster.

Managua's options emphasize one finding of our studies. There are relatively few things that individuals and societies can do to cope with intensive hazards, short of leaving the hazardous zone entirely. Man can protect himself better against intermediate hazards such as floods because there is some warning. Pervasive hazards allow the widest latitude of adjustment. People can cope with drought, for example, by employing a wide range of actions—simply bearing minor losses, irrigating fields, planting drought-resistant plant varieties, using special cultivation techniques, shifting between herding and farming, or migrating to new areas.

Opportunities for adjustment to the same, or similar, hazards differ greatly among societies. For much of human history, those directly

How Hazard Grows With a City

Managua, Nicaragua, population in thousands   ● Major earthquake

As Managua's population increases, *above,* the potential for loss of life and property grows. The 1931 quake, *above right,* flattened a city of more than 50,000; the 1972 quake, *far right,* struck a city grown nearly 8 times larger.

affected by drought or flood had to cope as best they could on an individual or small-community basis. The nation, if it existed, had little in the way of resources or technical expertise to provide help. Our interviews with farmers in preindustrial societies reveal the enormous ingenuity of people under these circumstances.

Rice farmers facing flood in Sri Lanka, for example, have a wide range of folk adjustments. They may shift the time of planting rice, or the method of cultivation. They may choose different varieties of rice seed, or plant an entirely different plant crop. They may even change the layout of their fields in order to minimize the extent of the flooding. In all, our researchers found that the Sri Lankans had several hundred options to choose from.

We found similar adjustment patterns in our studies of drought in Tanzania and cyclones in Bangladesh. In general, folk strategies are numerous and widely adopted. They are flexible and economic and can be added gradually. The people tend to modify their behavior or agricultural practice to harmonize with natural rhythms rather than attempt to control or manipulate the environment. For example, when the rains fail, a Tanzanian farmer plants his crops in land that is normally too wet but is just right during drought. Or he may plant less

desirable but still useful food crops that require less moisture. He can do this because he has communal access to other land or because he grows the crops to feed his family and does not depend on the commercial demands or consumer tastes of a distant market. Similarly, farmers in Mexico's Oaxaca Valley will plant one of two available varieties of corn, depending on the rainfall. One variety survives drought better, but the other variety provides higher yields of corn when it has sufficient moisture.

Such adjustments are effective in coping with many hazards, but they fail to prevent major disasters. More government or social organization may be needed when a major disaster occurs, such as an earthquake or widespread, long-term drought. Most developing countries can at best offer only emergency food supplies and some relief assistance after a disaster.

The governments of industrialized nations have become increasingly involved in major adjustments designed to protect citizens from hazards. In addition to extensive relief, they employ their impressive technological capacity to manipulate and manage the environment by building huge dams, major irrigation projects, sea walls, and complex monitoring and warning systems.

Measuring Risks: How Lethal for How Long?

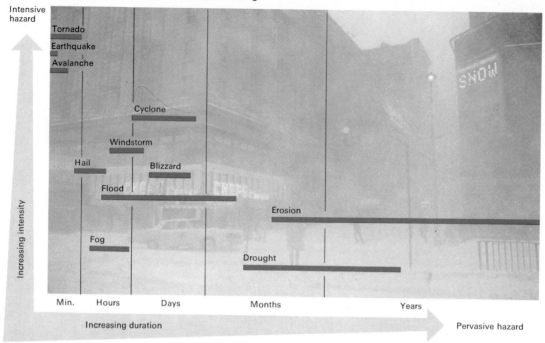

Intensive hazard

Increasing intensity

Tornado
Earthquake
Avalanche
Cyclone
Windstorm
Hail
Blizzard
Flood
Erosion
Fog
Drought

Min.　Hours　Days　Months　Years

Increasing duration

Pervasive hazard

Intensive hazards are sudden, brief, and affect small areas. Pervasive hazards develop slowly last longer, and generally affect larger areas

How a person adjusts to a hazard depends on his appraisal of its probability and likely intensity. The best-appraised events are the frequent, the recent, and those that greatly affect everyday life and livelihood. To test people's notions of probability, we included a simple story about the recurrence of hazard in our questionnaire. At each site, we used the common local hazard in the story. In a hurricane-prone area the story would go like this: "Once, after a hurricane, four men spoke about a hurricane coming again. The first said that a hurricane 'would come again soon' because when a hurricane happens, more are soon to come. The second thought that a hurricane 'would come again but he did not know when' because hurricanes can happen in any year. The third said that 'he knew when' the hurricane would come, for there is a regular time and that time must pass before it comes again. The fourth thought that a hurricane 'would not come again.' Which man has the best idea about the coming of hurricanes?"

While some people saw order in the random events, in most places, more than 60 per cent agreed with the second man in the story, who chose the proper random explanation of the probability of a hazard occurring. All in all, their judgments were surprisingly close to the record of past events and the judgments of trained observers about the likelihood of future events.

But, in a special Canadian study of London, Ont., a town of more than 200,000, we found out something else. We compared the perceptions of a large random sample of the population with scientific esti-

One person's hazard can be another's fun, and a snowbound commuter's feelings about snow may differ from those of his children at home. Such different perceptions may influence the way people prepare for and cope with hazards.

mates on how often five different hazardous events will occur. The people's estimates for tornadoes, hurricanes, and floods agreed closely with the scientific estimates. But, curiously, the estimates for the hazardous events that occur more commonly in Ontario—ice storms and blizzards—differed from citizen to scientist and citizen to citizen. We think the people viewed the seriousness and frequency of ice storms and blizzards differently because they are more familiar with them. Some do not think of them as hazards at all. Thus, individual judgments, and therefore the action taken concerning common hazards, may have much to do with an individual's psychological makeup.

Overall, each community develops a variety of ways to cope with a given hazard. Each person in a Tanzanian village may suggest one or two adjustments for drought, such as planting a hardier crop or going to live with relatives elsewhere until the drought ends. But if you canvass the entire village you may compile a list of 8 or 10 different actions. Similarly, each government office can tell about its own activities, such as building dams or providing insurance, but one must canvass all offices to determine the total governmental response to hazard. Many people know several actions to take to reduce damage in a hazardous area. But few act much in advance of a disaster. Some do nothing. So we find yet another question at the end of one long road of inquiry. Taking into account the important differences between hazards and societies, and the differences in individual appraisal, why do some people act to reduce their vulnerability while

Many citizens' groups are calling for closer study of the disaster potential of nuclear power and other new man-made hazards.

their neighbors do nothing at all? Why does one Puerto Rican farmer buy food, matches, candles, and lanterns and store water when a hurricane is predicted while another does not? We cannot adequately explain these individual differences. Psychologists, who are more concerned with individual behavior and thought processes, are still exploring these important questions.

White and Ian Burton of the University of Toronto and I are moving in two somewhat different directions. First, we are working to explain our major findings—that natural hazards are inevitable in the use of the earth but natural disasters are not. Both the rich and the poor are vulnerable to hazard. The poorest nations face a greater risk to life and property than do rich nations. In poor countries, the folk wisdom that helped people survive for so long is disappearing faster than the new governmental and social institutions can effectively replace it. In 1969, for instance, a small local drought struck the few areas of Tanzania where grain was grown commercially to feed that country's expanding urban population. This increased urbanization, coupled with the modernized farming methods and the consequent reduction of flexible family-style farming, magnified a localized natural event into a national problem. The drought was minor as measured by rainfall, but major as measured by production losses and relief costs to feed Tanzania's people.

Introducing technology too rapidly without making needed social adjustments at the same time can make such problems worse. The poor nations cannot escape completely the human toll of hazard, but their losses can be substantially reduced if these nations can combine the best of their own folk wisdom with the proper advanced technology of the developed nations.

Peculiarly, we have found that the developed nations are vulnerable, too, victims of their own technological success. Theoretically, their engineering achievements should enlarge the storehouse of useful actions for coping with hazard. Instead we find that for a long time technology tends to supplant all alternative adjustments until, belatedly and at very high cost, it is recognized that dams and levees are not the only answers to floods, or irrigation and water transfer the sole answers to drought. In time, of course, the lesson is learned. The United States is now paying greater attention to changing and restricting the use of flood plains than to simply protecting them with levees. Israel is developing crops that require less water rather than simply increasing irrigation systems for its farmers. Our goal is to encourage more of these solutions which compromise between man's needs and his environment.

We are also trying to apply our scientific techniques to a new and potentially more serious range of environmental hazards—those created by the technological environment. We are studying problems related to water supply and treatment, nuclear power, and man-made climatic change, and we have found one significant difference from our work on natural hazards.

Human experience played a central role in coping with an uncertain environment. But we have no pool of experience to draw on in dealing with most of the hazards of technology. They are too new and shrouded in great uncertainty. They affect us before we can detect and control them. For example, injecting particulate matter and heat from burning fossil fuels into the atmosphere certainly affects weather and climate, but we could be well on our way to a little Ice Age or, conversely, to a melting of the polar icecaps by the time that we experience the effects or are certain about them. Radiation or industrial chemicals may severely damage human genes before there is enough visible evidence of mutation for us to recognize the danger. There is a special and necessary role for scientific and technical assessors of such hazards, and for public and governmental guardians of the environment. Our new studies are especially focused on these groups. We hope that our techniques and findings may help these people to discover, measure, and deal with man-made hazards before the hazards become great—perhaps global—disasters.

But this new aspect of environmental risk worries us. Over the last 20 years, we learned to respect greatly the thousands of people living and working in the hazardous areas we have studied. We have confidence in their skills and common sense. But no such pool of common wisdom exists to deal with the new hazards created by man. Because of this, the world is a more threatened place.

The !Kung's New Culture

By Richard B. Lee

Studies of a changing Stone Age people are providing clues to the origin of many modern social problems

When I first went to the Kalahari Desert near the southern tip of Africa to study a group of primitive hunter-gatherers, I had no idea that my work would produce a link to such present-day problems as overpopulation, sexism, and social aggression. But my 12 years of study, beginning in 1963, coincided with a crucial change in the group's way of life. The observations of this change have convinced me that our own social problems are not rooted in any biological short-comings of the "naked ape," as some people claim. Instead, they result from changes in human economic relationships, sex roles, and child-rearing practices that first began to emerge more than 10,000 years ago, when our ancestors stopped hunting animals and gathering roots and berries and started farming.

The group I studied is the !Kung San (the exclamation point denotes a clicking sound in their language). Also known as Bushmen, these people were then the most traditional hunters and gatherers, unchanged by other cultures, that I could find. I worked with anthropologist Irven DeVore of Harvard University, Cambridge, Mass., and a team of scientists, studying !Kung eating habits, tribal customs, family size, and ways of organizing work schedules. Among the details we recorded were the distances and direction they traveled month by month, the quantities of food they ate at meals, and the day-to-day activities of each member of several family groups.

About 800 !Kung live in the Dobe area, which covers about 2,700 square miles (7,000 square kilometers) in the northwestern corner of Botswana, plus a part of Namibia (South West Africa). It is separated from the rest of Botswana by a belt of waterless, uninhabited country. Because of their isolation in this hostile environment, the !Kung have remained hunters and gatherers for at least 11,000 years. We know this because John Yellen, an archaeologist with the Smithsonian Institution in Washington, D.C., has been able to date spearpoints and other tools he found in their ancient campsites.

The Kalahari is a semidesert area, covered with wind-blown red and white sand. Temperatures can reach 115°F. (45°C) in the shade during the daytime, and can fall below freezing on winter nights. The drought-resistant vegetation includes grass, brush, and small nut-bearing mungongo trees. Water is the most important item for subsistence. No !Kung tribesman can ever settle far from one of the scattered water holes, because rainfall comes only in summer thunderstorms.

Until recently, the Dobe !Kung did no farming and owned no domestic animals except dogs. They lived in groups of about 25 persons, and moved from place to place within their territory, building temporary shelters of grass and branches. The women gathered berries, nuts, melons, roots, and seeds, while the men hunted animals ranging from rabbits to antelope, with poison-tipped arrows. There were no stores, schools, or missions in their area, and government presence was limited to one patrol by Land Rover every four to six weeks.

But by the time of my arrival in 1963, the !Kung were no longer living strictly as hunters in a world of hunters. They shared eight of their nine water holes with about 340 Bantu-speaking Tswana and Herero herders who had settled among the !Kung 40 years earlier with their cattle, goats, and donkeys. Some of the young !Kung men work for the newcomers herding cattle, and, at one time or another, members of the !Kung tribe have owned their own livestock. The !Kung told us that most of their people had tried to raise crops for several years before we arrived. When we returned to the Dobe area for our second field trip in 1967, we found that about one-third of the people in the various Dobe groups had planted crops at the start of what—for the Kalahari—was the rainy season.

A gradual shift was taking place, but it was not a clear-cut move from one way of life to another. When rainfall was heavier than normal, the Dobe !Kung planted crops. But during periods of drought, they had little choice but to return to full-time hunting and gathering.

The author:
Richard B. Lee
is professor of
anthropology at the
University of Toronto.
He is the author of
a new book, *Kalahari
Hunter-Gatherers.*

Range of the !Kung

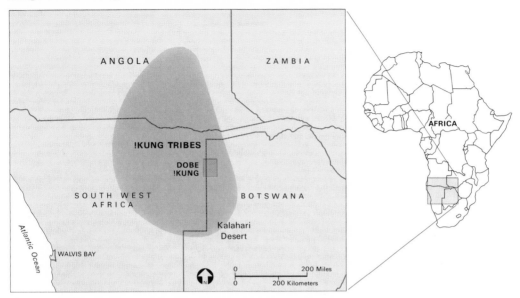

For example, during our first field work in 1963 the !Kung were in the midst of the worst drought of the century, and hunting and gathering was the only way they could survive. In 1967, the rains had been excellent, and they planted crops. More good rainy seasons followed, and by 1974 only a few of the !Kung in the Dobe area still lived entirely by hunting and gathering.

We noted in 1967 that the trend toward agriculture varied with location. The !Kung that lived around the eastern water holes had adapted more to farming and herding than those around the western water holes. We also saw that the eastern group was developing family and behavior patterns quite different from those of their hunting-and-gathering neighbors to the west.

This variation provided us with a natural laboratory for studying the !Kung as both foragers and farmers. It was also a window to our own past, to the time when mankind first began to cultivate wild plants and domesticate wild animals. These ancient experiments, probably lasting over many centuries, initiated the greatest change in the history of the human species—what some archaeologists call the Neolithic revolution. Out of this change flowed settled village life, population growth, urbanization, the rise of states and empires, and the development of the highly stratified militaristic societies that we refer to as "civilization" today.

The difference between the roving !Kung who live entirely by hunting and gathering and the more settled !Kung farmers is striking. The hunters and gatherers live surprisingly well. Their life is by no means the constant struggle for existence that some anthropologists describe as the lot of primitive hunters. My studies during the worst of the 1964

183

Nomadic !Kung men and women share
the responsibility for gathering food.
Two men, *top,* pick sour plums while
another sets a trap, *above,* to catch
a small animal, such as a hare. A
woman, *right,* carrying her baby and
a bundle of edible roots, uses an
ostrich-egg shell to collect scarce
water from the trunk of a tree.

Nomadic !Kung children prefer such sports as "riding the donkey" on tree branches to rough and competitive games.

drought indicate that they spent only from 12 to 19 hours a week searching for food. They enjoy an abundant and varied diet of over 150 species of wild plants and animals. They have no milk or grains, but nutritionists, geneticists, and biochemists who have studied their diet agree that it is nutritionally balanced and provides an adequate number of calories. These nomadic !Kung are exceedingly healthy and free of many of the diseases that continually plague people in more advanced societies.

In contrast, the !Kung farmers and herders consume a great deal of cow's milk and grain. They have more food, but nutritionists say that their diet is not as balanced as the hunters' and gatherers' diet. However, the physical effects of this altered diet are not yet well established because the !Kung have not been on it for long. But we have found that the children of these !Kung are taller, fatter, and heavier on the average than their nomadic cousins.

Mobility is one of the keys to the success of the hunting-and-gathering families. Lightly equipped, they move frequently from area to area, following the ripening of plants or the movements of game. Five or six moves during the summer rainy season are normal for them, though they tend to settle at one water hole for the dry season.

Farming, obviously, limits mobility. Crops and animals must be tended regularly, and so the farming !Kung have to limit their foraging to one-day or overnight trips. Instead of moving five or six times a season, they may move only once a year, or less often. One !Kung farming-and-herding group has occupied the same village site for 12 years. They abandon it only during times of drought.

The trend toward a more settled life has triggered a sharp rise in the farming !Kung birth rate. Nomadic !Kung women give birth only every four or five years. The fact that they constantly carry their infants as they move from place to place may contribute to this spacing of children. But a more likely reason is that most nomadic mothers

When a !Kung hunter kills an animal, he shares the meat with all members of the tribe.

nurse their children until they are at least 3 years old, though they supplement this with a few solid foods. Sociologist Nancy Howell of the University of Toronto, who has studied these people extensively, believes that the prolonged nursing has a contraceptive effect. The same contraceptive effect has been observed by scientists in many other tribes and population groups.

My research indicates that the women in the !Kung farming families bear children every two or three years. Howell suggests that the composition of their diet, not its quality or quantity, is the key to the higher birth rate. The hunting-and-gathering diet, though rich in proteins, vitamins, and minerals, lacks soft foods that babies can eat, so the mothers have to nurse them longer. The farmer's wife can give her children milk and porridge, foods that are easily assimilated by infants, allowing her to wean them earlier. Over a 20-year reproductive span, a !Kung farmer's wife could have up to 10 children compared with only 4 or 5 for a nomadic !Kung woman. If this theory is

The relationship between parents and children in the families of the nomadic !Kung is one of great joy. The children are carefree and show no signs of aggression.

Nomadic !Kung, who once moved from camp to camp, *above,* have now settled in one place to grow crops and herd cattle. A vital necessity for this life is water, which they obtain from deep wells, *opposite page.*

correct, it may explain why the Neolithic revolution triggered a sudden population growth some 10,000 years ago.

Our studies show that growing up in a hunting-and-gathering camp is pleasant and lacking in pressure. Boys and girls of all ages spend most of their time playing together. They do not develop distinct games and roles for boys and girls, and they do not have to join the hunt for food until they are teen-agers.

By contrast, farm life requires the child to contribute to the household economy. Anthropologist Patricia Draper of the University of New Mexico in Albuquerque, who has studied the effects of the more settled form of life on !Kung child-rearing practices, notes that the children of !Kung farmers play in groups of the same sex and similar ages. The boys and girls are put to work at an earlier age, but at different tasks. The boys, as they grow older, must herd cattle, so they go to the pasturelands, where they are away from adult influence. The girls are given household tasks, including caring for younger brothers

and sisters. They are closely supervised by adults. Draper concludes that this division of childhood labor may contribute to the loss of freedom and status of women that is evident among the farming-and-herding !Kung today.

Adult males and females in the hunting-and-gathering groups have equal roles. The males do not totally own the family property, as they do in the farming groups. !Kung women enjoy higher status, more freedom, and have a greater voice in group decisions in the nomadic bands than do the women in farming families. One reason for this is that the women, by gathering plant foods while the men are hunting, contribute more than half of the food that the bands consume. The women, of necessity, are as mobile as the men and leave the camp three or four days each week to gather food and to visit other camps, taking their infants with them. During this time, the men may care for the older children.

Similarly, couples do not always live in the husband's camp, as some observers of hunting-and-gathering societies have suggested. In fact, a man is more likely to move into his wife's camp. Yet the hunting-and-gathering life is in no sense a matriarchy. Men wield as much power in making decisions and in resolving conflicts as do women. Their life is one of give and take, rather than a situation of domination of one sex by another.

When !Kung settle down to farming and herding, their work patterns change. Women are confined to the home more, and their sphere of activities is restricted. They remain in the village to prepare food and take care of their homes. They become more homebound and more dependent on their husbands and male kin for subsistence. The men move far afield herding cattle for the Tswana and Herero. Because of this, more !Kung men than women learn the Bantu languages, and some younger men consciously emulate the Bantu men, whose culture is male dominated.

There is little competition among the members of !Kung hunting groups. The children do not even play competitive games, perhaps because the wide age range in most play groups would make competitiveness difficult. In addition, the children are constantly watched by adults, who quickly stop aggressive behavior.

The strong ethic of sharing and communality that characterizes their life style is particularly evident in the way the hunting-and-gathering !Kung share food. Each day, several men go to hunt with bows and poisoned arrows. If they kill a large animal, such as an antelope or a springbok, they share the meat equally with all members of the camp. Vegetables are more readily available and so are shared less widely than meat. Small quantities of nuts and berries are consumed by the gatherer and her family and by close relatives within the group. Because the camp is a sharing unit, members take great pains to maintain group harmony. If an argument breaks out, a family may leave the group and join another, rather than cause sharing to break down

When nomadic !Kung settle down and learn how to cultivate plants and herd cattle, they begin to put a value on personal possessions, such as donkeys, tools, and clothing. This, in turn, leads to a change in ethics from sharing to saving and sometimes even to taking property of others. One result is a !Kung herdsman on trial for stealing cattle, an act unheard of in the old !Kung hunting-and-gathering way of life.

within the first group. Months later, after the cause of the argument that split up the group has been forgotten, the family may leave its new group and return to the original one.

The farming !Kung cannot handle arguments this way because they cannot easily pick up and leave. Therefore, they rely on their Bantu neighbors to mediate their disputes. They no longer have the kind of life that involves community sharing of the spoils of a hunt, and aggressive behavior among young children occurs more often because the children are cared for by older sisters rather than adults.

Saving and the careful managing of one's own affairs are the key features of the farmers' and herders' life. Their goal is to increase the crop or the herd, not to squander it by giving it away. Hospitality is important and visitors are always cared for, but this is quite different from the hunters' total sharing. The !Kung tribesmen have to contend with the contradiction between the sharing ethic and the saving ethic as they shift to the new way of life.

When an energetic !Kung man builds up a small herd of goats or cattle he is behaving in an eminently rational way, from the herder's point of view. But what does he do when his kinsmen say, "You have meat and we have none; slaughter a goat to fill our hungry bellies?" If he grants their request, he fails as a farmer. But if he refuses, he will be accused of being "stingy and far-hearted," one of the worst accusations a hunter-gatherer can make.

Kxoma, a !Kung farmer, put it this way, "If someone asks me for food, I just give it to him." Kxoma is 35 years old and very proud of his crop of melons, beans, sorghum, corn, and sunflowers. I pushed the question further. "What if many people come, none of them farmers, and they want all your food. Will you then refuse?"

"No, I won't refuse," Kxoma answered. "If people ask, you must give it to them. After all my food is gone, I can hunt for bush foods."

Group pressure is difficult to overcome and many animals are undoubtedly slaughtered to fulfill such obligations. This may help to explain why so many !Kung who become livestock owners eventually lose their cattle. Those few !Kung who succeed in establishing themselves as herdsmen may have to pay a price for their success in terms of loss of kinship ties and even partial ostracism from the group.

When we compare the two life styles the question naturally arises: Why do the !Kung switch from hunting to farming—for that matter, why did our Neolithic ancestors switch? It had to be more than just a change in climate. I have painted a pleasant picture of the hunter-gatherer life—plenty of food, complete equality, and an ethic of sharing. By contrast, the settled village life has brought the !Kung higher fertility but probably lowered nutrition, the subjugation of women, and the replacement of sharing by saving for the future.

I think two kinds of answers are in order. First, the picture of the hunter-gatherer way of life is somewhat overdrawn. It is not all a bed of roses. There is hardship and higher mortality rates, perhaps because

A cattle herder and his wife, with three young children and a fourth on the way, contrast with nomadic !Kung families that have fewer children, more widely spaced.

the people have to move so often. Old and sick people might live longer if they could settle in one place permanently.

It also could be that in the Neolithic transformation, as in any other historical process, people really had no choice. The Neolithic revolution was a process that took many centuries. People began raising plants and breeding animals because of large-scale forces of which they were only dimly, if at all, aware. Just as climatic changes play a part in determining whether the !Kung plant crops or continue to hunt, the Neolithic revolution must have grown from such uncontrollable forces as population pressure, the slow rise of individual work effort, and the imperceptible decline of game and fruits that could be procured by the old hunting-and-gathering methods.

Kxoma, one of the forward-looking members of the !Kung group, welcomes the changes that are taking place. "I don't like the bush," he said. "You can die out there. It's work, work, work from morning till night. This village is the place I love most. When you live in the bush, as soon as the sun comes up, you are up and off with your spear under your armpit, looking for something to kill. By nightfall, you come home so tired you just fall asleep. When the sun comes up again, you are off again, your spear under your armpit, walking and searching for something to kill. What a life!

"Now, here in the village it's a different story. You wake up in the morning, and not too early. You sit up and say to yourself, 'Now what shall I do today? I think I'll stroll over to that Herero village over there.' You ask for milk and get a nice long drink, and then you sit and talk, and then you come home and talk some more. The next day you sit up and say, 'What shall I do today?' Then you hear, 'Kxoma, come here. A cow has died in the bush, let's go and eat it.' So you stroll over to the Herero village and eat some meat. I like the village. I'll just stay here from now on.''

Kxoma's statement is interesting because it expresses some of the uncertainties about where the next meal will come from that are a constant concern in the hunting way of life. However, the actual time spent by hunters and gatherers in the quest for food, as my 1964 studies indicated, was far less than Kxoma seems to remember. There are, of course, other !Kung farmers who don't particularly like the new life because of the social problems it is bringing them, but Kxoma

Sending their children to government schools and living in permanent Bantu-style homes, the !Kung are undergoing social change that may prove to be irreversible.

represents one of the !Kung who are readily adapting to it. How long the adjustment will take for the whole group depends somewhat on how many !Kung adopt Kxoma's positive view of the situation.

When I began to study the !Kung, they provided a particularly good natural laboratory for studying human change and adaptation because they were changing of their own accord, and were not being forced to change by the encroachment of civilization. Forced changes have occurred in many parts of the world with disastrous results to the affected tribal culture. Examples of this are the past destruction of various American Indian tribes in the United States and the present cultural emasculation of several primitive tribes in Brazil. Forced changes give us few if any clues as to how early man developed.

Yet, "civilization" has moved in on the Dobe area since we began our studies. In 1965, the South African government built a fence along the Botswana-Namibia border that bisects the Dobe area. The fence was part of South Africa's effort to maintain its control over Namibia in the face of opposition from the United Nations and African liberation movements. But this fence has cut the Dobe !Kung off from their western hunting lands and hastened their shift to agriculture. In 1967, a store was built in the area, and for the first time cash became a part of the !Kung economy. Several young !Kung men now work for wages at the store. Also, a primary school opened in 1973.

It became increasingly clear to the members of our research group that a slow and orderly Neolithic transformation was not in the cards for the !Kung. We could envision an invasion of the Dobe area by cattle dealers, prospectors, missionaries, tourists, counter-insurgency forces from the Republic of South Africa, and other outsiders like ourselves. At a meeting of our research group near Boston in early 1973 to discuss the future of the !Kung, we decided that our responsibility to the people we study goes beyond simply reporting our scientific results. It includes helping the !Kung in their struggle to determine their own future.

At this meeting we established the Kalahari Peoples Fund, a nonprofit organization to provide technical and financial assistance for the development of !Kung communities. This organization works closely with the Botswana government and with voluntary agencies. For example, the fund has provided scholarships for 22 !Kung children to attend the new primary school set up in the Dobe area.

No one claims that the changes taking place in !Kung society exactly duplicate those that took place when earlier hunter-gatherer societies turned to farming and herding. Any such attempt to reconstruct prehistoric society must be inexact. Yet, the record of the !Kung offers clues about early social forms that can be applied to other societies by skillful use of analogy. These should make us view with suspicion theories that seek to prove that male dominance in our present social order is part of an evolutionary heritage, or that man is by nature uncooperative and aggressive.

The Case of the Baffling Bursts

By Ian B. Strong

Astrophysicists suggest that space goblins, dwarf stars, speeding BBs, or burping black holes may be the sources of mysterious bursts of gamma rays

For the past three years I have puzzled over unexpected signals relayed to earth by a network of surveillance satellites. The brief messages occur 9 or 10 times each year–triggered by an intense wave of electromagnetic energy that sweeps across the earth and rumbles on into space. Where and how the waves start are parts of a continuing mystery that challenges many astrophysicists.

The mystery began about eight years ago when the United States launched the fourth pair of Vela satellites into a circular orbit about 75,000 miles (121,000 kilometers) above the earth. Deployed on opposite sides of the earth and armed with sensitive detectors, they watch for a sudden burst of energetic radiation streaming from a nuclear weapon blast in space. Such a burst would indicate a violation of the limited nuclear test-ban treaty.

These and later Velas carry special instruments that can even detect nuclear blasts on the far side of the moon. The moon would block direct radiation that triggers the primary detectors, but radioactive

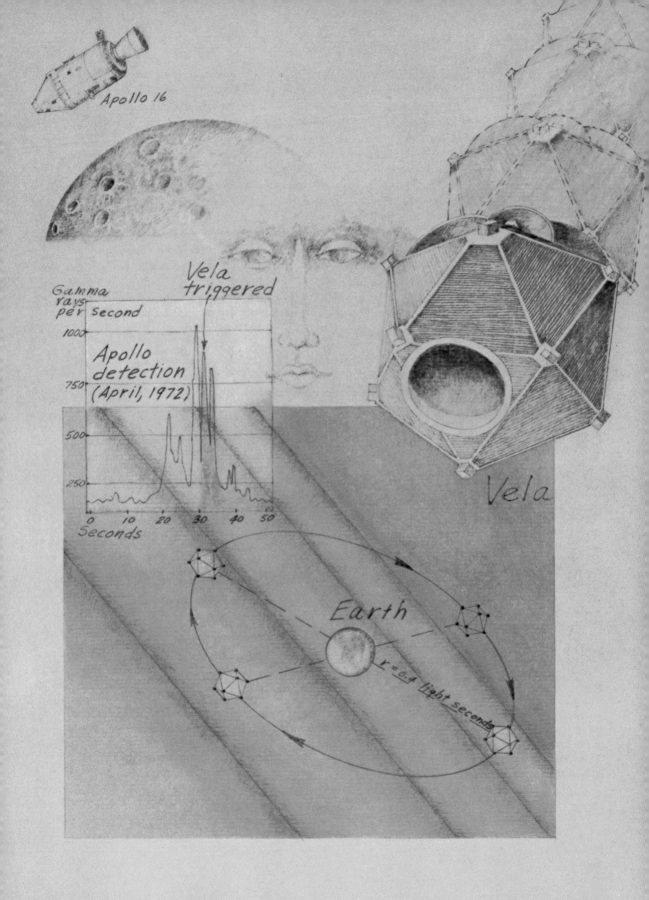

Apollo 16

Gamma
rays
per second

1000

Vela
triggered

Apollo
detection
(April, 1972)

750

500

250

0   10   20   30   40   50
Seconds

Vela

Earth

r = 0.4 light seconds

debris would emerge from behind the moon seconds after the blast, emitting gamma rays—electromagnetic radiation of higher frequency than X rays. The new detectors activate when the number of rays hitting them rises abruptly. They record and transmit to earth the burst's intensity along with the exact time of activation.

One of the Velas occasionally sent a message saying it had been hit. Until the summer of 1972, we ignored such messages because energetic charged particles from the sun or cosmic rays from space often trigger a detector. Besides, we know when spacecraft large enough to carry bombs are behind the moon.

Then, my colleague Ray W. Klebesadel began searching recent records to determine how often two Velas relayed signals at the same time. We had calculated that two satellites would activate at the same time by chance once every few hundred years at the most. But Ray found several coincidences each year. Sometimes, three, or even four, Vela satellites recorded hits within a second of each other.

I remembered that Stirling A. Colgate, then president of New Mexico Institute of Mining and Technology in Socorro, had occasionally called to ask, "Did your bomb satellites just see any gamma rays?" He had predicted in an article that an enormous amount of energy in the form of gamma rays would be emitted by a supernova, an exploding star that suddenly appears brighter. Having exhausted its nuclear fuel, the star's spent core collapses violently under the pull of its own gravity and its cooler outer layers explode, forming a wispy expanding cloud of dust and gas.

Could gamma rays from supernovae have caused the coincidental activations? I raced to the library and found Colgate's article. Much, but not all, of our data agreed with his predictions. The Vela bursts lasted much longer than he had predicted, and we knew of no supernovae seen at the times the bursts were recorded. Well, the detailed theory might be slightly wrong, but good theories have been nearly right much more often than they have been absolutely right. So I telephoned Colgate and told him of our discovery.

My co-workers and I began digging through the old Vela records, convincing ourselves that there were no errors. Data from different detectors ruled out radiation other than gamma rays as the cause of the coincidental activations. But where were the bursts coming from if not from supernovae?

The difference in time between activations gave us a way of roughly determining the direction from which each burst came. Gamma rays, moving at the speed of light, would take 0.8 second to cross the orbit and activate the opposite Vela satellite. Gamma rays traveling at right angles to the line joining opposite satellites would arrive simultaneously. Those coming from other directions would require travel times between zero and 0.8 second.

Conversely, we could calculate where a wave came from by using the exact difference in activation times. This triangulation method

The author:
Ian B. Strong is
an astrophysicist at
the Los Alamos
Scientific Laboratory
in New Mexico.

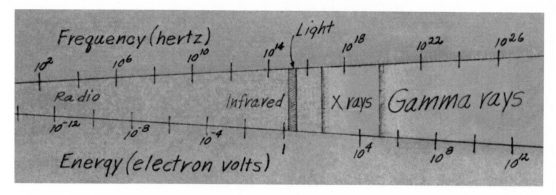

Frequency (hertz)

10^2 10^6 10^{10} 10^{14} Light 10^{18} 10^{22} 10^{26}

Radio   Infrared   X rays Gamma rays

10^{-12} 10^{-8} 10^{-4} 1 10^4 10^8 10^{12}

Energy (electron volts)

yielded two possible source directions for each burst. We quickly ruled out objects within the solar system as possible sources, but we found no correlations with other possible sources.

By this time, we were quite sure that the bursts were gamma rays—electromagnetic radiation that is like radio waves, light, or X rays, but has much higher frequencies (much shorter wave lengths). At times, electromagnetic radiation acts as though it is made up of photons, or bundles of localized energy. Each photon carries an amount of energy equal to its frequency multiplied by a constant. We measure this energy in electron volts (eV). Visible photons carry about 2.5 eV of energy. At higher energies (frequencies), we no longer call photons light. Those above about 100 eV are called X rays, and those above about 100,000 eV (100 keV) are called gamma rays. Gamma rays of about 150 keV trigger the Vela detectors.

We mailed a letter on our findings to the editor of the *Astrophysical Journal*, and it was published in June, 1973. Excited to have discovered something so unusual, we were anxious for confirmation from other investigators. What other satellites might also have detected the bursts? Could ground-based optical telescopes or radio telescopes have seen anything at the same times coming from roughly the same places? We decided to call other astrophysicists.

We had scarcely begun when Thomas L. Cline of the National Aeronautics and Space Administration Goddard Space Flight Center in Greenbelt, Md., called us. He and his colleague Upendra D. Desai had found what they called superevents using a detector on the Interplanetary Monitoring Platform-6 (IMP-6) satellite. They also were looking for confirmation. Cline heard of our discovery at a meeting on supernovae in Trieste, Italy. Unable to contact us from Europe, he called as soon as he returned home. We compared dates and times for several activations, which agreed to the nearest second. The IMP-6 detector positively identified the bursts as gamma rays, each of which carried about 150 keV of energy.

We next heard from Lawrence E. Peterson's X-ray astronomy group at the University of California at San Diego. Their X-ray telescope on the OSO-7 satellite, an orbiting solar observatory, had meas-

Unseen Light
Gamma rays occupy the high-frequency (high-energy) portion of the electromagnetic spectrum that also includes X rays, light, and radio waves.

Technicians assemble a pair of Vela satellites, *right,* later launched to watch for secret tests of nuclear weapons in space. The author and colleague Ray W. Klebesadel test the type of Vela gamma-ray detector, *below,* that discovered mysterious gamma-ray bursts.

ured radiation from a position that agreed almost perfectly with that of a burst we had detected at the same time.

Still more researchers reported prior detections during the next few months, including one by a detector on the Apollo 16 command module. It recorded, in the most detail yet, a burst on April 27, 1972, that was also detected by a Vela. By early 1975, about 15 teams had informally formed the Gamma-Ray Burst (GRB) Club. They included German, Italian, and Russian experimental teams, and even a Japanese balloon group joined us.

As membership in the club grew, we compared notes and gained confidence that these quite unexpected GRBs were real. They occur at an unpredictable time, in unpredictable parts of the sky and last no more than a few seconds. During 4½ years of nearly constant alert, we found 41 GRBs that had activated two or more distant satellites.

Not all these bursts were alike. About six of them lasted only 0.1 second. If we could listen to them with the loudspeaker of a Geiger-Müller counter, the very short ones would simply make a loud click. A typical complete GRB starts with a number of soft clicks. The main part of the burst, about 10 seconds later, sounds like a loud drum roll and lasts a second or two. Then there may be a muffled drum roll about a minute later. The loud drum-roll portion of the bursts stands out from the radiation that is always there about as much as an actual loud drum roll overpowers quiet conversation.

A Gamma-Ray Trap
When gamma rays with about 150,000 electron volts of energy collide with crystal atoms in the Vela detector, they knock electrons free. Color centers in the crystal capture the electrons and emit tiny flashes of light that a photomultiplier tube collects and converts to an electric signal.

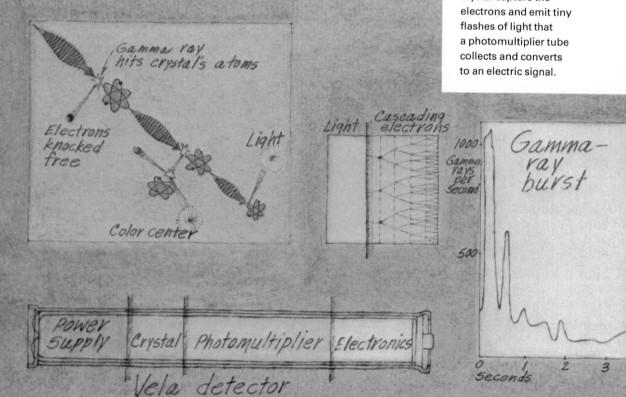

An optical reflector on Mount Hopkins in Arizona works like a searchlight in reverse, scanning the skies for faint visible evidence of gamma rays.

Discovery of these complicated GRBs breathed new life into the 10-year-old field of gamma-ray astronomy. This part of the electromagnetic spectrum is the latest to be used by astronomers as a window to space. The earth's atmosphere stops the otherwise lethal rays, so they cannot be detected directly from the surface. However, once physicists developed gamma-ray detectors for other experimental work and rockets were available to lift them above earth's atmosphere, gamma-ray astronomy began amidst great expectations. Theorists had predicted that all sorts of interesting gamma rays would be detected, but with few exceptions, none were.

The theorists based their predictions on well-known mechanisms for producing gamma rays: the relaxation of excited radioactive nuclei to a lower energy state; the breakup of elementary particles into pure energy; the spiraling motion of electrons or other charged particles about magnetic-field lines; or, even the heating of matter to about 1,000,000,000°C. Astrophysicists believe that conditions exist in space that can produce gamma rays in these ways.

The Compton effect, another gamma-ray production process, is the key to understanding how gamma rays can be detected. If a photon of perhaps only a few eV collides in billiard-ball fashion with a charged particle moving at nearly the speed of light, the photon will pick up some of the charged particle's energy and ricochet away as a gamma ray. Conversely, a gamma-ray collision with a relatively slow-moving outer atomic electron knocks the electron free from the atom at a high speed. The gamma ray ricochets away at a slight angle with less energy. In this way, a gamma ray from space soon loses its energy through numerous collisions with atoms in the earth's atmosphere. Or, it may even disappear near an atomic nucleus, producing a high-speed electron and its positively charged antiparticle, a positron, in a process called pair production.

The only way to detect gamma rays from space with detectors on the earth's surface is to look for the shower of particles produced when gamma rays with more than about 10,000,000 keV collide with atoms in the upper atmosphere. These particles move so fast that they create a kind of visible shock wave called Cherenkov radiation, which appears as a very faint flash in the upper atmosphere and lasts only a few billionths of a second. Jonathan E. Grindlay of Harvard University and Giovanni G. Fazio of the Smithsonian Astrophysical Observatory (SAO) used an SAO optical reflector atop Mount Hopkins, near Tucson, Ariz., to look for such Cherenkov radiation. So far, this device has found only one celestial object that appears to be a source of such high-energy gamma rays.

The Vela detectors use a crystal of cesium iodide that stops 150 keV gamma rays in a few inches or centimeters. The electrons that are scattered by the gamma rays emit light when they rejoin the crystal at sites called color centers. The net result is a single pulse of light from the crystal that a photomultiplier tube changes into an electric signal.

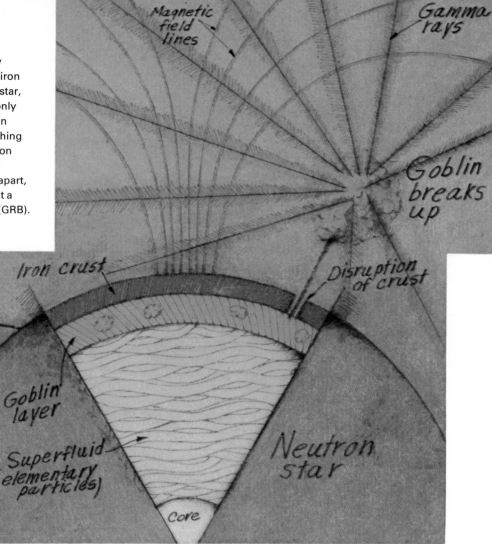

A Goblin Getaway
From beneath the iron crust of a neutron star, a nuclear goblin, only 10 feet (3 meters) in diameter but weighing more than 10 million billion tons, might escape and break apart, madly spewing out a gamma-ray burst (GRB).

Magnetic field lines

Gamma rays

Goblin breaks up

Iron crust

Disruption of crust

Goblin layer

Superfluid (elementary particles)

Neutron star

Core

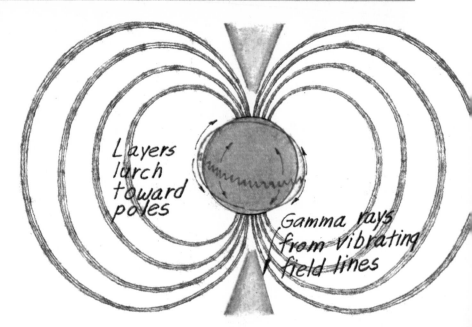

Geriatric Glitches
As it ages and slows, a pulsar loses some of its equatorial bulge when thin surface layers lurch toward its poles. In a young pulsar, this may cause a glitch, a sudden change in the precise period of its radio-wave pulses. In an old pulsar, shaking magnetic-field lines may create a GRB.

Layers lurch toward poles

Gamma rays from vibrating field lines

Soon after these detectors discovered GRBs, theorists rushed to explain their origins. By mid-1975, 24 different theories vied to account for the 41 GRBs. To theoreticians, the data is like a big sandpile to a child—there are almost no rules about what to build. We have enough information in the GRB signals to make rough models of their sources, but we do not yet know enough to prove or disprove the theories or describe in detail exactly how the gamma rays are produced.

From the signals, we know the GRBs appear to come from many directions in space. They are certainly not bunched at the center of our Galaxy, the Milky Way, but are slightly concentrated along its equator. This tells us that the sources are probably close to us, like the brightest and nearest stars, as opposed to very far away, like the distant galaxies that appear to be evenly distributed throughout the sky. Klebesadel and I concluded that GRBs probably originate less than 1,000 light-years away, in our sun's spiral arm of the Milky Way.

The GRB signals also give us important clues about the size and energy of the astronomical objects that produce them. Within each burst, the drum roll-like peaks rise from background to their summit in about 0.03 second. An object, or portion of an object, that turns on that quickly can be no bigger than the earth. For example, if the sun were to turn off completely, then turn back on all at once, we would first see radiation from the middle of the solar disk—that is, from the surface of the sun closest to us. It would take about 2½ seconds for it to expand until we could once again see the entire solar disk, because light from the edges would take that much longer to reach us.

Knowing how much energy hit a Vela detector, we can calculate how much must have left the source, assuming it traveled outward as an expanding spherical wave of radiation. If the source were as close as the nearer stars, about 100 light-years away, the energy released as gamma rays must be greater than that of 100 billion billion tons of TNT. Exploding a blanket of atomic bombs entirely covering the earth would produce about this much energy. But no astrophysicists believe that gamma-ray bursts come from nuclear wars on distant planets.

Bursts from Boiling BBs
Iron BBs, hurled from a neutron star, *bottom*, would move at nearly the speed of light until they met sunlight photons, *center*. The BBs would heat up and emit X rays characteristic of iron atoms. To observers in the BBs' path, *top*, the X rays would appear as gamma rays.

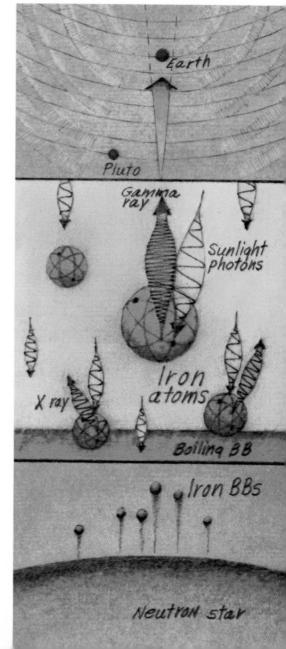

Like the other theories, Colgate's supernova model can account for this enormous energy release from a compact object. In a supernova, the star's core collapses to a neutron star—a dense ball of matter equivalent to packing all the sun's mass into a sphere only about 10 miles (16 kilometers) in diameter. Colgate suggested that the collapse creates a shock wave that moves outward through 50 million miles (80 million kilometers) of the expanding remnant to its surface. The farther the wave travels, the less dense the material becomes, and the wave speeds up. It is moving at nearly the speed of light when it breaks through the surface. Shaken violently by the wave, the charged particles at the surface radiate gamma rays.

Supernovae are perhaps the most spectacular examples of stars that collapse to compact astronomical bodies during their violent death throes. Ordinary stars like the sun eventually collapse under their own weight when the nuclear fires burning in their cores run out of fuel and stop emitting the radiation pressure that counterbalances the ultimately lethal gravitational pull. What happens when the nuclear fuel is gone depends on the star's mass. Those stars with less than about 1.4 times our sun's mass collapse to become white dwarfs about the size of the earth. At first white- or even blue-hot, the white dwarf cools to a dull red glow that is invisible to the eye even at only a few light-years distance from the once bright star.

Many models relate GRBs to white dwarf stars. Donald D. Clayton of Rice University in Houston and British astrophysicist Fred Hoyle have proposed a white dwarf model that is a kind of nuclear explosion. They imagine a white dwarf orbiting with a blue supergiant star about their common center of gravity in a kind of waltz that astrophysicists call a binary star system. If the partners in a giant-dwarf waltz are close enough, material from the giant funnels along the white dwarf's intense magnetic-field lines and piles up on its magnetic poles in a process called accretion. Clayton and Hoyle suggest that if enough matter piles up for nuclear fusion reactions to begin, then a nuclear explosion may follow, hurling the material back into space and radiating gamma rays in the process.

Although white dwarfs weigh an incredible thousand tons per cubic inch, they are far less dense than the neutron stars formed in a matter of seconds when supernovae cores collapse. Neutron stars weigh about a billion tons per cubic inch. Their surfaces are made incredibly smooth by the strong gravitational grip that shrinks mountains smaller than pinheads. Beneath their smooth surfaces, neutron stars have solid crusts of iron, under which layers of elementary particles and neutrons boil. Some neutron stars, called pulsars, spin as rapidly as 30 times per second, beaming periodic pulses of radio waves into space as their intense magnetic fields whip particles to near the speed of light.

Like white dwarfs, neutron stars can attract matter from a binary partner. This action is believed to account for most of the intense X-ray sources detected by the Uhuru satellite in the early 1970s and it

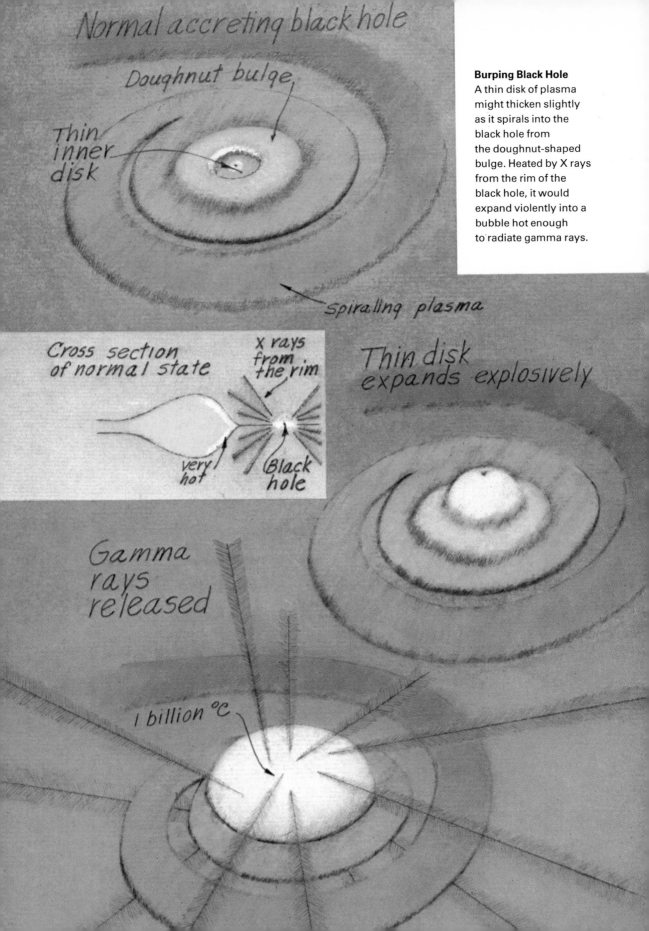

Normal accreting black hole

Doughnut bulge

Thin inner disk

spiralling plasma

Burping Black Hole
A thin disk of plasma might thicken slightly as it spirals into the black hole from the doughnut-shaped bulge. Heated by X rays from the rim of the black hole, it would expand violently into a bubble hot enough to radiate gamma rays.

Cross section of normal state

X rays from the rim

very hot

Black hole

Thin disk expands explosively

Gamma rays released

1 billion °C

has also been suggested as an explanation for GRBs. Hundreds of billions of tons of matter would fall in globs along intense magnetic-field lines to the hot polar surfaces, heating and spewing out gamma rays as it picked up speed.

Other models relate the GRBs to glitches—sudden changes in a spinning pulsar's precise period, the time interval between pulses. Franco Pacini of the Astrophysics Laboratory in Frascati, Italy, and Malvin A. Ruderman of the Columbia University Physics Department in New York City, suggest that GRBs are caused by glitches in old pulsars. Like a spinning ball, a pulsar's rotation causes its equator to bulge and its poles to flatten. After millions of years, a pulsar slows down, quits beaming radio waves, and adjusts to become more spherical. A crack may develop near the pulsar's equator and thin slivers of the surface, like the outer layers of an onion, might lurch toward the poles. These readjustments may disrupt the magnetic-field lines that are anchored in the surface and the vibrating lines would excite the surrounding charged particles, sending out gamma rays. Or, the lurching layers might generate frictional hot spots from which gamma rays would be radiated.

Another neutron star theory invokes a state of nuclear matter that stretches the imagination. Before his death in 1974, astrophysicist Fritz Zwicky suggested that nuclear goblins cause GRBs—a theory independently developed by Russian astronomer G. S. Bisnovatyi-Kogan and his co-workers. Nuclear goblins are temporarily stable giant cores, about 10 feet (3 meters) in diameter and weighing 10 million billion tons. They might form beneath the iron crust of a neutron star. If one escaped through a volcano or a crack in the crust, it would act like an enormous radioactive nucleus, madly sputtering neutrons, atomic nuclei, and gamma rays as it shot upward.

To astronomers, the most ingenious GRB source theory involving a neutron star appeared in print in early 1974. Grindlay and Fazio envision a neutron star emitting a few iron particles the size of BBs during a readjustment of its surface. The iron BBs would zip through space at the speed of light until they encountered a star like our sun. At a distance well outside the orbit of Pluto, they would heat up as they began colliding violently with sunlight photons. The BBs would ionize and then emit characteristic X rays. Those X rays emitted in the forward direction would appear to observers in their path as gamma rays, because of the high-speed relative motion.

The Grindlay-Fazio theory is one of the few that can be verified experimentally. A detector on a distant space probe that was not in line with the BBs' motion toward earth would see a burst of X rays at about the same time an earth-orbiting GRB detector activated.

If speeding BBs do not sound plausible, perhaps black holes do. A black hole is the ultimate fate of a collapsed star core containing more than about twice the mass of the sun. It has so much mass that it continues collapsing through the white dwarf and neutron star stages to infinite density and zero volume. It is called a black hole because nothing, not even light, escapes its complete gravitational grip.

There is mounting circumstantial evidence linking a number of GRBs to Cygnus X-1, a rapidly fluctuating X-ray source consisting of a massive blue star and an invisible companion that many astronomers suspect is a black hole. This link inspired me to propose that GRBs may originate in the distorted regions of space near black holes.

In a blue star-black hole binary, the powerful black hole would pull matter from its giant companion. The matter is hot enough to be a plasma—a mixture of ionized atoms and free electrons. The plasma spirals into the black hole, gaining speed and forming an accretion disk. At a radius of about 6,000 miles (10,000 kilometers) from the hole, the pressure of X-radiation from the edge of the hole counterbalances the intense gravitational pull and the plasma particles form a doughnut-shaped bulge in which they spiral slowly in nearly concentric circles. But a few plasma particles near the midplane of the disk are sucked inward from the bulge and quickly accelerate until they reach nearly the speed of light. They radiate the counterbalancing X rays before plunging into the abyss.

These terminal particles form a thin inner accretion disk within the doughnut-shaped bulge. It is only about 0.0001 as thick as the bulge. This inner disk may not be stable. If it thickens even a little, the X rays will quickly heat the disk, causing it to expand explosively into a hot sphere of thin plasma that may reach a temperature of 1 billion °C—hot enough to release vast numbers of gamma rays. Once the thin disk became unstable, the explosive release of gamma rays would follow in less than about 0.1 second.

To account for the GRB drum roll-like peaks, several hundred such expansions would have to occur in such rapid-fire succession that they often appear as a single signal. The detailed Apollo 16 GRB signal supports this view. During the largest peak, when the detector was registering about 1,000 counts per second, the counting rate suddenly dropped to zero for 0.015 second, then rose again. This is difficult to explain unless we assume that the GRB peaks are actually built of many much briefer spikes that our detectors do not always distinguish.

Do GRBs actually consist of hundreds of tiny spikes? We cannot say yet. Unlike the fast-flashing pulsars, which play a kind of periodic peekaboo that observers can monitor at leisure, the unpredictable GRBs storm across the earth never to be seen again. If we could pinpoint their sources, we could then train giant optical telescopes and radio telescopes on those spots and look for other radiation, even though a burst had passed. Now that we have accidentally discovered a little information about GRBs, we can design and build tailor-made gamma-ray detectors to tell us even more. We hope that a scientific gamma-ray satellite will be built to look in greater detail at GRBs. For now, their origins remain a major cosmic mystery whose solution is anyone's guess.

Solar Systems In Miniature

By Joseph Veverka

**The strange icy moons of the outer
planets are helping astronomers test
theories of how our solar system evolved**

Out beyond the orbit of Earth, beyond Mars, beyond the asteroid belt lies what appears to be a miniature replica of the solar system—the giant planet Jupiter and its satellites. The four largest, called the Galilean satellites after their discoverer Galileo Galilei, bear an uncanny resemblance in many ways to the planets of the solar system.

After Galileo's discovery in 1610, and the discovery of Saturn's major moons a few decades later, satellites were largely ignored for over 300 years. But since 1970, improved instruments on Earth, plus those carried by spacecraft, have allowed astronomers much closer looks at these objects. They have come to realize that the satellites are unusual objects, with environments that are quite bizarre. They also realize that the satellites, as well as the planets, must be studied closely if we are to get a comprehensive view of how our solar system formed.

The 33 natural satellites known to circle various planets of the solar system can be conveniently divided into two categories, mostly on the basis of size. The large, or regular, satellites—those with diameters of 400 kilometers (250 miles) or more—probably still occupy orbits close to the regions where they originally formed billions of years ago.

Seen from Europa, Jupiter's satellite Io sports a tail of glowing sodium atoms that have been chipped from its surface by radiation from Jupiter.

About half the known satellites belong to this group, including Jupiter's Io, Europa, Ganymede, and Callisto (the Galilean satellites); Saturn's Titan and Iapetus; the five satellites of Uranus; and Neptune's Triton. Studying them is likely to yield information about physical and chemical conditions in well-defined areas of the original solar nebula out of which all the planets and satellites formed.

The remaining satellites are much smaller, some only 10 kilometers (6 miles) in diameter. Most of them are probably stray pieces of interplanetary debris that wandered close enough to a large planet to be captured by its gravity. Small satellites have their own fascination, but they are difficult to fit into the history of the solar system because we cannot be sure where they formed.

The largest of the regular satellites are planets in every sense except that they orbit a larger planet and not the Sun. Several of them are larger than Mercury and nearly as big as Mars.

The large satellites of Jupiter, Saturn, and Uranus have another intriguing property—their orbits form miniature copies of the planetary orbits, with the parent planet replacing the Sun at the center. The orbits are nearly circular, they lie in or near the plane of the parent planet's equator, and, like the planets, the satellites travel in a direct orbit—counterclockwise, as seen from above the north pole. The smaller, stray satellites, on the other hand, tend to have irregular orbits that may be highly tilted to the planet's equatorial plane, and some even travel backwards, in retrograde orbits. The large satellites also may share another parallel with the planets of the solar system—a progressive decrease in density as their orbits move outward from the parent body. This is true for Jupiter's Galilean satellites and it may also be true for others.

Astronomers generally agree that the solar system formed out of the solar nebula—a large, flattened disk of gas and dust that surrounded the Sun soon after its formation. The roots of this idea go back to the work of the German philosopher Immanuel Kant and the French astronomer and mathematician Pierre Simon Laplace in the late 1700s. However, only since 1955 have we made any significant progress in understanding the chemical process involved.

There is still considerable dispute over the details, but the basic idea is this: The original solar nebula had the same chemical composition as the Sun—about 75 per cent hydrogen, almost 25 per cent helium, and less than 1 per cent heavier elements, by weight. Close to the Sun, the temperature of the nebula was so high that only those materials with extremely high boiling points, such as most metals and rocks, could condense into solids. Volatile materials—those that boil at much lower temperatures, such as water, ammonia, and methane—could only condense much farther out in the nebula, where the temperatures were very cold. At the distance of Pluto, for instance, temperatures even then were probably as low as $-255°C$ ($-425°F.$), cold enough to condense almost every material except helium and hydrogen.

The author:
Joseph Veverka is assistant professor of astronomy at Cornell University. The satellites, especially Titan, are among his major research interests.

One specific model, or theory, constructed to expand upon this general explanation of solar system formation is the chemical equilibrium model developed in 1971 by geochemist John Lewis of the Massachusetts Institute of Technology. His model makes precise predictions about the chemical composition of planet-sized bodies, based on the temperatures at various distances from the Sun when they were forming. The predictions match tolerably well what we have learned of the composition of the inner planets—Mercury, Venus, Earth, and Mars—as well as the composition of the asteroids, fragments of which we can study when they fall to Earth as meteorites. But what about his predictions for the region beyond the asteroid belt, where large amounts of water, ammonia, and methane should have condensed?

Unfortunately, the outer planets are not much help in answering this question. The giant planets, especially Jupiter and Saturn, are so massive that they were able to hang on to large quantities of the original nebula gases, which just blew off the smaller planets and satellites. Thus, their composition was determined more by their size and gravity than by the temperature conditions in their part of the nebula; it is probably identical to that of the Sun. They could have formed almost anywhere in the original nebula.

This is where the satellites come in. Even the largest ones have too little gravity to be able to hold large amounts of uncondensed solar nebula gas. Their composition had to be determined by the conditions in the part of the solar nebula where they formed. Because we believe that they formed close to where they are now, they provide a test for solar system models.

For instance, Lewis' model predicts that satellites at the orbit of Jupiter should consist mostly of water and ammonia in the form of water ice and ammonium hydrate solution; at the distance of Saturn, some frozen methane would be mixed in with the water ice and ammonium hydrate; and out near Uranus and Neptune, satellites could be mostly frozen methane, possibly with some frozen argon. Our limited knowledge of satellites provides some support for these predictions. Callisto may well consist of an outer crust of frozen water and a liquid interior of ammonium hydrate solution, while Titan may also contain frozen methane and methane hydrate. As yet, we know too little about the satellites of Uranus and Neptune to know whether they could be mostly methane as the model requires.

There may be a complicating factor, however. Just as the hot temperatures near the Sun prevented any volatile materials from condensing nearby, so the heat generated by a giant planet as it contracted may have affected the condensation of materials in its immediate vicinity. This would cause local variations in the predictions of the simple chemical equilibrium model. For example, it appears that Jupiter was still contracting appreciably when its satellite system began forming some 4 or 5 billion years ago. During this period of contraction, Jupiter's surface temperature was several thousand degrees, hot enough to keep volatile materials from condensing in abundance anywhere close by. This could explain why the two inner Galilean satellites, Io and Europa, appear to be dense and rocky, while the outer two, Ganymede and Callisto, are light and probably icy.

A Catalog of Moons
The major satellites
of the solar system
resemble planets in
many ways, *opposite
page.* Some are larger
than Mercury and
nearly as big as Mars.

In order to test the accuracy of Lewis' predictions, we need to know the chemical composition of each satellite. But until we can obtain actual samples, our only definite information about average composition is based on measurements of a satellite's density—that is, the average mass of one cubic centimeter of its material. Rocky satellites should have densities of about 3 grams per cubic centimeter; those made mostly of water and ammonia should be close to 1.7; and objects consisting mostly of frozen methane should be near 1.0.

To determine the average density of a satellite, we must know both the satellite's mass and its volume. Neither is easy to measure. For a distant satellite, the radius, from which we can figure the volume, may be difficult to determine precisely because our Earth-based telescopes cannot make a sufficiently sharp image of the satellite. Satellite masses are also difficult to measure because this usually involves studying how the satellite deflects the paths of other satellites, and this is generally a small effect. It is much easier to determine the mass from the way a satellite deflects a passing spacecraft. Pioneer 10 provided this information, as well as refined data on size, for the Galilean satellites; we hope to get the same type of data for some of Saturn's satellites when Pioneer Saturn (formerly Pioneer 11) approaches Saturn in 1979.

The average densities of the Galilean satellites decrease progressively as their distance from Jupiter increases. Io, the closest large satellite, has a density of 3.5 grams per cubic centimeter. The density of Europa is 3.4; of Ganymede, 2.0; and that of Callisto, the most distant of the four, is 1.7. These measurements give us a crude clue to the average chemical composition of each satellite.

In addition to measuring density, we can also learn about a satellite by studying its spectrum and by measuring the way it polarizes, or alters the vibrations of, incoming sunlight. Solids absorb radiation at characteristic wave lengths determined by their chemical composition. By comparing the spectrum of a satellite with laboratory spectra of known materials, we can learn something of the satellite's surface composition. Polarization measurements help us determine whether a surface is solid, liquid, or gas, because each of these forms of matter polarizes light in a different way.

Astronomical observations, other than density measurements, tell us only about a satellite's surface, not about its average chemical composition. In other words, we must beware of surface camouflage. For instance, spectra of Europa at infrared wave lengths show that the surface is covered with water frost. Yet the mean density (3.4 grams per cubic centimeter) indicates that Europa must be primarily rocky.

On the other hand, Callisto has just the right mean density (1.7 grams per cubic centimeter) to be the kind of water-ammonia object

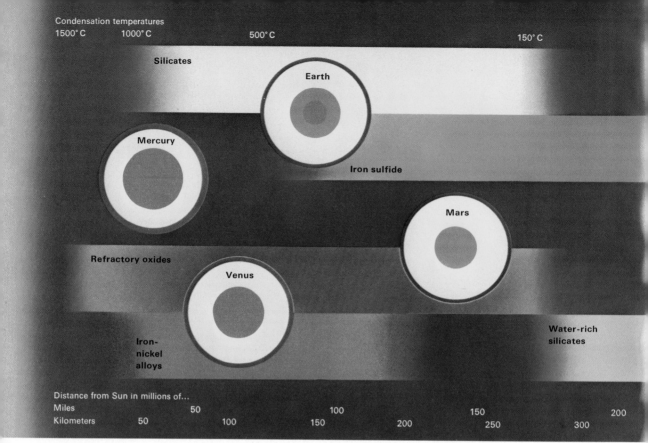

Condensation temperatures

1500°C   1000°C   500°C   150°C

Silicates

Earth

Mercury

Iron sulfide

Mars

Refractory oxides

Venus

Water-rich
silicates

Iron-
nickel
alloys

Distance from Sun in millions of...

Miles   50   100   150   200
Kilometers   50   100   150   200   250   300

Where Chemicals Condensed

Temperatures in the emerging solar system determined where various materials condensed out of the solar nebula's gases. Predictions about the composition of bodies that formed as temperatures fell off fit what is known of the inner planets, but still must be verified for the outer regions.

that Lewis predicts should be found near Jupiter. Yet its surface is covered with a peculiar dark material that looks like rock powder. The interior of Callisto may also be quite strange, consisting of a liquid solution of ammonia and water, with only a thin outer crust of ice. Imagine the bizarre geologic processes on such an object—any crack in the outer crust would produce water-ammonia volcanoes.

Perhaps more than anything else, the Galilean satellites are profoundly influenced by Jupiter's intense radiation belts, the high-energy protons and other charged particles trapped in the planet's magnetic field. Io, the closest to Jupiter, moves through the thickest part of the radiation belts and its surface is continually bombarded by high-energy protons and electrons.

One sign of the effect of Jupiter's radiation on Io is the cloud of glowing sodium atoms that envelops the satellite. Such sodium clouds are not seen around any other satellite. Presumably, protons from Jupiter's radiation belts are chipping off, or sputtering, sodium atoms from the surface of Io. About 10 million protons of suitable energy (1,000 electron volts) hit 1 square centimeter (0.155 square inch) of Io every second, sputtering off about 100,000 atoms. Most of these atoms are knocked off with speeds fast enough to escape the gravitational pull of Io, but not fast enough to escape the neighborhood of Jupiter. As a result, Io is surrounded by a cloud of sputtered atoms that trails nearly one-third of the way around its orbit. Hydrogen is the only

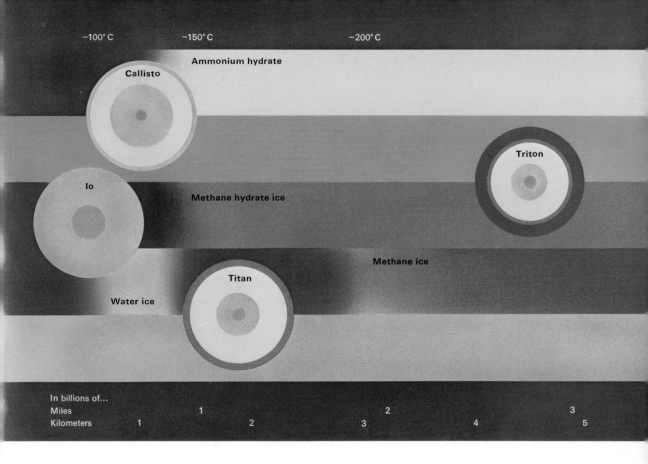

other sputtered atom found around Io so far, but it is likely that other atoms are also being chipped off.

Exposure to intense bombardment by protons may also explain why the closer to Jupiter a satellite is, the redder it is. Callisto, the most distant of the four, appears gray when viewed through a telescope, but Io is definitely reddish-orange. It seems likely that charged particles in Jupiter's radiation belts interact with the satellite surfaces to produce substances that strongly absorb ultraviolet and blue light, giving some satellites a reddish hue.

Perhaps we can understand how such substances might be produced if we look at a model of Io proposed in 1974 by Fraser P. Fanale, Torrence V. Johnson, and Dennis L. Matson of the Jet Propulsion Laboratory (JPL) in Pasadena, Calif. They devised their model after a study involving proton bombardment of various substances in a laboratory, chemical and optical analyses of meteorites, and spectral observations of Io. They suggest that Io originally formed from the same materials that make up the most primitive meteorites, called carbonaceous chondrites. In addition to containing the silicon compounds usually found in stony meteorites, carbonaceous chondrites are rich in carbon and in volatile materials such as water. They also contain radioactive elements such as thorium, uranium, and potassium. When a solid body the size of Io—3,640 kilometers (2,260 miles) in diameter—is made of such material, the radioactive elements generate heat faster than it can be conducted to the surface. The inside temperature rises until the interior melts, producing large quantities of water that is rich in dissolved materials such as sulfates of magnesium, calcium, and sodium. Such salts are occasionally found in some carbonaceous chondrites, lending credibility to this idea.

In the case of Io, the JPL scientists suggest that the solution containing the dissolved sulfates somehow found its way to the surface. The water then evaporated, leaving the salts that now cover the surface. These deposits, rich in sodium compounds, would provide an ample source of atoms for Io's sodium cloud.

However, to match Io's spectrum, we need a surface material that is very bright at red wave lengths, but dark in the blue and ultraviolet. Sodium is no help here; the element is colorless, and so are most common sodium compounds, including sodium sulfate. But sulfur, one of the few materials whose redness resembles that of Io, can be efficiently produced by proton bombardment of sulfates and other sulfur compounds. This process would produce the most sulfur near the poles of Io, where most of the protons from Jupiter's radiation belts hit, and may account for Io's dark-red polar caps, first noticed in 1884 by astronomer Edward Emerson Barnard.

It seems likely that sulfur is one of the substances that makes Io red. But is it the only one? To try to answer this, we must turn to Titan, Saturn's largest satellite.

Titan, the largest satellite in the solar system with a diameter of 5,800 kilometers (3,600 miles), is an intense red color. From the way Titan polarizes incoming sunlight, it is clear that what we see at visible wave lengths is not a solid surface, but merely the top of a thick cloud layer. These clouds, then, must contain some substance that absorbs ultraviolet and blue light, reflecting back only red.

So far, Titan is the only satellite known to have an extensive atmosphere. As early as 1940, astronomer Gerard P. Kuiper found evidence of methane in Titan's spectrum. But methane alone cannot account for the redness, nor can any of the other gases that have been detected in Titan's atmosphere, such as ethane and other simple hydrocarbons.

The most likely explanation seems to be that ultraviolet radiation from the Sun acts on the methane high in Titan's atmosphere to produce larger, more complex molecules in a kind of photochemical smog. This process has been demonstrated in a number of laboratory simulations that typically produce a reddish goo, rich in large organic molecules, whose color closely matches that of Titan.

Such materials may not be limited to Titan. There is a striking similarity between the color of Titan and that of the reddish regions on Jupiter and Saturn, suggesting that the same materials are probably produced in the upper atmospheres of these planets. But on the giant planets, large accumulations of these compounds are unlikely. As the organic goo settles deeper into their atmospheres, the temperature will eventually rise high enough to break down complex molecules. In

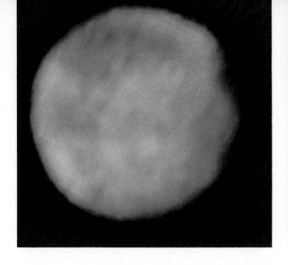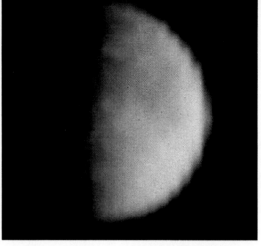

Titan's atmosphere, however, temperatures never get high enough to break down complicated molecules and many of these probably settle out of the atmosphere and accumulate on the surface. Titan's surface could be littered with large organic molecules similar to those thought to have played prime roles in the creation of life on Earth–an exciting prospect to those studying the possibility of life elsewhere.

Saturn's rings also have a puzzling red color. The rings are made up of myriads of very small satellites, each in its own orbit around the planet. Most observations suggest that these particles range in diameter from a few centimeters to perhaps several meters (from about an inch to several yards). Infrared spectra of the rings reveal strong absorption bands characteristic of water ice, proving that the particles are ice balls. But the ice must be contaminated, because the rings are dark at blue and ultraviolet wave lengths, unlike pure water ice.

But what is the material that absorbs ultraviolet and blue light to give the rings their red tinge? Is it related to the red organic goo produced on Titan by ultraviolet light, or the red sulfur produced on Io by Jupiter's radiation? Is there some universal process that produces reddish materials whenever mixtures of ices–water, methane, ammonia, hydrogen sulfide–are exposed to energetic radiation? A vigorous program of laboratory research is now underway to answer these questions.

Meanwhile, the production of the red photochemical smog that absorbs ultraviolet light high in Titan's atmosphere may explain another peculiarity of this satellite. At Titan's distance from the Sun, its temperature should be between −195° and −185°C (−315° and −300°F.). But the actual temperature, determined by studying Titan's infrared spectrum, is higher than expected by about 40°C (72°F.).

There are at least two possible explanations for this. The simplest theory is that as the red hydrocarbon particles in the atmosphere greedily absorb ultraviolet light, they heat up to what are high temperatures by Titan standards. It is the heat from these particles at the top of the cloud layer that accounts for the high observed temperatures. The surface temperature could be as low as −195°C (−315°F.). This model requires only enough atmosphere to support the particles,

Volcanoes break through a frozen surface to spew forth their gases in an imaginative view of Titan. Sunlight may act on methane in the thick, red clouds to form complex molecules that gather in rusty pools of organic goo.

Heating the Particles
The unexpectedly high temperature observed for Titan may be caused as sunlight heats up hydrocarbon particles high in the atmosphere. Temperatures could be much lower on the surface. Atmospheric pressure might be only 2 per cent of Earth's.

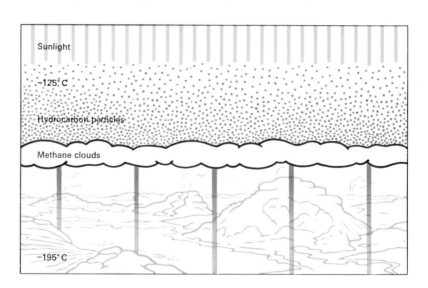

Sunlight

−125°C

Hydrocarbon particles

Methane clouds

−195°C

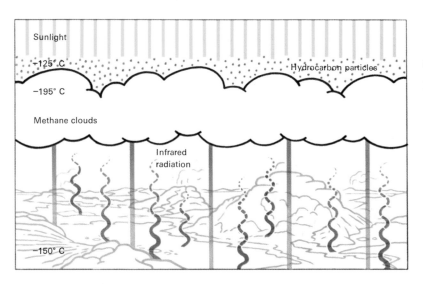

Sunlight

−125° C

Hydrocarbon particles

−195° C

Methane clouds

Infrared
radiation

−150° C

Stoking the Atmosphere
An alternative theory of
Titan's high temperature
is that the atmosphere
allows sunlight through,
but traps the infrared
radiation from the
surface. This requires
substantial atmospheric
pressure and could
produce relatively high
surface temperatures.

and the surface atmospheric pressure could be only 20 millibars, 2 per cent of the sea-level pressure on Earth.

A more complicated model employs the greenhouse effect, in which the atmosphere allows sunlight to penetrate, but traps much of the infrared radiation subsequently emitted by the surface, causing the temperature to rise. On Earth, the greenhouse effect is caused by carbon dioxide and water vapor; on Titan, it could be due to methane, or perhaps hydrogen. To trap heat in this way, Titan would need enough atmosphere to give a surface pressure of at least 400 millibars, 40 per cent of that on Earth. The surface temperature in this model could be as high as $-125°C$ ($-190°F.$) or warmer.

The true situation on Titan probably falls between these two extremes. A weak greenhouse effect may keep the lower atmosphere warm, while the red hydrocarbon particles may heat up the upper atmosphere. Continuing observations indicate that the surface temperature is probably about $-150°C$ ($-235°F.$), but it could be higher. As yet, we cannot be sure how deep the red clouds of Titan are, whether there are other clouds below them, or what the surface pressure is. It is certainly more than 20 millibars, because the amount of methane seen above the cloud tops alone is enough to create this pressure.

The possible presence of molecular hydrogen on Titan, announced in 1972 by Laurence Trafton of the University of Texas, poses another interesting problem. Titan, with its relatively small mass, should long ago have lost all the hydrogen in its atmosphere. What replenishes the supply? Perhaps ultraviolet light breaks down complex atmospheric molecules containing hydrogen, or the hydrogen may be produced from volatile surface materials. A volatile-rich surface is quite likely on Titan. Judging from its low mean density (about 1.4 grams per cubic centimeter), Titan is mostly ice, probably frozen methane and water on the outside with a liquid interior of ammonium hydrate—just the sort of composition predicted by Lewis' chemical equilibrium model. One theory envisages ammonia gas escaping from ammonium hydroxide volcanoes on the surface to be broken down into hydrogen and nitrogen by ultraviolet light.

But there may be another way of keeping hydrogen in Titan's atmosphere—recycling. Gas molecules escaping from Titan, like those leaving Io, are not completely lost. Many of them go into orbit around Saturn. Thomas R. McDonough and Neil M. Brice of Cornell University in Ithaca, N.Y., calculated in 1973 that these atoms could form a large doughnut-shaped cloud that completely envelops Titan's orbit. As Titan moves through this cloud, it may pick up as many hydrogen atoms by collision as it loses.

Titan's atmosphere and surface should be relatively easy to explore by spacecraft. Unlike Saturn or Jupiter, Titan is not massive enough to accelerate an incoming probe so much that it would burn up as it enters the atmosphere. And Titan has a solid surface to land on, something that Saturn and Jupiter probably cannot offer. Thus, Titan will likely provide our first detailed look at an object with a hydrogen-rich, oxygen-poor atmosphere similar to those prevalent on the giant planets and the kind thought to have existed on the early Earth. In this kind of atmosphere occur the complex chemical processes that produce the large organic molecules that are the forerunners of life.

One of the strangest aspects of many satellites is the change we observe in their brightness as they travel around their orbits. The difference is most pronounced for Iapetus, the ninth satellite out from Saturn. Like most satellites, Iapetus always keeps the same side turned toward its parent planet. This means that one half is always facing ahead as it moves along its orbit, while the other is always facing behind. When it is on the eastern side of Saturn and its leading side faces us, Iapetus appears very faint. But it becomes six times brighter as it swings to the west, turning its trailing side toward us. This difference cannot be explained by an irregular shape, as is the case with some small asteroids. Iapetus is too big—about 1,600 kilometers (1,000 miles) in diameter—to be anything but round. Internal pressure forces such a large object into a ball.

One explanation for the variation in brightness is that interplanetary dust may have chipped off a thin layer of bright ice from the leading side, revealing a dark surface underneath. But if this is so, why is the other side unaffected? Another idea is that dark material is being deposited on the leading side, but not on the trailing side. But what is this dark material, where does it come from, and why is it deposited only on one side? In either case, why is the effect much more pronounced on Iapetus than on the other satellites? Is Iapetus a freak? Or is the striking difference between its two faces a reminder that we are still unaware of some key process in satellite development?

We know enough about the larger satellites of Jupiter and Saturn to know that they warrant close exploration. But so far, we know almost nothing about the satellites of Uranus and Neptune. The five satellites of Uranus have beautifully regular orbits, suggesting that they too may have formed in much the same way as the planets. If so, their properties may change significantly with distance from the planet, as do those of Jupiter's Galilean satellites. Triton, the larger of Neptune's two satellites, is of interest because it may have an atmosphere. And, unlike the other large satellites, it moves in a retrograde orbit. Then there is the strange case of the planet Pluto. According to Kuiper, Pluto may be an escaped satellite of Neptune. It would be interesting to see if any similarities exist between Pluto and Triton.

As yet, we have only begun our study of the satellites. Many questions remain, and new ones are constantly arising. A major task of future space exploration will be to test models of the formation of the solar system. Unfortunately, we have only one solar system at our disposal, but the large satellite systems may provide the best laboratories to test our speculations. Beyond that, we will continue to explore the solar system simply out of a spirit of adventure.

Science File

Science Year contributors report on the year's major developments in their respective fields. The articles in this section are arranged alphabetically by subject matter.

Agriculture

Anthropology

Archaeology
Old World
New World

Astronomy
Planetary
Stellar
High-Energy
Cosmology

Biochemistry

Books of Science

Botany

Chemical Technology

Chemistry
Dynamics
Structural
Synthesis

Communications

Drugs

Ecology

Electronics

Energy

Environment

Genetics

Geoscience
Geochemistry
Geology
Geophysics
Paleontology

Immunology

Medicine
Dentistry
Internal
Surgery

Meteorology

Microbiology

Neurology

Nutrition

Oceanography

Physics
Atomic and Molecular
Elementary Particles
Nuclear
Plasma
Solid State

Psychology

Public Health

Science Support

Space Exploration

Transportation

Zoology

Agriculture

Plant breeders at Cornell University, Ithaca, N.Y., working with researchers in Mexico, announced in January, 1975, that they have found varieties, or strains, of corn that are resistant to two of the world's most destructive corn pests — the sugar cane borer and the European corn borer. Through crossbreeding, the scientists hope to incorporate the genes for the resistance into commercial varieties of corn within three years. Corn is the world's third most important food grain, after wheat and rice.

Vernon E. Gracen of Cornell directed the research in collaboration with scientists at the International Center for Maize and Wheat in El Batan, Mexico. With graduate student Sue Sullivan, he tested most of the 12,000 corn strains in the world maize collection housed at the Mexico center during 1973 and 1974. First they planted the corn strains in upstate New York during the two summers and deliberately infested them with the European corn borer. Seeds from the corn that survived were then sent to Mexico for winter planting where the resulting plants were exposed to the sugar cane borer. The scientists thus got two crops a year and found several varieties — all from among wild types of corn that originated on islands near Antigua in the West Indies — with resistance to both insects. According to Gracen, the mechanism behind the resistance in these plants is as yet unknown, but it is different from and more effective than the resistance to the corn borer found in some commercial American varieties.

Sorghum hybrids. Bruce Maunder of the U.S. Department of Agriculture (USDA) experiment station at Lubbock, Tex., successfully incorporated resistance to greenbugs into two outstanding hybrids of grain sorghum. Limited quantities of seeds of the new hybrids were available for 1975 plantings. Sorghum is the world's fourth largest cereal grain crop.

USDA experiment stations in the South have developed another variety of sorghum, called Theis, which is superior for syrup (molasses) production. The new variety is more resistant to

A Beefalo, produced by crossbreeding buffalo and beef cattle at Tracy, Calif., thrives on grass alone and produces tender, less costly meat.

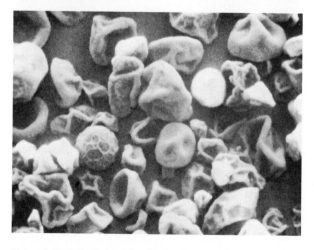

Pinhead-sized plastic capsules filled with a powerful insecticide, *top,* became the first microencapsulated pesticide approved in the U.S. As the capsules release insecticide gradually, they deteriorate, *center and above.* Farmers spray less often with this product.

disease than traditional varieties and it produces more syrup. Theis is adapted to the climate and soils of the Southeastern states.

Purdue University scientists produced a high-protein variety of sorghum in 1974. The new variety is high in lysine, a substance that increases digestibility. Plant researcher John D. Axtell and doctoral student Daya P. Mohan produced the new variety in the laboratory by chemically treating some commercial American strains of sorghum to induce mutations. Because it combines the high-protein trait with the heavy-yielding trait of American strains, the new variety is expected to be welcomed by farmers. It is also adapted to temperate climates.

Other new crop varieties. Three new early-maturing varieties of rice — IR28, IR29, and IR30 — were developed in 1974 by the International Rice Research Institute at Los Baños in the Philippines. The new varieties are highly resistant to most of the insects and diseases that attack rice, and they grow much faster than other varieties developed at the institute. Farmers using these plants can produce more than one rice crop in a year, or they can plant other crops on the land after they harvest the rice.

Larry N. Baker, a plant breeder at the Michigan Agricultural Experiment Station at East Lansing developed four hybrid supercarrots in 1974. He named them Spartan Sweet, Spartan Fancy, Spartan Delite, and Spartan Bonus. They have more vigor, uniformity, color, sweetness, and carotene, a pigment that is converted into vitamin A in the human body. They also produce a higher yield of seeds.

Animal breeding. Harold Hafs and his students, working with dairy cows and horses at the East Lansing experiment station, found that a hormone-like substance known as prostaglandin $F_{2\alpha}$ effectively controls estrus and makes artificial insemination more efficient.

Estrus is the breeding period during which a female animal is most receptive to the male. It coincides with ovulation, and therefore the female is likely to conceive during estrus. The new hormonelike substance controls ovulation and thus enables artificial insemination of cows and horses when detection of estrus is difficult or impossible. The new process should prove of great value in cattle breeding.

Scientists for the first time used embryo transplants in dairy cattle to produce living offspring at a farm in Mt. Vernon, Wash. Herbert Johnson of Mt. Vernon performed the transplant surgery in 1974. The object of such embryo transplants is to allow an animal that normally has only one calf a year (and probably a maximum of from 5 to 10 calves in a lifetime) to "incubate" from 20 to 30 calves a year. This would greatly enhance the reproductive role of genetically superior cows.

Instead of producing their offspring through the long-term route of pregnancy, genetically superior cows can become living laboratories for producing eggs. Once fertilized, their eggs can be taken from them surgically and placed in common cows for incubation to the full 9-month term.

William N. Day, professor of physiological reproduction at the University of Missouri, Columbia, is using embryo-transfer techniques to develop a new kind of "piggy bank." He intends to start embryo pigs from the eggs of "supersows" and from the sperm of "superboars" in a test tube, then implant them in other sows where they can develop until birth. Animal breeder John S. Roussel of the Louisiana State University Agricultural Experiment Station at Baton Rouge has developed a procedure for freezing rabbit embryos and later transplanting them into foster mother rabbits.

Fire blight. Plant pathologist Robert N. Goodman of the University of Missouri identified a bacterial toxin that produces fire-blight symptoms in apples and pears. Fire blight is the most devastating disease for pears and frequently for apples. The toxin quickly identifies the levels of resistance in small seedlings of promising pear and apple varieties. This will save time by eliminating the wait for experimental trees to grow to maturity and produce fruit.

Machinery. Food scientists Justin T. Morris and Ahmad A. Kattan and agricultural engineer Glen S. Nelson of the University of Arkansas at Fayetteville developed a harvester for strawberries in 1974. Agricultural engineer Clarence M. Hansen of Michigan State University in East Lansing invented a decapper, also for strawberries. The two machines can be used in combination to harvest and process strawberry crops rapidly. The machines are better adapted to work with some varieties of strawberries than others, and special care must be exercised in planting the strawberry beds to maximize the efficiency of the harvester. The fruit is removed from the plant by air suction, which has the advantage of protecting the fruit from being bruised by rough machine surfaces.

Hjalmar D. Bruhn and Richard C. Koegel of the University of Wisconsin at Madison designed a new system for processing alfalfa for livestock feed. After the alfalfa is mechanically pressed to remove its liquid, the fibrous parts of the plants are put into a silo for fermentation and the liquid portions are heat-coagulated and prepared as a protein concentrate. Yields of up to 1,000 pounds of protein per acre (454 kilograms per hectare) have been achieved.

Agriculture in China. Ten U.S. agricultural scientists, sponsored by the National Academy of Sciences and led by Sterling Wortman, vice-president of the Rockefeller Foundation, toured the farms of China in September, 1974. The group was surprised at China's great agricultural development during the past 25 years.

Chinese farmers now have the largest irrigated land area in the world, more than twice as much irrigated land as U.S. farmers. They began using dwarf high-yielding rice varieties in the early 1960s, five years before such plants were developed for use in the Philippines and Southeast Asia by scientists at the International Rice Research Institute. The Chinese also have achieved high specialization in multiple cropping, harvesting as many as 12 vegetable crops per year on the same land in some southern provinces.

Lunar soil. Botanist Ralph Baker and soil chemist Willard L. Lindsay of Colorado State University in Fort Collins dispelled previous optimism that soil brought back from the moon was more fertile than that found on earth. Their carefully controlled experiments in 1974 showed that lunar soil has the same growth-promoting properties as the earth's.   [Sylvan H. Wittwer]

Anthropology

The fossil remains of an early relative of modern man were found in the Awash River Valley in central Ethiopia in 1974 by C. Donald Johanson of Case Western Reserve University and the Cleveland Museum of Natural History. The 75 bone fragments, representing about 40 per cent of a complete skeleton, are estimated to date from more than 3 million years ago. The find was by far the most important of the season.

The skeleton, referred to as "Lucy" by Johanson, appears to be that of an 18- to 20-year-old female who stood 3½ feet (106 centimeters) tall. She had lived and died on the shore of an ancient lake. Fossilized turtle and crocodile eggs and crab claws found with her may represent part of her diet. Lucy's significance lies in her completeness and the relatively modern appearance of her skeleton, considering her great antiquity.

Diet and teeth. New information on the probable diet of *Gigantopithecus*, a giant fossil primate with small front teeth and massive back teeth, has been obtained from an unlikely source – the giant panda, whose tooth structure is similar. The lower jaws of three *Gigantopithecus* fossils were found in China between 1956 and 1958.

Paul E. Mahler of Queens College, City University of New York, compared the teeth and jaws of the giant panda to those of bears. The shape of the panda's small front and large back teeth separate it from bears. Its diet of hard bamboo shoots that have to be crushed by the teeth during mastication is unique. Mahler suggested that the panda's dental similarities to the fossil *Gigantopithecus* imply dietary similarities, and that the giant primate also ate hard foods that required crushing.

Language. Attempts to communicate with chimpanzees have included the use of plastic symbols, the sign language of the deaf, and, most recently, a computer. Lana, a chimpanzee at Yerkes Primate Research Center in Atlanta, Ga., has learned enough of a modified English, called Yerkish, to converse with people through a computerized keyboard. Duane M. Rumbaugh of Georgia State University re-

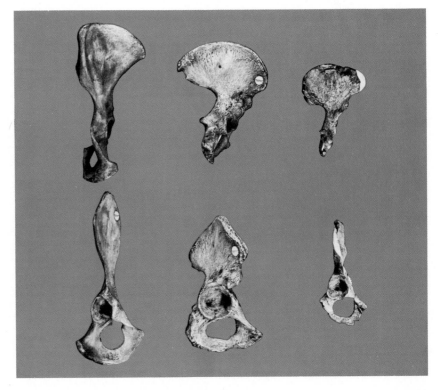

Comparing two views of the pelvic bones of, left to right, a chimpanzee, modern man, and fossil man, shows the difficulty that anthropologists have in distinguishing ape from human bones. In the top row, the fossil bone resembles the human one. In the bottom row, the fossil is like the ape.

ported in December, 1974, that Lana had moved to a stage similar to that of a human child who suddenly discovers that everything has a name and eagerly begins to request the names.

While working with Timothy V. Gill, her teacher and friend, Lana wanted a box filled with candy, an object whose name she had not yet been taught. Lana pushed buttons on her computer console to read, "Tim give Lana name of this?" Gill replied on the computer console that the object was a box, whereupon Lana said, again by console, "Tim give Lana this box."

Lana has since requested the names of other objects. Scientists continue to argue about whether Lana and her fellow apes have really learned language, but all researchers agree that the chimpanzees' linguistic feats are truly amazing.

Zaire chimp. The pygmy chimpanzee (*Pan paniscus*) became the object of new anthropological interest in 1975. The pygmy chimpanzee lives in the remote forests of Zaire in central Africa, and it is seldom seen in zoos.

Geoffrey H. Bourne, director of the Yerkes center, reported in May, 1975, that three recently captured pygmy chimpanzees are being studied there. He said that a number of the small chimpanzee's anatomical features, especially its higher forehead and more dome-shaped skull, are somewhat similar to features found in humans. He suggested that the pygmy chimpanzee occupies an evolutionary position closer to that of man than its larger, more common relative (*Pan troglodytes*). The teeth of the pygmy chimpanzee are also said to be more like those of humans, and researchers have been impressed by its apparent intelligence.

Walking. A large number of primate fossils were recovered from the Fayum deposits in Egypt by Yale University expeditions between 1961 and 1967, and some interesting results of this work were announced in April, 1975, by physical anthropologist Glenn C. Conroy of New York University. The fossil apes *Apidium* and *Aegyptopithecus*, which date from the Oligocene Epoch about 26 to 40 million years ago,

Anthropologist Donald Johanson displays the remains of "Lucy," a 3-million-year-old fossil woman he found in Ethiopia in 1974.

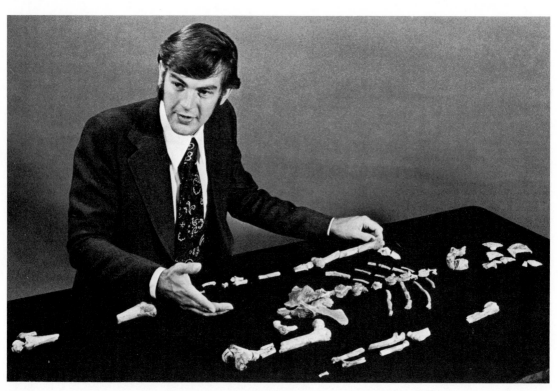

showed greater similarity in limb structure and implied habits of walking to some of the living South American monkeys, such as the squirrel and howler, than to any present-day African or Asiatic apes and monkeys. This suggests that students of primate evolution may obtain more useful information about the evolution of walking upright by studying New World rather than Old World monkeys.

Old age. An analysis of 125 skeletons from 500-year-old burial urns at Hacienda Ayalan on Ecuador's southern coast disclosed that the prehistoric Indians of this area lived much longer than Indians living at the same time in Charles County, Maryland. The study was carried out by Douglas H. Ubelaker of the U.S. National Museum of Natural History.

The average life span of the Ecuadoreans was 60 years, nearly twice the 33-year average age of the Maryland Woodland Indians. Ubelaker attributes the Ecuadoreans' longevity to good nutrition and a reliable food supply, and possibly also to genetic factors. Some

present-day Ecuadoreans, living in remote Andean villages, have also been found to live long lives and still be healthy and vigorous in their old age. Ubelaker's study shows that this phenomenon of long life among these Indians is not just a recent development.

Earwax and cancer. Variations in earwax have been noted by anthropologists for many years. They have found that Asiatics and American Indians have dry earwax, while Europeans and Africans have a wet, or sticky, variety.

Nicholas L. Petrakis of the University of California at San Francisco reported in April, 1975, that wet earwax appears to be associated with a higher incidence of breast cancer than the dry variety. The cancer incidence, which Petrakis thinks may be associated with secretion activity in a nonlactating breast, is probably not directly related to the type of earwax. But the glands that produce earwax and the female breast are both apocrine glands, and the fluids that they secrete may well be conditioned by the same genetic factors. [Charles F. Merbs]

Archaeology

Old World. Three Egyptian tombs in the cemetery of Khufu in Giza (Al Jizah), near Cairo, were excavated by archaeologist Alexander M. Badawy of the University of California at Los Angeles in 1974. They were dated to between 2723 and 2423 B.C. One, a rock-cut tomb, contained paintings and carvings showing scenes of people dancing and tending cattle. The other two had paintings of agricultural scenes but no carvings.

Also in 1974, archaeologist Geoffrey Martin of London's University College led a team that excavated an elaborate crypt at Saqqarah, 18 miles south of Cairo. The crypt was prepared for Horemheb, a general who became pharaoh of Egypt. The 250-square-foot (22-square-meter) tomb was only half completed. Horemheb was a general under the boy-king Tutankhamon, and ruled as pharaoh during the eighteenth dynasty. He was credited with rebuilding a decayed administrative system and helping to restore Egypt's ancient religion after the sun-worship cult of Tutankhamon's father-in-law, King

Akhenaton, was overthrown. Horemheb died about 1320 B.C.

Ancient metallurgy. What may be the world's oldest underground copper mine was discovered in the Timna Valley near the port of Elat in southern Israel. The mine, which was found by Beno Rothenburg of Tel Aviv University and Ronald Tylecote of the University of Newcastle in England, dates to about 1400 B.C. Chisel marks showed that the Egyptians once used stone hammers and metal chisels to mine ore there in a highly complex system of some 200 shafts and galleries that went deep into the earth.

Each tunnel had its own narrow air channel, and along with the shafts had footholds and handholds cut in the walls. The archaeologists also found what appears to be a smelting plant near the copper mine. Analysis of the slag found there indicated that the methods used to separate copper from crude ore as early as 1200 B.C. were as efficient as those known today.

China. Archaeologists excavating near Wuhan in central China's Hupeh

A boomerang fragment about 10,000 years old was found in a South Australian peat bog by archaeologists from Australian National University.

Archaeology

Continued

An 8-day-old Soay lamb was born at the Butser Ancient Farm Project in England. Archaeologists study these rare sheep because they are the descendants of sheep raised by Iron Age farmers about 300 B.C.

Province announced in February, 1975, that they had unearthed the foundations of a palace built at least 3,400 years ago. They also excavated a nearby tomb that dates from the same period. The coffin found in the tomb, though rotten, has the oldest known woodcarvings in China. They were described as "an exquisite animal-mask design and thunder and cloud decorations." There were many other objects.

The find is historically important because it shows that the Shang dynasty culture, which lasted from about 1500 B.C. to 1027 B.C., had spread south from its origins in the Yellow River region to the Yangtze River region when the dynasty was still young.

The palace foundations measured 125 by 36 feet (38 by 11 meters), about half again as big as a tennis court. The building had four rooms leading off an outer corridor, and at least 43 columns supported its roof.

In August, 1974, Chinese archaeologists found a large group of manuscripts, philosophical treatises, and medical texts in a tomb in the southern

city of Changsha. These Han dynasty documents, dating back more than 2,000 years, included two military maps, the oldest found in China.

Little Butser farmstead. Under the direction of Peter J. Reynolds, members of the Council for British Archaeology are farming a reconstruction of an Iron Age farmstead at Little Butser, near Petersfield in Hampshire, England. They are studying how people living in Great Britain about 300 B.C. raised stock, cultivated and stored crops, and erected buildings. The council began the experiment in 1972 and published the first results in late 1974.

The project, called the Butser Ancient Farm Project, has proved that Iron Age people could have preserved their grain in small underground pits, the type that has been found in excavations of Iron Age farms throughout England. Reynolds reports that over 90 per cent of the grain that was stored wet over the winter in such unlined pits germinated in the spring. The group planted emmer and spelt, varieties of wheat known to Iron Age farmers.

Archaeology

Continued

They also raised wild Soay sheep, a type that has survived for over 2,000 years on the island of Soay off the northwest coast of Scotland and is identical with the sheep skeletons found at Iron Age sites. The archaeologists built an Iron Age mud-and-straw roundhouse that easily survived winter weather and strong winds.

The scientists planted the emmer and spelt wheats both in the winter and in the spring so that they could compare the yields of the two crops. Some archaeologists have suggested that Iron Age farmers planted two crops a year in order to obtain enough grain to insure against partial crop failure.

The Little Butser experimenters harvested their yield by using sickles of the Iron Age type. Studies have shown that emmer and spelt have twice the protein value of modern bread wheats. However, the structure of the plants makes them much more difficult to thresh than the modern varieties. The yields of the two types of wheat at the project compared well with the yields of modern wheat fields.   [Judith Rodden]

New World. Tools and spearpoints used by four different groups of Stone Age peoples, discovered in 1974 in a cave in Argentina, have allowed archaeologists to determine how long ago people reached the southern tip of South America. The find also enabled them to date the earliest rock paintings found in the New World.

Augusto Cardich of the University of La Plata, Argentina, found the tools and weapons while digging into a 12-layer deposit of dirt and rock debris that covered the floor of one of the Los Toldos caves in Santa Cruz province. The oldest collection of tools, all stone scrapers, were found in the lowest layer. They were dated at about 10,650 B.C. As yet, the people that used these tools has not been named.

A later group of people left spearpoints and tools two levels above. These have been identified as belonging to the Toldense, a Stone Age people known to have made the paintings found in nearby caves. Archaeologists had not previously been able to date these people, but Cardich's find shows that the

Remains of the oldest house of public worship yet found in Mesoamerica, a structure dating to 1350 B.C., were uncovered near Oaxaca, Mexico, by Kent Flannery of the University of Michigan.

Archaeology

Continued

The Indian stone medicine wheel in the Bighorn Mountains of Wyoming may have been a giant calendar used to mark the coming of summer. Stone piles and the wheel's spokes are aligned with the sun at the summer solstice and to other astronomical events preceding it.

Toldense people lived between 9000 and 7000 B.C. Tools of a third group, the Casapedrense people, were found above the Toldense spearpoints. The Casapedrense tools were dated at 5500 to 3500 B.C. Traces of more recent Patagonian hunter-gatherers were found in the top level.

Central America. Also in 1974, archaeologist Kent V. Flannery and his associates from the University of Michigan at Ann Arbor found the remains of what apparently was a public house of worship. It once stood at San José Magote in the Valley of Oaxaca, Mexico. The oldest structure of this type yet found in Mesoamerica, it was dated to 1350 B.C. The small building once contained an altar for worship and a storage pit filled with powdered lime that was used to make stucco for the walls. The building was made of pine posts and wattle (woven twigs) and clay daub. After it collapsed, other ceremonial structures were erected there.

Archaeologist Peter J. Schmidt of the German Foundation for Scientific Investigations found portions of a large,

stone, human effigy while excavating a pyramid in Mexico in 1974. The pyramid is at San Luis Coyotzingo, in Puebla state. The stone figure, which belongs to the middle postclassical period of Toltec culture and dates to about A.D. 1000, was found deep inside the Coyotzingo pyramid. It is a reclining figure similar to others of that period found in Mexico, the best-known examples of such stone statues—called chacmool figures—are found at Chichen Itzá in Yucatán and at Tula in the state of Hidalgo.

An archaeological team from Southern Methodist University in Dallas, Texas, announced on April 24, 1975, that they had found the largest and oldest Mayan pyramid yet discovered. The pyramid is in Belize, formerly British Honduras, and it is approximately 12 stories high. The archaeologists also found evidence of two similar pyramids that once stood near this one.

North America. Charles E. Holmes and his co-workers at the University of Alaska in College found a prehistoric campsite south of Fairbanks at Dry

Creek, near the Nenana River. They found stone knives, scrapers, pointed cutting tools, and bits of charcoal. Radiocarbon dating of the charcoal placed the age of the site at 8740 B.C. The tools were of types called Denali and Akmak, known from other prehistoric Alaskan sites, as well as Diuktai, a type found in northeastern Siberia.

The first systematic excavation of a Stone Age hunting camp in Ontario, Canada, was conducted in 1974 by William Roosa of the University of Waterloo, Ontario. The site, called the Brophy site, lies northwest of London. Roosa found many flint spearpoints, scrapers, drills, knives, and other tools there. As yet, no date has been established for the campsite, though it was found in soil just above fossil animal deposits that have been dated at 9200 B.C. Many of the spearpoints were Folsom points, the type first discovered at Folsom, N. Mex.

James W. Porter of Loyola University in Chicago excavated a small Illinois village site in Monroe County, near St. Louis, dating to 500 B.C. or earlier. The site was occupied by the first pottery-producing people of the area. Porter found a collection of rather crude, thick, flat-bottomed, and straight-sided jars similar to some previously found in excavations in eastern Iowa and a number of Midwestern and New England states. There were also many spearpoints, scrapers, drills, and knives.

The household remains of four different groups of primitive people were found in 1974 and 1975 in the Tuttle Creek Reservoir area of Pottawatomie County, Kansas, by Larry J. Schmits of the University of Kansas in Lawrence. Three of the groups were dated between 3700 and 2900 B.C. The fourth was older, but has not yet been dated. The diggers found food remains that included bones of bison and deer, as well as the remains of smaller mammals, fish, and amphibians. Grinding slabs and a collection of seeds indicated that the three dated groups also ate plant foods, including bulrush and goosefoot seeds. There were also spearpoints among the remains of the earliest group. [James B. Griffin]

Astronomy

Planetary Astronomy. After 506 days in space, Mariner 10 was switched off on March 24, 1975. The spacecraft had passed the planet Mercury three times in its flight. The first time, on March 29, 1974, it reported the discovery of Mercury's magnetic field, radiation belts, and helium atmosphere, and obtained detailed television pictures of more than 40 per cent of its surface. The second pass, on September 21, extended television coverage into the south polar region.

Mercury's magnetism. On March 16, 1975, Mariner passed only 327 kilometers (200 miles) above the surface, near Mercury's north pole. According to Norman F. Ness of the Goddard Space Flight Center in Greenbelt, Md., this final pass unequivocally proved that the magnetic field is a property of the planet and not a superficial effect caused by interaction with the magnetized solar wind. The field is a magnetic dipole, like the magnetic fields of the Earth and Jupiter, with the magnetic axis tilted about 7 degrees from Mercury's rotational axis. The strength of the field at the equator is about 100 times less than that of the Earth.

The origin of the field is still unclear. It could be caused by motions in a fluid planetary core, or it could be a field that was induced from outside into magnetically susceptible rocks during some primitive epoch. Whatever its origin, it is certain to play a major role in understanding planetary magnetism and the structure of planetary interiors.

What the pictures show. Bruce C. Murray of the California Institute of Technology in Pasadena and his colleagues on the Mariner 10 imaging team propose that the obvious similarity of the surface features on the Moon and Mercury argues strongly that both have a similar crust material and also had a similar history of meteorite bombardment. They also argue that the presence of this low-density, lunarlike crust material on Mercury, which has a high average density, means that the planet must have a dense central core rich in iron. Because differentiation—separation and concentration of denser material into a core—could hardly

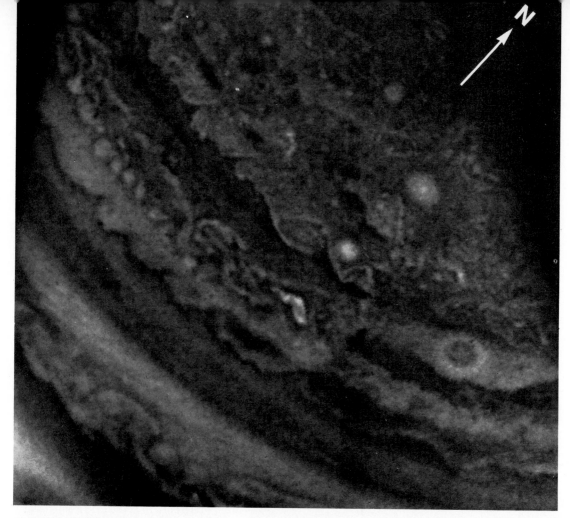

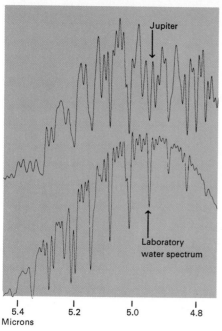

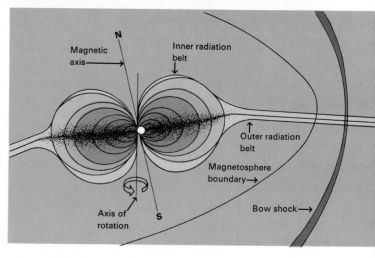

The first look at Jupiter's north pole, *top,* shows that the normal cloud bands give way to a stormy, bubbling cauldron. Dips in a spectrum of the upper atmosphere match those in a laboratory spectrum of water, *left,* proving there is water on Jupiter. Details of Jupiter's radiation belts, *above,* show that high-energy, ionized particles, trapped in the huge magnetosphere, stretch out in a thin disk around the planet.

occur without tremendous disturbance to the surface material, the presence of craters dating back to the planet's earliest history implies that differentiation took place while the planet was still forming and was completed before the bombardment ended.

Television pictures taken during the first Mercury fly-by revealed an enormous crater about 1,500 kilometers (930 miles) in diameter. Newell J. Trask of the U.S. Geological Survey, Washington, D.C., and John E. Guest of the University of London prepared a geologic terrain map from the pictures showing that the crater, called Caloris Basin, and its extended debris dominate almost 25 per cent of the planet's surface. In addition, a region of peculiar, jumbled landforms has been found on the other side of the planet exactly opposite to Caloris Basin. This terrain may have formed when shock waves released during the tremendous collision that formed the crater were focused on the opposite side of the planet.

Doubts about Uranus. Uranus is unique among the planets because the orbital plane of its satellite system is essentially perpendicular to the plane of its orbit around the Sun. Because the rotational axes of the Earth, Mars, Jupiter, and Saturn are all nearly perpendicular to the orbital planes of their major satellites, astronomers believed that Uranus' rotational axis must be tilted over so far that it is almost parallel with the planet's orbital plane. This means that Uranus, rather than spinning like a top, would seem to roll along certain portions of its orbit. Spectroscopic measurements made in 1930 seemed to support this view and also indicated a short rotation period of about 10 hours.

These measurements are now being questioned. Morris Podolak and Robert E. Danielson of Princeton University reported in March, 1975, that they encountered grave difficulties in constructing structural models of Uranus that fit its observed oblateness, or flattening at the poles, and its currently accepted rotation period. Either the assumed direction of the rotational axis is incorrect, or the rotation period must be very much longer than astronomers thought. The validity of the theoretical models is supported by the 1973 discov-

ery that Miranda, Uranus' closest satellite, has an eccentric orbit that is also tilted to the orbital plane of the other four satellites. This discovery, by Ewen A. Whitaker and Richard J. Greenberg of the Lunar and Planetary Laboratory in Tucson, Ariz., suggests that our knowledge of Uranus' rotation is much less firm than was previously thought.

The spectrum of Uranus shows strong, dark bands caused by methane gas in the planet's atmosphere. Early estimates suggested that methane made up roughly 10 per cent of the total atmosphere, which was thought to be mainly molecular hydrogen. Two new, independent estimates announced in March by Jay Bergstralh of the Jet Propulsion Laboratory (JPL) in Pasadena, Calif., and by Michael J. S. Belton and Sethanne Hayes of Kitt Peak National Observatory near Tucson now suggest that the proportion of methane relative to hydrogen may be 10 to 20 times less than previously supposed. In addition, both investigations indicate that Uranus' atmosphere contains another unidentified gas in amounts comparable to molecular hydrogen. This means that the atmosphere may be roughly half molecular hydrogen and half the unidentified gas, and less than 0.5 per cent methane.

A magnetosphere on Saturn? Larry Brown of the Goddard Space Flight Center announced in March that he had tentatively identified radio bursts from Saturn. His conclusion is based on analyzing satellite data for radiation bursts synchronized with Saturn's rotation period. Detection of these bursts, despite a high noise background caused by radio emissions from Earth, implies a strong nonthermal source. This source is probably similar to the one for Jupiter, where radio waves are caused by energetic particles accelerated in the planet's magnetosphere.

Water on Jupiter. Because oxygen and hydrogen, the atomic constituents of water, are both abundant in the universe, astronomers have long supposed that Jupiter's atmosphere should contain water molecules. From the Earth's surface, however, it is extremely difficult to detect the characteristic signature lines of water in Jupiter's spectrum because of the strong overlying lines caused by water in the Earth's

Astronomy

Continued

atmosphere. Also, in most parts of the spectrum where water absorption lines would be seen, they can be overlaid by the stronger lines of methane, hydrogen, and ammonia gas, all of which exist in Jupiter's atmosphere. And finally, the deeper, warmer parts of Jupiter's atmosphere, where water should be present, are obscured by clouds of condensed ammonia higher up.

Overcoming all these difficulties, Harold P. Larson, Uwe Fink, Richard Treffers, and Thomas Gautier of the Lunar and Planetary Observatory announced in March that they had definitely detected water vapor on Jupiter. From a plane flying 12 to 15 kilometers (7.5 to 9 miles) above the ground, far above the Earth's water vapor, they observed a region of the infrared spectrum that probes "hot spots" in Jupiter's clouds and found a series of strong absorption lines characteristic of water. The relative strengths of the lines indicate that the experimenters were probing a region of Jupiter's atmosphere where temperature levels were above 27°C (80°F.).

Jupiter's corona. Gunnar Fjeldbo and his associates at JPL completed an analysis of Pioneer 10 radio-occultation data for Jupiter's ionosphere in mid-1974. Such data is obtained by studying how the radio signal from a spacecraft fades as the craft passes behind a planet. They found that the ionosphere is more than 3,000 kilometers (1,850 miles) high and consists of at least seven separate layers of ionized atoms, with peak electron densities of about 300,000 electrons per cubic centimeter.

By studying the rate at which electron density decreases with height in the top layer, the experimenters deduced that the temperature of Jupiter's upper atmosphere is about 625°C (1160°F.), a surprising result. Theoretical calculations, based on the assumption that the Sun's ultraviolet radiation is the sole source of energy for the upper atmosphere, rarely predicted temperatures above −125°C (−190°F.). Astronomers now speculate that the heat in Jupiter's upper atmosphere is caused primarily by atmospheric waves generated in the turbulent lower atmosphere, which break and dissipate their energy at high altitudes.   [Michael J. S. Belton]

Stellar Astronomy. Astronomers in 1974 and 1975 proved that the sun is truly round, obtained the first pictures of the surface of another star, discovered two remarkable infrared nebulae, and made major pulsar discoveries.

A round sun. Measurements made in 1966 by Robert H. Dicke and H. Mark Goldenberg of Princeton University seemed to indicate that the sun is not a true sphere, but is slightly flattened at the poles. This implied the interior might turn much more rapidly than the surface.

But Henry A. Hill of the University of Arizona announced very different results in June, 1974. His conclusions are based on precise measurements obtained in 1972 and 1973 that suggest the difference between the sun's polar diameter and its diameter at the equator is no more than the 0.016 second of arc — about 11 kilometers (7 miles) — that is expected if the sun rotates as a whole in the same 27-day period observed for its visible surface.

The central armpit. Astronomers at Kitt Peak National Observatory in Arizona announced in December, 1974, that they had obtained the first photographs showing the surface of a star other than our sun. The subject was the red star in Orion called Betelgeuse (possibly from the Arabic *Ibt al Jauzah*, meaning Armpit of the Central One).

Roger Lynds, John W. Harvey, and S. Peter Worden used the 4-meter (158-inch) Mayall telescope at Kitt Peak to take a series of rapid-exposure photographs of the red supergiant in March, 1974. By processing the photographs through a computer, the observers could eliminate the distorting effect of the earth's atmosphere and produce a disk image of the star's surface. Previous photographs showed stars only as points of light.

The photographs show a characteristic darkening toward the edge of Betelgeuse's disk, just as conventional photos of the sun show it to be darker at the edges than at the disk center. Computer comparison of photos made at two different wave lengths yielded a map of the surface that shows large regions of differing temperatures. The warmer regions may be gigantic convection cells, areas where hot gases are circulating between the interior and the star's sur-

New 400-centimeter (158-inch) telescope
at Chile's Cerro Tololo Inter-American
Observatory, *above,* is largest in Southern
Hemisphere. Its first photograph, produced
in October, 1974, was of 47 Tucanae,
the brightest globular star cluster, *right.*

Astronomy

Continued

face. Martin Schwarzschild of Princeton predicted that such cells might possibly exist in the atmospheres of red giant stars in June, 1974, before the Kitt Peak data were analyzed.

Strangers in infrared. The Egg Nebula and the Red Rectangle are the names astronomers have given to two strange infrared sources whose discovery was revealed in early 1975. The infrared emission was originally found in a series of rocket flights directed by Russell G. Walker and Stephan D. Price of Air Force Cambridge Research Laboratories. When ground-based astronomers photographed the positions where strong emission was found, they discovered the two objects and recognized them as members of a new class of phenomena – the biconical nebulae.

At the heart of the Egg Nebula is a star, and a binary, or double, star is in the center of the Red Rectangle. The stars are surrounded by dense dust clouds, possibly doughnut-shaped. According to interpretations proposed by Edward P. Ney of the University of Minnesota and Martin Cohen of the

University of California at Berkeley and their colleagues, the central stars shine through a hole in the middle of the doughnut-shaped clouds, illuminating nebular clouds of gas and dust surrounding them. The nebular clouds reflect the light and thus appear in the visible-light photos. The central stars also heat the doughnut dust clouds, producing the infrared radiation.

The Egg Nebula and Red Rectangle may each be a new solar system in the making. Planets may be condensing from the dust clouds and comets may be forming in the reflection nebulae.

Star measurements. A team of scientists led by Stephen P. Maran of the Goddard Space Flight Center in Greenbelt, Md., in August, 1974, presented the first report on infrared star measurements made by U.S. Air Force satellites orbiting the earth. They found infrared emission associated with the slow, rhythmic pulsations of giant red variable stars. Such infrared radiation could cause the radio waves emitted by water vapor and the hydroxyl radical ($-OH$) that have been detected in

Astronomy
Continued

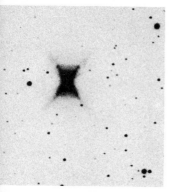

A strange nebula, called the Red Rectangle, has been identified as a source of infrared emission, created when a star heats surrounding dust.

molecular clouds that surround the star R Aquilae. The radio waves could be produced by a process in which an increase in the infrared radiation stimulates a proportional increase in the radio emission from molecules.

Interstellar molecules. Radio astronomers confirmed the detection of a number of new interstellar molecules in late 1974 and early 1975. Of special interest was the detection of methylamine, announced in August, 1974, by a group of Japanese observers. They used the 6-meter (20-foot) radio antenna at Mitaka, Japan, and the 11-meter (36-foot) National Radio Astronomy Observatory antenna on Kitt Peak. Another group used the 64-meter (210-foot) radio telescope at Parkes, Australia. Methylamine is similar in chemical structure to the amino acid glycine. Its discovery in interstellar space increases the confidence of theorists that amino acids, the building blocks of life, can form in space by natural processes.

Another discovery was a form of the ethyl alcohol molecule known as transethanol. A group led by Benjamin M. Zuckerman of the University of Maryland detected it in October, 1974, in the interstellar cloud Sagittarius B2. The alcoholic content they measured in the cloud is the equivalent of about 10-billion billion billion fifths of liquor at 200 proof, more than "the total amount of man's fermentation efforts since the beginning of recorded history."

Pulsar advances. The most important pulsar discovery since they were first detected in 1967 occurred in July, 1974, when Russell A. Hulse and Joseph H. Taylor of the University of Massachusetts discovered the first pulsar in a binary star system. Long-term monitoring of the binary pulsar should allow astronomers to test basic predictions of gravitation and relativity. See ASTRONOMY (High Energy).

In January, 1975, F. Curtis Michel of Rice University in Houston suggested that a pulsar seems to fade out as its rotation slows down because the object's magnetic field can no longer control the escape of the plasma, or ionized gas, that emits the radio signals. Apparently the radio emission can occur only when the plasma escapes in an organized manner, guided by the magnetic field.   [Stephen P. Maran]

High-Energy Astronomy. Major advances occurred in 1974 and 1975 in understanding astronomical objects in which high-energy events occur. Astronomers are beginning to understand the place of binary X-ray sources — those found in double-star systems — in the scheme of stellar evolution and they can ask more sophisticated questions about the nature of the objects and the systems in which they are found.

Binary pulsar. Russell A. Hulse and Joseph H. Taylor of the University of Massachusetts found the first radio pulsar in a binary system in July, 1974. Their discovery establishes that a neutron star can form as one of a relatively low-mass binary pair without disrupting the system.

When the first radio pulsar was discovered in 1967, the experimental evidence led astronomers to identify the source of the pulses as a neutron star. A star becomes a neutron star after it has exhausted its nuclear fuel so that the gravitational attraction pulling its surface inward is no longer balanced by the pressure exerted by internal heat. When this happens, the star collapses.

Astronomers believe that the excess mass is blown off in a supernova explosion at the moment of collapse, leaving behind the core of the original star transformed into a neutron star. Such a star is very dense and may be only about 10 kilometers (6 miles) in diameter. It has a strong magnetic field and rotates rapidly. The radio pulses are thought to be caused by radiation from electrons that are accelerated in the rotating magnetic field and flung out from the star.

Before the Hulse-Taylor discovery, all known pulsars were isolated stars, even though at least half of all stars are found in binary systems. Astronomers speculated that the supernova explosion creating a neutron star would blow apart a binary system, leaving behind only the pulsar. However, the 1971 discovery of the pulsating X-ray sources Hercules X-1 and Centaurus X-3 — also believed to be neutron stars — in close binary systems casts considerable doubt on this view. Either the X-ray sources were not neutron stars, or neutron stars could form without a supernova explosion, or the explosion would not always disrupt the binary system.

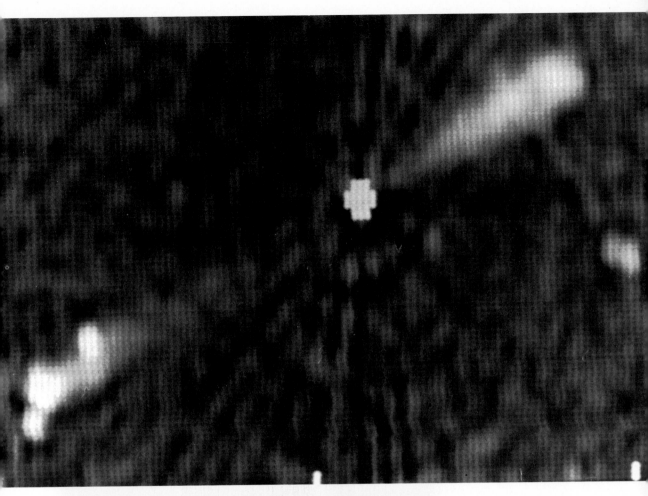

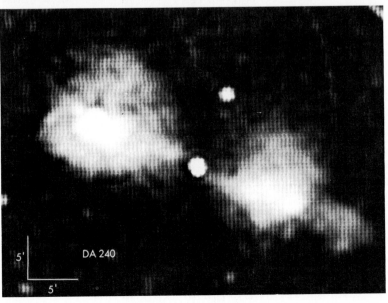

DA 240

5'

5'

Radio source 3C 236, *above,*
18.6 million light-years wide,
is the largest known object
in the universe. DA 240, *left,*
the second largest object, is
6.6 million light-years wide.
The photos were derived by
recording radio emissions from
swirling clouds of gas flung
out from a bright central core.

Astronomy

Theoretical calculations by Craig Wheeler of the University of Texas and Myron Lecar of the Smithsonian Astrophysical Observatory (SAO) showed that the last suggestion was indeed possible. The discovery of a pulsar in a close, low-mass binary system clinches the argument. The pulsar and its companion revolve around each other every six hours in a highly eccentric orbit, with a radius about equal to that of the sun. The discovery permits detailed study of the motions of a pulsar in a strong gravitational field, which it experiences because its partner is so close.

Hercules X-1. Another important contribution to the study of neutron stars came in December, 1974, when Jerry E. Nelson and John Middleditch of the University of California at Berkeley reported measuring the optical pulsation of Hercules X-1. The optical pulses are caused when the X-ray beam from the neutron star is reflected off the atmosphere of the companion as visible light.

The new data, obtained by carefully following changes in the optical pulses, allowed Nelson and Middleditch to refine previous measurements of the neutron star's mass to 1.3 times the mass of the sun. This value can be used to evaluate different theories that describe how matter behaves under the extremely high pressures inside such stars.

Streaming gas. The X-ray emission observed in a binary system is caused as gas flows from one star to the other. How this happens was explained much more clearly in December, 1974, after Polish astronomer Wojciech Krzeminski discovered the optical counterpart of Centaurus X-3 – a blue supergiant about 20 times as massive as the sun. N. V. Vidal of Tel Aviv University in Israel observed streams of matter leaving the star at speeds of 800 kilometers (500 miles) per second. Using data from the Uhuru satellite, Ethan Schreier of SAO discovered that the intensity of the X-ray source was changing by factors of 10. This made it possible to determine how the gas is distributed in the system.

These results open up a new field of X-ray investigations dealing with how the neutron star interacts with the gas in the system. Shock waves, created as the gas travels at supersonic speeds, can be studied in the X-ray light. Because of the large X-ray luminosity, the level of X-ray emission determines the ionization and temperature of the gas. These also can be measured by studying X-ray effects.

Black hole. Continuing observations of Cygnus X-1 have revealed further details of its X-ray emission that fit quite well with theories based on the idea that Cygnus X-1 contains a black hole. This gives astronomers greater confidence that the first black hole has been found.

In a black hole, gravity would be so strong that even light could not escape, but matter falling into it could become hot enough to radiate X rays before being swallowed up. In 1974, Richard E. Rothschild of Goddard Space Flight Center in Greenbelt, Md., found fast, random, X-ray pulses from Cygnus X-1, as brief as 1 millisecond (0.001 second). These pulses precisely match predictions for the disk of inflowing matter around a black hole. Theoretical work by Kip S. Thorne of the California Institute of Technology and Richard H. Price of the University of Utah on the energy distribution of the X-ray radiation also shows striking agreement between the new observations and black hole predictions.

An earlier theory that Cygnus X-1 might be a triple star system – an attempt to explain the X-ray emission without suggesting a black hole – was ruled out with the discovery of absorption dips – changes in intensity – that coincide with the binary period. The work was reported in late 1974 by Paul Sanford of the University of London and George Clark of the Massachusetts Institute of Technology.

Radio galaxy. P. J. N. Davison at the University of London discovered in 1974 that the X-ray emission from the radio galaxy Centaurus A has quadrupled in two years. This confirms that the emitting region is only a few light-years in diameter, as suggested earlier on the basis of spectral measurements by Uhuru. The production of these enormous amounts of energy in such a small region in the nucleus of a galaxy means that processes other than high temperatures must be involved. The cause of this violent activity is not yet understood.   [Riccardo Giacconi]

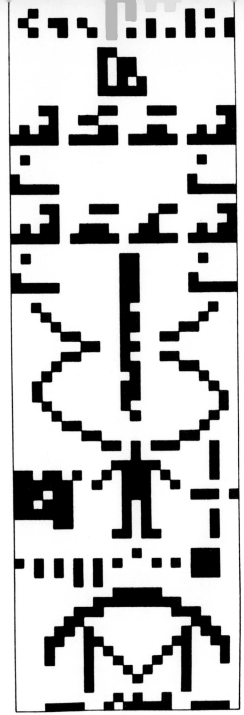

Talking to Other Worlds
Message radioed to space in 1974 starts by describing binary numbers, which are then used to give chemical formulas for the molecules in deoxyribonucleic acid, the human genetic code. A double helix winds down to a crude human form, to indicate the connection with human life. Below it is a diagram of the solar system, with Earth out of line and centered under the figure. The transmitting telescope is at the bottom.

Cosmology. The big bang theory of cosmology received an important boost in April, 1975, when Paul Richards and his group at the University of California at Berkeley reported measuring the spectrum of the microwave background radiation at submillimeter wave lengths.

The discovery of the cosmic microwave radiation in 1965 ranks as one of the principal observational achievements of modern cosmology. Its microwave spectrum seemed to match the spectrum of a black body, or perfect radiator, at a temperature only 3 degrees above absolute zero. This indicates that it was the remnant of the cooled-off radiation that filled the universe just after the big bang.

Each temperature from a black body has its own characteristic spectral curve that peaks at a different wave length. Astronomers can determine the temperature of an emitting source by measuring its emission intensity at various wave lengths and comparing the resulting curve with those already known.

Measurements of the background radiation in the microwave range were not conclusive because astronomers could not measure the entire curve. To be sure that the radiation resulted from the big bang, they needed measurements at submillimeter wave lengths.

To obtain the submillimeter spectrum, Richards sent a helium-cooled spectrophotometer aloft in a balloon. He found that the radiation reaches its maximum intensity at a wave length of 1.7 millimeters, which is the peak for a 3° black body, and the two spectral curves match exactly. The measurements conclusively prove that the radiation is a remnant of the big bang.

Expanding universe. Several developments in 1974 and 1975 have reinforced the idea that the universe is open and will continue to expand indefinitely.

James Gunn and J. Beverley Oke of Hale Observatories in California published new data on distant clusters of galaxies in January, 1975. Nearby galaxies are moving away from us at speeds that increase proportionately with distance. But the most distant galaxies seem to be receding more slowly than they would if the expansion were completely uniform. A closed universe would require such an effect. The gravity of the universe would cause deceleration until the expansion would eventually be reversed, causing a catastrophic collapse of the universe some tens of billions of years from now. Gunn and Oke, however, reported that the deceleration rate appears too small to reverse the expansion. See JAMES GUNN.

A new yardstick. Allan R. Sandage, also of Hale Observatories, and Swiss astronomer Gustav Tamman published a series of papers in 1974 and 1975 that take an alternative approach to determining whether expansion will ever reverse. By carefully remeasuring the distances to both nearby and distant galaxies, they obtained a new, lower value for the Hubble constant, the rate at which the universe is expanding. The new value indicates that expansion has been going on for about 20 billion years if the universe is open, and for less than 14 billion years if it is closed. (A closed universe slows down faster and thus reaches the lower expansion rate at a younger age.) Sandage and Tamman also point out that the oldest objects in our Galaxy, the globular cluster stars, are known to be at least 14 billion years old. The universe must be even older, coinciding with the expansion time of an open universe.

In another study in early 1975, Sandage mapped the variations from the uniform expansion rate found among nearby galaxies. He found that the nearby galaxies are receding from us in all directions in an extremely uniform manner, as would be expected in an open universe.

The missing mass. In order for eventual recollapse, the average density of all the matter in the universe must be above a certain critical value. Direct measurements of matter we can see, such as stars and nebulae, fall far short of this value. However, in late 1974, two separate studies—by Estonian astronomers Jaan Einasto, Ants Kaasik, and Enn Saar; and by Jeremiah Ostriker, James Peebles, and Amos Yahil of Princeton University in New Jersey—suggested that a considerable amount of invisible mass may exist as extended halos around galaxies. This increases previous density estimates, but still falls short of what is needed.

Another way to determine average density is to study the primordial gas falling into clusters of galaxies. Susan Lea of the University of California at Berkeley found in 1974 that the gas would heat up to hundreds of millions of degrees as it flows into a cluster's gravitational field. This produces excessive X rays which can be studied to determine the density of the unseen material. Space-borne telescopes have observed X rays from several galaxy clusters, but the density of the incoming gas appears to be less than 20 per cent of what is needed to close the universe.

Interstellar deuterium, an isotope of hydrogen, provides another important cosmological clue. Theoretical work published in May, 1974, by Richard Epstein, David Arnett, and David Schramm of the University of Texas and in September by Thomas Weaver and George Chapline of the Lawrence Livermore Laboratory in California indicates that deuterium is unlikely to be synthesized in a supernova explosion.

This reinforces the idea that all the deuterium now found in the universe originated in the big bang. And, if this is so, the amounts observed today require that the universe be open. In a closed universe, matter would have been much denser in the early stages and much of the fragile deuterium would have been destroyed in collisions with other atoms.

Background uniformity. At some point in the past, radiation and matter in the primordial universe were intimately coupled. The failure so far to discover any localized "hot spots" in the background radiation indicates that primordial matter was evenly distributed.

But Joseph Silk of the University of California at Berkeley theorized in late 1974 that matter must have started clumping together very early because galaxies today are not uniformly distributed. The radiation also must have been uneven, guaranteeing the existence of hot spots. The intensity of the hot spots—which Silk believes are on the verge of detection—would depend on whether the universe is open or closed. An open, low-density universe would have had much larger fluctuations in primordial matter and the associated hot spots would also be much stronger.   [Joseph Silk]

The Inconstant Constant

Gravity may be weakening. In April, 1974, Thomas C. Van Flandern of the U.S. Naval Observatory in Washington, D.C., presented evidence of this based on a careful analysis of the orbits of the moon around the earth, and of the earth around the sun. These orbits are slowly expanding in a way that Van Flandern believes can be explained only by a small weakening of gravity.

Gravity is the weakest, yet the most far-ranging, of the forces of nature. Sir Isaac Newton was the first to describe gravity as a universal force governing planets, stars, and the universe itself. According to Newton, gravity is an attraction that exists between any two masses. It is equal to the product of the two masses divided by the square of the distance between them, all multiplied by the fundamental constant of nature called G.

By 1915, Albert Einstein had developed his general theory of relativity, a theory of space and time that is also a law of gravity superseding Newton's. The constant G still represented the force of gravity, though. Most physicists continue to accept Einstein's theory of relativity, with G as a fundamental constant.

G is small. Lift this book, and you are defying the gravitational pull of the entire earth. G is so small that it cannot be measured directly with any great accuracy. Van Flandern's evidence is therefore indirect, based on timing when the moon passes in front of various stars, using atomic clocks. As a result of these precise timings, Van Flandern declares that G is decreasing by 8 parts in 100 billion per year.

Previous indirect evidence indicated otherwise. Irwin I. Shapiro and his co-workers at the Massachusetts Institute of Technology in Cambridge deduced in 1971 that G was constant, based on a radar examination of planetary orbits. In early 1974, David S. P. Dearborn and David N. Schramm, then at the University of Texas at Austin, argued that certain clusters of stars could not exist unless G were constant. Moreover, Einstein's general theory of relativity has survived many tests, and this theory is built on a constant value for G. But none of this evidence is conclusive enough to conflict with Van Flandern's results.

The value of G has cosmic implications. In the 1920s, Edwin P. Hubble discovered that the universe is expanding in accord with Einstein's theory. This expansion and the theory of relativity together imply that the universe is from 10 to 20 billion years old. A cosmic coincidence has also been found. The average density of matter in the universe — about one atom of hydrogen per 100,000 cubic centimeters (6,100 cubic inches) — times the square of the age of the universe is about equal to the reciprocal of G, or 1/G.

If this rough equality is more than accidental, then perhaps the influence of the stars on the earth is decreasing as the universe expands. This lessening influence could be reflected in a decrease in G. Moreover, since the universe is from 10 to 20 billion years old, G should decrease by one part in 10 or 20 billion each year. This is roughly the amount Van Flandern found.

A decrease in G is part of several theories of gravity put forward over the years. Three of the most influential are those of Paul A. M. Dirac, Carl H. Brans and Robert H. Dicke, and Fred Hoyle and Jayant V. Narlikar. Indirect arguments supporting these theories relate to the ages of stars, the temperature of the sun, and continental drift.

The most exciting prospect is that properties of elementary particles and atomic nuclei may also be involved. In the 1930s, British astronomer Sir Arthur S. Eddington listed several numerical coincidences relating elementary particles and the universe. A decrease of G close to that predicted from cosmic expansion could thus yield further understanding, not only of gravity, but also of the universe and the submicroscopic world of nuclear physics.

Van Flandern's results must yet be verified. Shapiro's radar measurements are continuing. Another experiment being conducted at the University of Texas uses a laser to determine the distance to the moon with extreme accuracy. It, too, should provide a clue as to whether the moon's orbit is actually expanding. But until these experiments and others can be refined, physicists will continue to wonder if Einstein's general theory of relativity must be modified and whether gravity is truly weakening.   [Lawrence C. Shepley]

Biochemistry

Biological Water Wheel

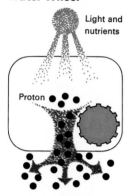

Light and nutrients

Proton

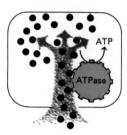

ATP

ATPase

Energy from light or from nutrients can drive protons out of a cell. When protons flow back in, they trigger the enzyme ATPase to help form ATP, the cell's energy molecule.

All living cells, from the simplest bacteria to specialized nerve cells in the brain, must carry out at least two basic functions — produce energy for their life processes, and recognize and respond to signals from the outside world. Biochemists moved closer in 1974 and 1975 to an understanding of the chemical mechanisms by which cells accomplish these functions, called energy transduction and signal transduction. What we have learned indicates that both may be much simpler than we thought.

A biological water wheel. Living cells store energy in the chemical bonds of adenosine triphosphate (ATP). When converted to another substance, ATP releases the energy for muscle contraction, nerve function, and a host of other processes. So cells must continually synthesize ATP. To do this, plant and animal cells use energy they obtain from sunlight or from nutrients such as carbohydrates. In October, 1974, Peter C. Maloney, Eva R. Kashket, and Thomas H. Wilson of Harvard Medical School provided evidence that supports one theory of how cells use the energy to produce ATP.

The theory, proposed several years ago by Peter Mitchell of Glynn Research Laboratories in Cornwall, England, suggests that the energy produces a gradient, or difference in concentration, of hydrogen ions (H^+), or protons, across a cell's membrane. The proton gradient then provides the energy for ATP synthesis.

Just as energy is required to pump water from a valley to a mountaintop, so energy is needed to pump ions across a membrane from a region of low concentration to one of high concentration. Conversely, when ions flow down a concentration gradient, energy is released, just as energy is released when water rushes down a mountainside. A water wheel or a turbine placed in the mountain stream captures some of this energy, and can convert it into useful mechanical or electric energy. Mitchell's theory proposes a similar mechanism to drive the synthesis of ATP.

The Harvard researchers demonstrated that ATP synthesis could be driven by a proton gradient across the membranes of the bacterium *Escherichia coli*. Other investigators had observed that as *E. coli* "burns" nutrients,

protons are pumped out of the bacterium. This causes a proton gradient to form across the cell membrane, and ATP is synthesized.

The Harvard group suspended cells in a nutrient-free medium containing a higher concentration of protons and a lower concentration of potassium ions (K^+) than the cells contained. The scientists added the antibiotic valinomycin, which specifically causes cell membranes to become much more permeable to K^+, to induce the outward flow of the positively charged ions. This left the inside of the cell with a slight excess of negative charges, which drew in H^+ from the medium. The cells began to synthesize ATP as soon as the valinomycin was added. Since there were no nutrients present, the experiment proved proton influx alone provides enough energy for ATP synthesis.

The researchers repeated the experiment with cells that lacked functional ATPase, an enzyme normally found in the cell membrane. These cells made no ATP. Apparently, ATPase plays the crucial role of the water wheel, capturing the energy released in the downhill (inward) flow of protons by synthesizing ATP.

Similar observations have been made in other laboratories with energy-transducing membranes from both animal and plant cells. In one experiment, bacterial cell membranes containing a substance capable of capturing the sun's energy to produce ATP in a whole cell were isolated from the rest of the bacteria and converted into closed spheres. Supplemented with ATPase and the raw materials of ATP, these spheres produced ATP when light was turned on them. Taken as a whole, all these findings show that in a wide variety of organisms, a proton gradient driving a relatively simple mechanism involving a single enzyme may be central to energy transduction.

A single signaler. The nervous system of complex animals consists of an enormous number of long, thin, interlocking nerve cells that carry signals very rapidly. When a nerve cell is stimulated, its membrane becomes more permeable to sodium ions (Na^+) at the stimulated end. This change sweeps the length of the nerve cell carrying an electric impulse to its far end, where the

Biochemistry

Continued

impulse must be passed either to the next nerve in a circuit, or to a target tissue, such as a muscle. There is a small gap between nerve and nerve, or between nerve and muscle, which the electric impulse does not normally bridge. Instead, the impulse triggers the release of a chemical substance called a transmitter at the far end of the stimulated nerve. The transmitter diffuses across the gap between cells and binds to a protein called a receptor in the membrane of the second cell, initiating a new electric impulse.

In November of 1974, Jean-Pierre Changeux and his colleagues at the Pasteur Institute in Paris and Michael A. Raftery and co-workers at the California Institute of Technology in Pasadena used a new technique to obtain a relatively pure quantity of such a receptor. Then they inserted this material into very simple membrane spheres. When the scientists added chemical transmitter, they found that the Na^+ permeability of the membrane increased strikingly. This suggests that a single protein, the receptor, is responsi-ble for the entire signal transduction, both recognizing the transmitter and initiating the new impulse. Future experiments with the purified receptor should help scientists to explore further the molecular details of how nerve cells function, and thus to move one step closer to an understanding of the thought process itself.

The structure of gene controllers. The genes of all organisms are part of a long molecule called deoxyribonucleic acid (DNA), which is essentially a chain of chemical units called bases. All cells have complex machinery for translating the genetic information in DNA into the thousands of different proteins and other substances they need to grow and function.

Not all of a cell's DNA translates into proteins, however. Certain regions contain signals that control nearby genes. In bacteria, the control mechanism involves interactions between these control regions and certain proteins called regulatory proteins.

Late in 1973, Walter Gilbert and Allan Maxam of Harvard University

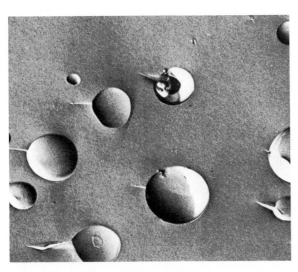

Tiny membranelike sacs, *above,* revealed by an electron microscope, were prepared in the laboratory, then added to a preparation of cell membranes. The arrows, *right,* indicate spots where the sacs probably fused with a membrane. Such sacs might someday be filled with materials to be carried into cells in experiments or for medical treatment.

Biochemistry
Continued

A model of a new shape proposed for some mitochondria was deduced through changes in the form of the structure's cross section as a series of slices were made through it. Most mitochondria, tiny cell structures involved in energy production, were previously thought to be sausage shaped.

identified and determined the sequence of 24 bases making up part of the region controlling genes in *E. coli* that make substances which help the organism metabolize the sugar lactose. In January, 1975, Robert C. Dickson and John Abelson of the University of California at San Diego, and Wayne M. Barnes and William S. Reznikoff of the University of Wisconsin established the sequence of this entire control region, containing some 122 bases. They allowed large numbers of *E. coli* to become infected by a virus called lambda. When this virus infects *E. coli*, it occasionally incorporates bits of the bacterial DNA into its own, much smaller, DNA molecule. Viruses that have picked up the lactose control region can be easily detected; they carry these genes into other bacteria which lack them, and the bacteria suddenly can metabolize lactose.

The DNA isolated from such lactransducing lambda is a much better starting material for the isolation of the lactose control region, because lambda DNA is 100 times shorter than that of

E. coli. Using recently developed techniques, the scientists isolated the region and determined the sequence of its bases, a feat that most biochemists would have believed impossible five years earlier.

Early in 1974, three of the regulatory proteins that interact with the lactose control region were isolated and their chemical structure determined. All that is needed now is the 3-dimensional structure of the regulatory proteins to give scientists their first glimpse of the chemical details of the protein-DNA interaction that controls gene activity.

Thomas A. Steitz and his colleagues at Yale University in New Haven, Conn., have obtained tiny crystals of one of the regulatory proteins, called the lactose repressor. Crystallization is the first step in determining the 3-dimensional structure of a protein by X-ray diffraction. So far, the repressor crystals that Steitz's team has produced are too small to use for determining structure, but it is reasonable to suppose that the research team can soon produce larger ones.   [David L. Nelson]

Books of Science

Here are 31 outstanding new science books suitable for general readers. The director of libraries of the Smithsonian Institution selected them from books published in 1974 and 1975.

Anthropology. *A Guide to America's Indians: Ceremonials, Reservations and Museums* by Arnold Marquis. This book is a brief, but detailed, account of the origins, societies, cultures, and histories of the major tribes of American Indians. In addition to describing the chief characteristics of the tribes and their past and current ways of living, it contains many maps and lists that locate today's tribes and reservations, as well as museums devoted to Indian lore. (University of Oklahoma Press, 1974. 267 pp. illus. $9.95)

Archaeology. *Atlas of Ancient Archaeology* by Jacquetta Hawkes contains hundreds of one-page descriptions in words and diagrams of the world's important archaeological sites. It lists the major findings at each site that contribute to understanding the development of civilization. The descriptions are arranged by region so that the reader can follow the history of a particular culture. (McGraw-Hill, 1974. 272 pp. illus. $19.50)

Peking Man: The Discovery, Disappearance and Mystery of a Priceless Scientific Treasure by Harry L. Shapiro. A noted anthropologist reconstructs the life, manners, and appearance of these precursors of modern man. He tells how the remains of Peking man were found and identified, and recounts the story of their disappearance in China during World War II. He also gives a brief review of other important fossil discoveries, and how they relate to the evolution of man. (Simon & Schuster, 1974. 190 pp. illus. $7.95)

Astronomy. *Comets, Meteoroids, and Asteroids: Mavericks of the Solar System* by Franklyn Branley and Helmut Wimmer is a clear and simple exposition of what is known about meteors, meteorites, asteroids, comets, cosmic rays, dust, solar wind, and tektites. The book tells how they were discovered and what their origins were. (T. Y. Crowell, 1974. 115 pp. illus. $5.50)

Biology. *The New Genetics* by Margaret O. Hyde explains the promise and the dangers of genetic engineering. The author tells how the function of genes was discovered, and explains the mechanism by which genes impart the messages that shape the development of all living creatures. The relationship of genes to diseases of the blood, and the results of errors in genetic structure are also described, along with the techniques used to repair faulty genes. The book also tells of artificial insemination and other ways of propagating life. (Franklin Watts, 1974. 134 pp. $6.95)

Botany. *Carnivorous Plants* by John F. Waters. This book describes over 500 plants that eat insects. The plants are divided into three categories according to the method they use to catch the insects. Details of how the plants use the method are given. The author also dispels the myth that there are man-eating plants. (Franklin Watts, 1974. 60 pp. illus. $3.45)

Trees Alive by Sarah Riedman. The life processes of trees are described in detail in this book. It explains how they make food, take up water, change through the seasons, breathe, and suffer diseases. It also tells how and where trees live together to create forests, and how they support plant and animal life. (Lothrop, Lee & Shepard, 1974. 124 pp. illus. $5.50)

Climate. *Weather Changes Man* by Ben Bova is an introduction to both the short- and long-range effects of weather on human life. It presents theories of how the atmosphere developed and the role it plays in fostering and sustaining life, and describes how climate affects survival of humans. (Addison-Wesley, 1974. 140 pp. illus. $5.50)

The Weather Machine by Nigel Calder details the effects of the sun, volcanoes, and variations in the earth's orbit in creating climatic and weather conditions. Calder explains the techniques of assessing climates of the past from tree rings, ocean mud, and ancient ice. He also theorizes on the approach of another Ice Age. (Viking, 1974. 143 pp. illus. $14.95)

Drugs. *Mind Drugs*, edited by Margaret O. Hyde, is a brief, succinct presentation of the scientific and social aspects of drugs. It presents the work of several authors outlining the effects and abuses of the major drugs, including alcohol, amphetamines, barbiturates, caffein, cocaine, heroin, LSD, and marijuana. The authors describe how the

drugs work and the processes of addiction and cure, citing cases and listing sources of assistance. (McGraw-Hill, 1974. 190 pp. $4.95)

Energy. *Earthbound: Minerals, Energy and Man's Future* by Charles F. Park, Jr., and Margaret C. Freeman is a thorough, advanced, but not technically difficult explanation of the conflicting needs for mineral commodities and the steps that must be taken to ensure their availability. It covers the technical, political, and economic aspects of the extractive (mineral) and energy resource industries. The authors include soil and water among the resources, and cite arguments for mineral conservation as well as use. (Freeman, Cooper, 1975. 279 pp. $8.95)

The Energy Trap by Daniel S. Halacy, Jr., traces the history of energy development from the discovery of fire to nuclear reactors. The author describes problems of overconsumption and assesses potential sources of power such as nuclear fusion. (Four Winds Press, 1974. 143 pp. illus. $6.95)

The Great Energy Search by Elaine Israel is an essay on the creation, discovery, and use of oil, gas, and coal to produce energy, and on the growing shortage of these resources. It portrays the struggle between environmentalists and energy producers attempting to develop new sources of fuel. Nuclear, solar, and geothermal energy are briefly discussed. (Messner, 1974. 63 pp. illus. $5.95)

Entomology. *Good Bugs and Bad Bugs in Your Garden: Back Yard Ecology* by Dorothy C. Hagner contains short descriptions of harmful and beneficial bugs commonly found in gardens. Its theme is the balance in nature between plants and animals and the problems introduced when this balance is upset by insecticides and other control techniques. There is practical advice for the average gardener, including how to use companion plants to aid or hinder growth. (T. Y. Crowell, 1974. 88 pp. illus. $5.50)

Gypsy Moth: Its History in America by Robert M. McClung. The author details the life cycle of the gypsy moth, which denudes trees in its caterpillar stage. While describing attempts to control this moth, he covers the shortcomings and successes of chemical and biological methods of controlling insect pests in general. (Morrow, 1974. 96 pp. illus. $4.75)

Geology. *Continents in Motion: The New Earth Debate* by Walter Sullivan. *The New York Times* science editor explains how a number of disparate geologic discoveries were brought together into a new theory of continental drift and its effect on life forms. Plate tectonics helps explain how mountains, deserts, lakes, and seas were made and destroyed. Sullivan recounts the many stories of discoveries, their proponents, and the difficulties of achieving acceptance of new scientific ideas. (McGraw-Hill, 1974. 399 pp. illus. $17.95)

Measurement. *Metric Is Here* by William Moore brings the metric system to life by relating it to measurements commonly used in factory, home, laboratory, office, and sports. Equivalents and conversions between customary and metric systems are carefully and thoroughly described in the text and concluding tables. (Putnam's, 1974. 94 pp. illus. $5.95)

Medicine. *The Transplant: A True Account of a Modern Surgical Miracle* by A. Q. Moubray. In fine storybook fashion, the author weaves an absorbing account of the details of physiology, the technique of organ transplants, and the training of a surgeon, while describing an actual kidney transplant case from the onset of the illness through surgery to the patient's recovery. (McKay, 1974. 266 pp. $8.95)

Microphotography. *Zooming In: Photographic Discoveries Under the Microscope* by Barbara and Lewis Wolberg. This book features dozens of photographs taken through a microscope. They include microscopic animals found in water, human cells, crystals, plants, and parts of animals and minerals. It also includes a general description of photomicrographic techniques. (Harcourt, Brace, Jovanovich, 1974. 68 pp. illus. $7.75)

Natural History. *Yellowstone: The First National Park* by Ruth Kirk is a naturalist's description of the Yellowstone Park area. It features the interesting geologic history and describes the present structure of the area, including the park's geysers and remnants of ice from the Ice Age. (Atheneum, 1974. 103 pp. illus. $6.25)

The Great Barrier Reef by Isobel Bennett discusses the history, geography, and geology of the coral reefs and islands off the coast of Australia, as well as the many varieties of plants, animals, and fish found there. The book offers many insights into ecology. (Scribner's, 1974. 183 pp. illus. $17.50)

Natural Resources. *Water: The Fountain of Opportunity* by H. G. Deming. This is a comprehensive and wide-ranging book, including historical, cultural, and scientific lore and knowledge about a vital resource – water. The author emphasizes management and use of water in chapters on agriculture, industrial processes, sewage disposal, irrigation, purification, fresh water from the ocean, and flood control. He describes water as an energy carrier and the vehicle of life. (Oxford University Press, 1975. 342 pp. $12.50)

Physiology. *Conception, Contraception: A New Look* by Suzanne Loebl. This book is a straightforward examination of the history and science of birth control. The process of conception and how it was discovered are included, along with a brief biography of Margaret Sanger and discussion of the development of the birth control movement in the United States. (McGraw-Hill, 1974. 147 pp. illus. $6.95)

Your Body's Defenses by David C. Knight tells how the body responds to injuries and diseases. Besides describing the processes of natural and induced immunity, the author suggests elements of personal care to increase resistance to disease and injury. (McGraw-Hill, 1975. 95 pp. illus. $5.72)

Space Exploration. *The Galactic Club: Intelligent Life in Outer Space* by Ronald N. Bracewell is a somewhat sophisticated, but readable, examination of the many speculations about life in other parts of the universe. The author proposes ways to try to detect and communicate with other life, and various methods of space travel outside the solar system. (W. H. Freeman, 1975. 134 pp. illus. $6.95)

Technology. *Plastics* by James and Lynn Hahn briefly describes several chemical processes and the technology used in manufacturing plastics. The authors include a brief history of the development of plastics, outline some of the uses of plastics in medicine and

architecture, and speculate on future uses of plastics in transportation, communication, building, household goods, and clothing. (Franklin Watts, 1974. 65 pp. illus. $3.45)

The Scientific Breakthrough: The Impact of Modern Invention by Ronald W. Clark. This book examines six major technologies that have profoundly affected human life: atomic energy, electromagnetic communication, genetic engineering, manned flight, photography, and plastics. It demonstrates how important scientific phenomena are often discovered by chance, and how rapidly products may be created that capitalize on scientific theory. (Putnam's, 1974. 208 pp. illus. $14.95)

Zoology. *Animal Invaders: The Story of Imported Wildlife* by Alvin and Virginia Silverstein recounts some of the disasters that have occurred when animals are moved into new regions, upsetting the balance of the existing ecology. The authors tell of American muskrats undermining dikes in the Netherlands, African snails eating gardens in Florida, Asian walking catfish destroying fish in Florida ponds, and many similar problems. (Atheneum, 1974. 124 pp. illus. $5.95)

Diving Companions: Sea Lions, Elephant Seal, Walrus by Jacques-Yves Cousteau and Phillipe Diolé. This book is a mixture of science, description, and personal narrative by two famous marine biologists who focus on the cold, hostile Arctic environment. It is profusely illustrated with color photographs. (Doubleday, 1974. 304 pp. illus. $10.95)

Wild Travelers: The Story of Animal Migration by George Laycock. This story of animal movement focuses not only on the animals and where they go, but also on how migration has been studied. (Four Winds Press, 1974. 110 pp. illus. $5.95)

Mind in the Waters: A Book To Celebrate the Consciousness of Whales and Dolphins by Joan McIntyre. This book is a collection of articles about whales and dolphins from mythology, literature, and science. It contains several classic articles on the intelligence of whales, the life cycle of individuals and colonies of whales, and conservation of the species. (Scribner's, 1974. 240 pp. illus. $14.95)　　　　[Russell Shank]

Botany

Plant scientist K. N. Kao, of the Prairie Regional Laboratory, National Research Council, in Saskatoon, Canada, used polyethylene glycol to induce the formation of small clumps of plant protoplasts, plant cells with the cell walls removed. This work, reported in October, 1974, is a first step in fusing cell materials.

Plant breeders have always been limited by the fact that different species of plants will not crossbreed. Thus, it has been impossible to introduce desirable genes from one species into another. Cell fusion is a step in that direction, however, because it creates living cells with genes from two species.

Kao's work was followed by evidence from his colleague Oluf L. Gamborg that protoplasts of soybean and barley, of soybean and rape — an important oil seed plant — and of clover and rape had been fused in the laboratory. Although whole plants have not yet been developed from these tissue cultures, many scientists believe that, in time, hybrids of almost any plant combination can be produced at will. Devising plants that produce, say, both fruit and edible roots could benefit a hungry world.

Longer life for seeds. Equally important to agriculture is extending the length of time during which seeds retain the ability to germinate.

N. W. Pammenter, J. H. Adamson, and Patricia Berjak of the University of Natal in Durban, South Africa, have found that seed life can be extended by placing seeds on a negatively charged conductor. The conductor provides a source of electrons which seem to reduce attacks on the seed by free radicals, unattached atoms or groups of atoms in the atmosphere. The results support the theory that free radicals damage cells and age seeds.

Controlling rust fungi. Millions of dollars worth of crops are destroyed each year by rust fungi, even though many rust-resistant strains have been developed. Growing rust fungi in the laboratory for study has been difficult, but R. J. Bose and Michael Shaw of the University of British Columbia in Vancouver, Canada, produced luxuriant growth and spore formation of the flax

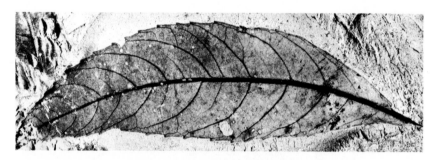

Fossil leaves that grew in a subtropical area in North Dakota 50 million years ago are helping botanists perfect a new system of classifying ancient plants by vein pattern.

Botany
Continued

rust fungus in 1974. They grew the organism in a liquid medium that they sterilized and seeded with rust spores.

The new technique will permit scientists to study damaging parasitic fungi more easily and to learn how to control the spore formation of fungi.

Fossils show climate change. In addition to revealing the evolutionary history of plants themselves, botanical fossils yield evidence about climatic conditions. John H. Troughton of the Carnegie Institute of Washington, P. V. Wells of the University of Kansas, and H. A. Mooney of Stanford University reported in 1974 that the ratio of the isotope carbon 13 to carbon 12 in fossil samples of a Nevada cactus (*Opuntia polyacantha*) went down during the period from 40,000 years ago to 10,000 years ago. This marks a change in how plants used carbon. The change occurs under dry conditions, indicating that the late Pleistocene rainy climate had become drier in Nevada at least 10,000 years ago.

Bacteria and plants. Botanists P. Christee, Edward I. Newman, and R. Campbell, working at the University of Bristol in Great Britain, reported in 1974 that microorganisms inhabiting the root surface and the soil around the roots can affect a plant's uptake of soil nutrients. Two common plants, perennial ryegrass (*Lolium perenne*) and the narrow-leaved plantain (*Plantago lanceolata*), were grown both in pure stands and mixed plots. When mixed, both species consistently grew more and had a higher density of root microbes than in pure stands. Apparently, the plants have a mutually reinforcing effect on the microbes which provide them with nitrogen and perhaps other nutrients. See THE NITROGEN FIX.

Trees and water supply. Conversion of a forest in the southern Appalachian Mountains from mixed hardwood trees to white pine has apparently reduced the waterflow in streams there. Wayne T. Swank and James E. Douglass of the Southeastern Forest Experimental Station in Franklin, N.C., reported in September, 1974, that 15 years of measurements show that the annual streamflow was reduced by 20 per cent. The sharpest drop occurred in the dormant period and early in the growing season. This indicates that pine trees use more

water and have shorter, less complete dormancy periods than hardwoods. Because many municipal watersheds in the Eastern United States have extensive plantings of white pine, it is likely that waterflow from these areas has also been reduced. Planting hardwood trees instead of white pine could increase water supplies in urbanized areas by reducing water use in the watershed.

The Sunken Forest on Fire Island, a barrier island in the Atlantic Ocean off Long Island, New York, is a small sand-dune ecosystem that gets much of its mineral nutrition from materials in the ocean spray. This apparently compensates for the lack of nutrients in the highly weathered sandy soil there. Henry W. Art of Williams College, F. Herbert Bormann and Garth K. Voigt of Yale University, and George M. Woodwell of Brookhaven National Laboratory reported in 1974 that the growth rates in the Sunken Forest are comparable to those in inland temperate forests where nutrients are derived through the slow weathering of soil particles. The Sunken Forest's growth rates contrast sharply with those in the sand-dune ecosystems bordering Lake Michigan, which have similar sandy soil but very little nourishment from wind-driven spray. Such an ecosystem requires thousands of years to develop trees, but radiocarbon dates show that the Sunken Forest reached tree stage in only 200 to 300 years.

Ancient flower types. In August, 1974, William L. Crepet, David L. Dilcher, and Frank W. Potter of Indiana University, Bloomington, found fossil catkins in the Southeastern United States that date back to the Middle Eocene Epoch, 50 million years ago.

Catkins are clusters of small florets that are easily carried by the wind for pollination. The flowers of willow trees and many other temperate trees are catkins. Paleobotanists have long wondered whether this flower form is an ancient survival or a recent development in the evolution of flowering plants. This new research indicates that catkins developed fairly early in the history of flowering plants. The structural details of these fossil catkins are much like those in the modern walnut family, Juglandaceae, and confirm the family's antiquity. [Howard S. Irwin]

Chemical Technology

Chemical technologists introduced many new ideas in 1974 and 1975 to solve industrial problems. They ranged from new sources of protein to better batteries and safer cigarettes.

Single-cell protein. The growing world food shortage increased the pressure to develop new sources of protein. Until now, soybeans have been the major source of these proteins. But proteins from alfalfa, corn germ, cottonseed, rapeseed, and wheat are entering the market. Researchers are also making "single-cell protein" — bacteria or fungi grown in cultures and harvested. This process has several advantages, because it is independent of agricultural or weather conditions. The organisms also multiply rapidly.

The single-cell protein processes in operation at present produce animal feeds almost exclusively. They are generally not used for human consumption because some of the materials used in the fermentation process are petroleum products. One, called gas oil, contains carcinogens. Single-cell proteins also contain high concentrations of nucleic acids, which determine genetic functions in cells. Scientists do not yet know if this would have any adverse effects on humans. Researchers are now testing processes that use materials such as starches and sugars. These also might be more readily available than petroleum in developing countries where the lack of protein may be most crucial.

Nevertheless, single-cell protein is probably the most attractive long-term approach, according to bioengineer Arthur E. Humphrey, dean of the University of Pennsylvania's College of Engineering and Applied Science. Humphrey claims that computer-control techniques for monitoring the activities of growing microorganisms are making possible a new era of growing them in quantity. Amoco Foods Company in Chicago plans to open a plant early in 1976 that will produce a yeast for human consumption that is rich in protein, vitamins, minerals, and fats.

Microorganisms also hold the key to a new process for recovering protein from whey, the nutrient-rich but generally unwanted by-product of cheese-

Drawing by D. Fradon; © 1975 The New Yorker Magazine, Inc.

"What's the opposite of 'Eureka!'?"

making. Processors recover the dry solids from only 1 billion of the 30 billion pounds (13.6 billion kilograms) of liquid whey produced each year. This is done by evaporation only, a very expensive operation because of whey's high water content. However, Milbrew, Incorporated, of Juneau, Wis., opened a plant in January, 1975, that treats whey by fermentation with yeast to yield a high-protein yeast. The fermentation process, which converts the lactose in whey to yeast, increases the protein content of the solids to 50 per cent, as compared to 12 per cent in the evaporation process. Water and ethyl alcohol, yielded as a by-product, can be evaporated and the solids used as an animal feed. Even though liquid still must be evaporated, the high protein content makes the process economical. The fermented broth can also be centrifuged to produce a yeast for human consumption.

Watered fuel. Scientists experimented with adding water to oil and gasoline to make those fuels burn better. Water apparently plays several key roles in the combustion of liquid fuels. The trick is to combine the water (constituting about 20 to 25 per cent of the total mixture) with the oil to form an emulsion.

Researchers use ultrasonic or chemical techniques to drive tiny water droplets inside small oil droplets. Then the emulsion is sprayed into a combustion chamber, where the water trapped in each oil droplet suddenly evaporates. This shatters the oil droplet into tiny fragments, providing more surface for contact with air. So the small fragments burn more quickly, completely, and efficiently than do big drops of oil. This more complete combustion reduces exhaust emissions, unburned particulates, and undesirable oxides of nitrogen.

Tymponic Corporation of Plainview, N.Y., and Elf Union, a French oil company, were selling ultrasonic emulsifiers for heaters and industrial diesel engines in 1974. However, the U.S. Environmental Protection Agency was still evaluating the devices.

Gasoline-water emulsions also work the same way in automobile engines. Industrial technologists claim the emulsions increase mileage and reduce nitrogen-oxide emissions. But major automakers are not enthusiastic about installing emulsion systems in cars. They point out that recirculating exhaust gas produces the same results.

Safer smoking. A man-made tobacco substitute called New Smoking Material (NSM), described as "safe," was developed and tested by Imperial Chemical Industries Limited of Great Britain and Imperial Tobacco Limited. An NSM manufacturing plant was scheduled to be opened late in 1975 in Scotland.

The tobacco substitute has a cellulose base, and it can be blended with real tobacco. It is gray, has relatively little flavor or odor, and contains no nicotine. The developers claim tests have demonstrated that NSM tar produces only one-fifth as many tumors in animals as does tar from tobacco and has only about one-fourth as much tar. NSM also contains none of the irritants that cause bronchial ailments.

NSM is made from sheets of bleached, paper-grade, wood-pulp cellulose that is treated with chemicals and heat to drive off tar and other undesirable materials. It is blended with special additives that make it burn like tobacco. The NSM plant will be able to produce 15,000 short tons (14,000 metric tons) of tobacco substitute a year.

Better batteries. Today's lead-acid battery will be the major commercial automobile and truck battery at least until the mid-1980s. But new types of batteries are being developed for use in electric vehicles and for energy-storage systems that can help electric utilities meet peak-period demands for power. See ENERGY ON THE SHELF.

All storage batteries produce electric energy as a result of a chemical reaction that creates positive and negative ions. When the battery is charged with electric current, this energy-storing reaction occurs. To release electricity, the chemical reaction reverses, and the battery discharges. Lead-acid batteries use lead plates as positive and negative electrodes in an acid electrolyte, or current-carrying solution. Lead-acid batteries can be recharged a limited number of times. The ideal battery would be able to store a large number of watts per pound of battery and be capable of innumerable recharges.

Bars coated with epoxy and set in concrete are examined for signs of corrosion after standing in salt water for two years. The coating is designed to protect reinforcing bars used in bridges from the corrosion caused by deicing salt.

Chemical Technology

Continued

Energy Research Corporation of Bethel, Conn., in 1974 produced prototypes of a battery with zinc and nickel oxide electrodes and a potassium hydroxide electrolyte. The batteries, which power small electric appliances, have a long life. More powerful versions may be suitable for electric vehicles.

High-powered batteries can be made from zinc-air and iron-air electrode combinations. The Compagnie Générale d'Électricité (CGE) in Marcoussis, France, has begun to develop a zinc-air model. A thick mixture of powdered zinc in a potassium hydroxide solution is stored in an external tank. The mixture flows through tubes in the battery that are exposed to air. The zinc acts as the negative electrode; the air, as the positive electrode. As the zinc oxidizes in the potassium hydroxide, electricity is generated. The range and speed of a vehicle powered by a zinc-air battery could be increased by simply enlarging the battery's tank and increasing the number of tubes.

A study published in October, 1974, by England's Electrical Research Asso-

ciation and the Electrochemical Unit of London's City University indicated iron-air batteries are particularly suitable for heavy-duty vehicles, such as buses. These batteries last 2.5 times longer than lead-acid batteries and weigh about 40 per cent less.

Ford Motor Company has a $3-million contract with the National Science Foundation to demonstrate the technical possibility of using sodium-sulfur batteries in vehicles and for energy storage in power plants. Most sodium-sulfur batteries are cylindrical, with an inner cylinder of aluminum oxide that acts as a current conductor and also separates the sodium and sulfur. These batteries operate at about 600°F. (315°C), presenting a major design problem of how to protect passengers from the hot contents in case of a motor-vehicle accident.

Plastics. A new polymer has been developed that can resist temperatures up to 1000°F. (538°C), according to a report presented at the annual meeting of the Society of Plastics Engineers in San Francisco in May, 1974. Code-named Air Force Resin (AF-R) 530, it was developed by Whittaker Corporation's Research and Development Department in San Diego under a U.S. Air Force contract. Its developers say it is also amazingly unaffected by harsh chemicals. AF-R 530 will first be used in the aerospace industry, but eventually, it may be used in turbine engines and heat exchangers.

Plastic wallpaper was introduced in Great Britain in the spring of 1975 by Imperial Chemical Industries. The wallpaper, made of polyethylene foam, is lightweight and water resistant and once hung on a wall, it can be straightened or removed without difficulty.

In early 1975, two companies — Kerona Plastics Extrusion in Stockton, Calif., and Centaur Manufacturing in Milpitas, Calif. — began manufacturing pipe made of foamed polyvinyl chloride (PCV). The pipe weighs 38 per cent less than pipe of the same diameter and thickness made of solid PVC. But the foamed material is not as strong as the solid PVC. However, if the wall thickness of the foamed PVC pipe is increased by about 35 per cent, it becomes as strong as conventional PVC pipe and also a better insulator.

In early 1975, Polyair Maschinenbau Ges.m.b.H. in Kittsee, Austria, announced it had developed a polyurethane automobile tire that lasts longer than any other tire. The plastic tire is virtually puncture-proof, but the company is developing a version that can run even if it is damaged and without internal air pressure.

Scrap wood can be reclaimed to give stiff competition to plastics, according to MK-RDA, an engineering firm in Portland, Ore. The firm helped develop a process in which chunks and pieces of timber are ground into particles and fibers that are arranged in layers, first in one direction, then in another. This mat of fibers can be stamped into shaped wood products, such as molding, siding, and veneer for furniture and paneling. MK-RDA also claims it is stronger and lighter than plastic and more fireproof. The fibers are treated with fire retardants and preservatives.   [Frederick C. Price]

Chemistry

Chemical Dynamics. Research in 1974 and 1975 took impressive strides toward the goal of inexpensive, plentiful isotopes, atoms of chemical elements that differ only in the number of neutrons in their nuclei. A research team reported using lasers to concentrate, or enrich, measurable quantities of uranium 235, the isotope that fuels nuclear power reactors and nuclear weapons. See Close-Up.

Although excitement focused on the possibility of easily enriching nuclear fuel, research into the enrichment of other elements with selected isotopes was welcomed by scientists who use these isotopes in agricultural, biological, and medical research. Scientists may soon be able to specify the relative abundance of isotopes in the chemicals they order and pay far less for them.

Laser-enrichment teams work with diverse elements and molecules and use different lasers. But all base their work on the fact that different isotopes, or molecules containing different isotopes, absorb light of slightly different colors to reach excited states. By precisely tuning a laser, the scientists can excite a single isotope in a mixture. Then, using another laser or some other physical or chemical process, they can concentrate or even remove the excited isotope.

During 1972 and 1973, Edward S. Yeung and C. Bradley Moore of the University of California at Berkeley used lasers to separate molecules containing hydrogen from those containing the hydrogen isotope deuterium. At about the same time, Russian physicists R. V. Ambartzumian and Vladilin S. Letokhov of the Institute of Spectroscopy in Moscow succeeded in enriching isotopes of nitrogen.

A flood of research findings was reported, beginning in June, 1974. A West German team led by Herbert Walther enriched the calcium 40 (Ca-40) isotope in a gaseous beam of calcium atoms using a two-step photoionization process. They excited Ca-40 with a precisely tuned dye laser. Then the team ionized the excited atoms with an argon laser, changing them into positively charged atoms that migrated to a negatively charged collector.

Lasers: A New Light on Nuclear Fuel

Intensive, worldwide research into the use of lasers to process nuclear fuel began shedding its cloak of secrecy in June, 1974. Physicist Benjamin B. Snavely and his co-workers at the University of California's Lawrence Livermore Laboratory announced that they had increased the relative abundance of highly fissionable uranium 235 (U-235) in a stream of hot natural uranium gas as it passed through a precisely tuned laser beam and the light of an arc lamp.

As first announced, the reaction enriched only minute amounts of uranium. But by June, 1975, the team had increased the production rate 10-millionfold and had visible evidence of their laser-enrichment work.

Like most elements found in nature, natural uranium is a mixture of isotopes – atoms with the same number of protons in their nuclei but a differing number of neutrons. Highly fissionable U-235 nuclei have 92 protons and 143 neutrons, while less-fissionable uranium 238 (U-238) nuclei have 146 neutrons. On the average, natural uranium ore contains only about 0.7 per cent U-235;

that is, only 7 of every 1,000 atoms are U-235. But nuclear-reactor fuel must be at least 3 per cent U-235.

To concentrate U-235, scientists have developed various physical separation schemes based on the small weight difference between the two isotopes, including the gaseous diffusion process now in use. But gaseous diffusion plants are expensive. A typical plant costs about $3 billion and uses at least 2 million kilowatts of electric power. The Energy Research and Development Administration (ERDA) operates only three plants, the largest of which is at Oak Ridge, Tenn.

Uranium separation by laser also depends on the isotopes' slight mass difference. Because of this difference, one isotope, or a molecule containing the isotope, absorbs colored light of a slightly different shade, or frequency, than the other isotopes. By tuning a laser to the absorption frequency of one isotope, scientists can selectively excite it and then separate it from the unexcited isotopes through a chemical or physical process.

Excited by a precisely tuned laser beam and ionized by a second laser beam, charged U-235 atoms from a natural uranium vapor collect on a negatively charged plate. This process yields visible amounts of enriched uranium with enough U-235 to be used as nuclear reactor fuel.

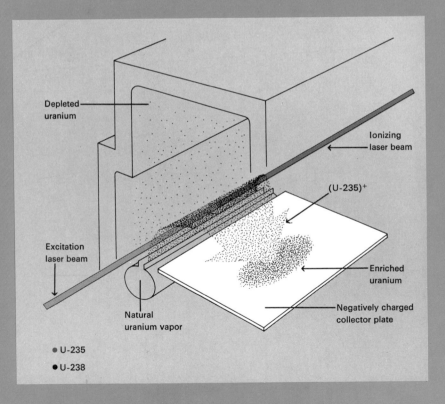

Depleted uranium

Ionizing laser beam

(U-235)+

Excitation laser beam

Enriched uranium

Natural uranium vapor

Negatively charged collector plate

● U-235
● U-238

The Livermore researchers used a scheme called two-step photoionization that was pioneered by Russian scientists. In their 1975 work, the Livermore group passed a gaseous beam of uranium atoms from an oven through a laser beam of yellow light precisely tuned to U-235's absorption frequency, and the laser excited outer electrons in the U-235 atoms. When these atoms passed through a second beam of light from a krypton laser, an electron was knocked free from each, leaving behind positively charged U-235 atoms that were attracted to a negatively charged collector plate. After two hours, about 0.004 gram of uranium had collected on the plate. The uranium was dissolved, dried to a powder, and found to contain about 3 per cent U-235.

To achieve this enrichment level in a gaseous diffusion plant requires thousands of sequential passes. Thus, the scientists showed the high enrichment possible in a single pass using lasers.

Snavely's colleague Sam A. Tuccio cautioned against assuming that the process can be duplicated on a large scale. "We are working with a technology in its infancy," he said. "The enrichment concept we have demonstrated is only one of the many that appear theoretically possible." Even if the process proves commercially feasible, large-scale laser-processing plants could not be in operation much before the mid-1980s. Their impact on the economics of nuclear-generated electricity would be "significant but not earthshaking," according to an ERDA spokesman.

Laser isotope enrichment may be more significant for countries that must now buy nuclear fuel from the United States because they cannot afford to build gaseous diffusion plants. The development may also have military implications. Following an announcement by Russian scientists in April, 1975, researchers at the Los Alamos Scientific Laboratory near Santa Fe, N. Mex. — where the world's first atomic bombs were developed — reported that they too had developed a laser separation process and had used it to enrich isotopes of boron, chlorine, and sulfur. Previously, their laser-enrichment work had been kept secret under the code name Project JUMPer.　　　　[Richard N. Zare]

Also in June, a research team led by Lowell L. Wood, at the University of California's Lawrence Livermore Laboratory, reported the separation of barium isotopes by shining a laser perpendicular to a beam of barium atoms. Atoms that absorb the light recoil sideways, creating a parallel portion of the barium beam that is richer in the selected isotope.

One of the most promising schemes for separating isotopes uses laser-controlled chemical reactions rather than physical means. Michel Lamotte, Harry J. Dewey, Richard A. Keller, and Joseph J. Ritter of the National Bureau of Standards (NBS) in Gaithersburg, Md., announced in January, 1975, that they had enriched isotopes of chlorine in the molecule thiophosgene ($CSCl_2$) by capitalizing on the excited molecule's greater reactivity with diethoxyethylene (DEE).

The NBS researchers could excite either $CSCl_2$ molecules containing chlorine 35 (Cl-35) or those containing chlorine 37 (Cl-37). For example, they used a dye laser tuned to 4705.5 angstroms to excite $CS(Cl-35)_2$. The excited molecule reacted with DEE and after four hours the amount of $CS(Cl-35)_2$ dropped from its natural abundance of 75 per cent to only 64 per cent.

Dirkson D.-S. Liu and Saswati Datta performed similar laser-controlled photochemical reactions at Columbia University using iodine monochloride (ICl). They also developed a laser-controlled isotope-exchange process in which the selected isotope of chlorine in the excited ICl replaced a natural chlorine isotope in the other reactant.

Other research teams used infrared lasers, which are more efficient than visible or ultraviolet lasers. Steven D. Rockwood and Sherman W. Rabideau of Los Alamos Scientific Laboratory (LASL) in New Mexico used a carbon dioxide (CO_2) infrared laser to separate boron isotopes. The laser light excites boron trichloride (BCl_3) containing the boron 11 (B-11) isotope, which is then ionized by ultraviolet light from a xenon lamp. The ionized BCl_3^+ reacts with a small quantity of oxygen to form a removable solid compound.

As an example of how swiftly the research progressed during the year, Rockwood's LASL colleague C. Paul

Robinson announced in April, 1975, that the group no longer needed ultraviolet light to enrich BCl_3. They used a new single-laser method to enrich boron and also to enrich microgram quantities of sulfur 34 (S-34) in sulfur hexafluoride gas by as much as 33 times its natural abundance. Earlier, Letokhov had written a letter to United States scientists describing even greater success using the same technique.

According to Robinson, S-34 is a nonradioactive isotope that would be a valuable tracer in food-chain cycles and in medical diagnosis. Its high cost—$1,000 per gram—has discouraged wide usage, but Robinson points out that the energy cost alone for their laser enrichment method is only about 40 cents per gram.

Reaction rates. Exactly how excited molecules affect the rate of chemical reactions between gas molecules continued to baffle researchers. To study this effect, they have devised clever ways of exciting molecules so they vibrate, rotate like tumbling dumbbells, or move in one of many possible excited states. Then the scientists react the molecules with other molecules before the excitation energy is able to spread to other energy states.

For example, Robert J. Gordon and M. C. Lin of the U.S. Naval Research Laboratory in Washington, D.C., irradiated ozone (O_3) with an infrared laser, causing it to vibrate. They quickly reacted the vibrating O_3 with nitric oxide (NO) to yield excited nitrogen dioxide (NO_2*) and oxygen (O_2). They found at least a tenfold increase in the reaction rate compared to a similar reaction in which O_3 does not vibrate.

Experiments in which vibrational excitation of the reactants did not increase reaction rate were also reported. In some cases, the rate decreased. For example, the reaction rate of cyanogen radical (CN) with O does not increase when CN is vibrationally excited, and the reaction of CN with O_2 decreases when CN is vibrating.

Richard B. Bernstein and his co-workers at the University of Texas in Austin probed the effect of rotational excitation on reaction rates. They found that energy of motion is more effective than rotational energy in boosting reaction rates.   [Richard N. Zare]

Structural Chemistry research during 1974 and 1975 focused on fluxional molecules, which switch back and forth from one structure to other equally likely configurations. For example, Kurt M. Mislow and his co-workers at Princeton University in New Jersey synthesized a highly symmetric methane derivative, trimesitylmethane (TMM), and then studied its structure. In a TMM molecule, three flat mesityl rings bond to a central carbon atom. Because the rings are too bulky for all three to fit in the same plane, they twist slightly to a form that resembles a three-bladed propeller. A solution of TMM contains an equal number of propellerlike molecules pitched in both left- and right-handed directions.

The Princeton chemists in 1974 explained that TMM molecules can be made to reverse their pitch in a process requiring 21.9 kilocalories of energy per mole. Each propeller reversal requires that two mesityl rings twist in a direction opposite to the third. The energy needed to do this is nearly twice that needed by related fluxional molecules. Because of this large difference in energy, the left- and right-handed molecules possibly can be separated.

The Princeton team used nuclear magnetic resonance (NMR) spectroscopy to study TMM in solution. They placed the solution between the poles of an electromagnet and monitored the patterns, or spectra, of radio-frequency energy absorbed by atomic nuclei.

Chemists use NMR studies because the spectra reflect the surroundings of the various atoms in the molecules. Because a fluxional molecule changes structure very rapidly in solution, its NMR spectrum represents an averaged structure just as a photograph of a rapidly spinning pinwheel shows only a blurred circle. But to provide an instantaneous view of molecules, chemists use X-ray crystallography in which a beam of X rays passes through a crystal and the diffraction pattern emerging from it is analyzed to establish the molecular structure.

F. Albert Cotton and his co-workers at Texas A&M University at College Station investigated organometallic molecules of the type $(COT)M(CO)_3$, in which COT represents cyclooctatetraene (C_8H_8); M represents an atom of

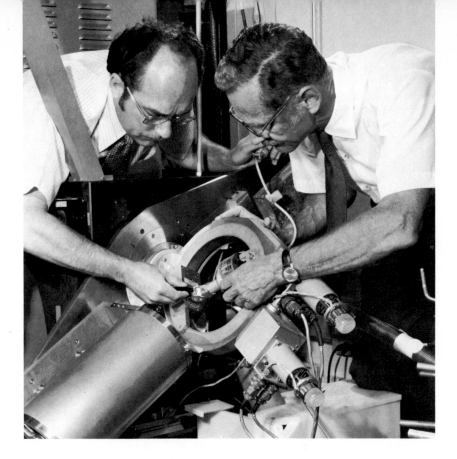

Chemists at Argonne National Laboratory adjust a neutron diffractometer that they use to study KCP. These platinum salts are one-dimensional conductors, which pass current easily in only one direction.

Chemistry

Continued

chromium, molybdenum, or tungsten; and CO is carbon monoxide. The scientists analyzed the NMR spectra of the compounds at different temperatures and found that M bonds symmetrically to all eight carbon atoms of the C_8H_8 ring during the molecule's rearrangement sequence in solution. But their X-ray studies showed that in the crystalline state M bonds to only six carbon atoms of the COT ring.

Opening and closing bridges. The scientists also investigated the fluxional behavior of metal-metal bonded molecules. They studied the cluster compound $Rh_4(CO)_{12}$, which, in the crystalline state, forms a pyramid structure with rhodium (Rh) atoms at its corners. The carbonyl (CO) ligands form two types of coordinate bonds with Rh atoms. Ligands are ions or groups of atoms that form coordinate bonds with metal ions. By attaching to one Rh atom, the CO ligands form terminal bonds, and by attaching to two at the same time, they form bridging bonds. In solution, however, all carbonyl ligands appear identical.

Cotton proposes that the rapid inter-conversion of terminal and bridging CO ligands in metal-metal bonded systems is an important and general phenomenon. With colleague Richard Adams, he demonstrated that metal-ligand-metal bridges open and close in pairs. They used the ligand methyl-isocyanide ($CNCH_3$), which is closely related to CO.

However, other researchers may have found an exception to this rule in $Pd_2(CNCH_3)_6^{2+}$, the unusual $CNCH_3$ complex with palladium (Pd) first synthesized in 1974 by Alan Balch and co-workers at the University of California at Davis. NMR studies show all ligands to be identical in solution at room temperature.

In April of 1975, however, Balch and Stephen Goldberg and Richard Eisenberg of Rochester University in New York determined the X-ray structure for crystals of $Pd_2(CNCH_3)_6^{2+}$. They found that two of the $CNCH_3$ ligands differ from the other four. The crystalline structure consists of two squares joined at a common corner by a

261

Chemistry

Continued

An IBM scientist points to a site in a plastic model of the one-dimensional organic conductor TTF-TCNQ at which a selenium atom replaces a sulfur atom. These substitutions increase conductivity along the axes of the ion stacks.

Technicians remodel a powerful heavy-ion linear accelerator (HILAC) located at the Lawrence Berkeley Laboratory. SuperHILAC later was used to synthesize element 106, now the heaviest element in the periodic table.

Pd-Pd bond, and the plane of one square is perpendicular to that of the other. The chemists believe that the six $CNCH_3$ ligands become equivalent in solution by a process in which the squares distort into pyramids joined at a common apex by a Pd-Pd bond.

1-D conductors. Chemists and physicists continued their studies of new materials that conduct electric current preferentially in one direction. These one-dimensional (1-D) conductors consist of stacks of cations, or positive ions, and anions, or negative ions, in the crystalline state. The regularly spaced ions overlap and form bonds along which electrons conduct electricity.

For three years, many independent research teams have been studying the 1-D conductor made by joining the cations tetrathiofulvalene (TTF) with the anions tetracyanoquinodimethane (TCNQ) to form tiny, brittle, needle-shaped crystals. Chemists Thomas J. Kistenmacher, Dwaine O. Cowan, and their co-workers at Johns Hopkins University reported in 1974 that X-ray crystallography studies showed TCNQ anions to be spaced 3.17 angstroms apart within each anion stack.

In September, 1974, Edward N. Engler and Vishnu V. Patel announced that they had doubled the conductivity by using selenium instead of sulfur in TTF in experiments at International Business Machines' Thomas J. Watson Research Center in Yorktown Heights, N.Y. The change altered the original TTF-TCNQ structure only slightly.

Some inorganic complexes, which are formed when groups or ligands bond to metal ions, are also conductors because of their stacked structures. A specific example is the material $K_2Pt(CN)_4Br_{0.3}3H_2O$ (KCP) first synthesized more than 100 years ago.

The true structure was first established in 1968 and was re-examined last year at Argonne National Laboratory near Chicago. A team of researchers led by Jack M. Williams used neutron beams rather than X rays to probe the structure. In August, 1974, Williams reported that bromine atoms play a key role in the conduction process and in the behavior of KCP at $-423°F.$ $(-253°C)$ when it becomes an electric insulator.   [Richard Eisenberg]

See PHYSICS (Solid State Physics).

Chemical Synthesis. Scientists synthesized a wide range of new products in 1974 and 1975. They include promising new antibiotics; organometallic materials that may allow vital industrial processes to take place at normal atmospheric pressures and temperatures; a combination of carbon and oxygen atoms that has defied synthesis until now despite its simple, stable triangular structure; and an element that has 106 protons in its nucleus. See PHYSICS (Nuclear Physics).

Burton G. Christensen and his research group at the Merck Sharp & Dohme Research Laboratories in Rahway, N.J., announced in November, 1974, that they had synthesized a new family of antibiotics that may surpass penicillins in effectiveness. The new antibiotics, oxacephalothin and carbacephalothin, are chemically similar to the cephalosporin antibiotics derived from a natural fungus. To synthesize the two new antibiotics, however, the Merck chemists do not need a naturally occurring starting substance.

Cephalosporins are chemically related to penicillin antibiotics; in fact, the cephalosporin antibiotic cephalothin contains a sulfur atom in the same position as that found in the penicillins. Starting with the chemical building blocks of cephalothin, the chemists replaced this sulfur atom with an oxygen atom to synthesize oxacephalothin and with a methylene group to synthesize carbacephalothin.

Quest for catalysts. Because of the high price of energy, many synthesis chemists have begun searching for new catalysts—materials that cause or speed up chemical reactions. Some vital industrial reactions take place only at high atmospheric temperatures and pressures. For example, the complex Haber-Bosch process used to manufacture ammonia fertilizer from atmospheric nitrogen and oil requires vast quantities of fuel to drive temperatures up to about 1022°F. (550°C) and pressures to about 200 times atmospheric pressure. A cheap, plentiful catalyst that would allow the reaction to take place under less extreme conditions would be a real breakthrough. See THE NITROGEN FIX.

Researchers have tried to develop organometallic catalysts. These are sta-

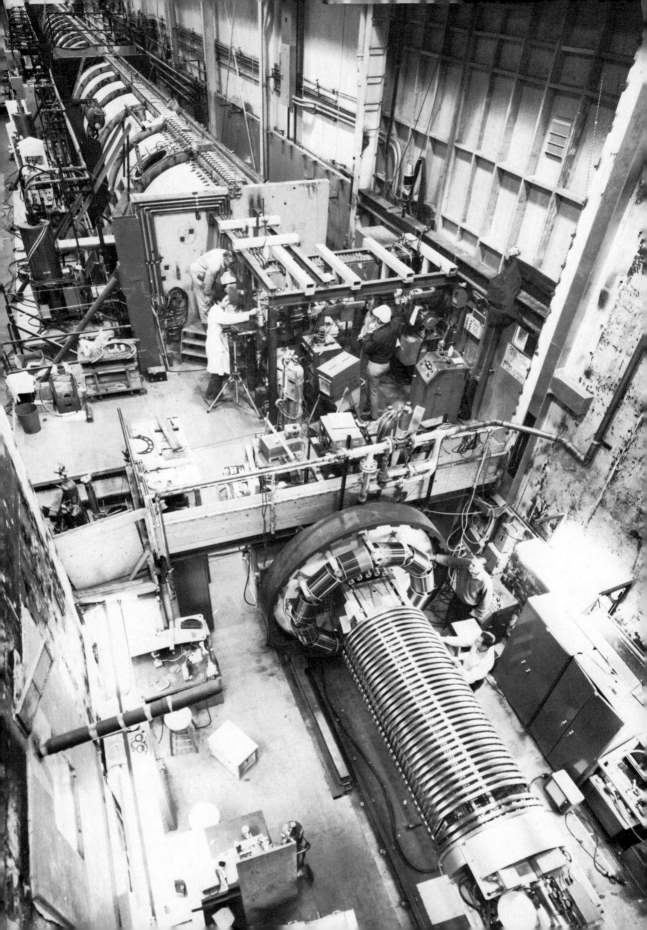

ble organic compounds containing a transition metal such as molybdenum in a very low or zero oxidation state — that is, the metal has all, or nearly all, of its normal complement of electrons. Conventional synthesis methods are of little use because the products invariably contain oxidized metal atoms.

But two independent research groups in 1974 developed a general synthesis procedure called hot atom chemistry for the direct formation of organometallics that contain an unoxidized metal atom. One group is headed by Philip S. Skell of Pennsylvania State University, the other by Peter L. Timms of Bristol University in England. Both teams found that hot transition metal atoms produced by heating the metal in a vacuum until it evaporates can be easily captured by solutions containing unsaturated hydrocarbons in inert solvents at low temperatures.

An ingenious but simple device keeps the normally solid metal in a vaporous state and the normally liquid hydrocarbon in a solid state. The chemists dissolve the unsaturated hydrocarbon or other reactant in an inert solvent such as liquid propane at $-301°F.$ ($-185°C$) and place this solution in an inclined flask with the bulbous part immersed in a cooling bath. Rapidly rotating the flask spreads a film of the solution on the bulb's inside surface. The chemists then pump the air out of the flask and turn on an electric heater that is suspended at the center of the bulb along with the metal. When the heated metal atoms strike the cold hydrocarbon film, they give up heat and form the organometallic product. Several ounces or grams of metal can be evaporated in an hour in the laboratory. Because the process is so simple, it could easily be scaled up to produce pounds or kilograms of the product per hour.

Timms's group prepared fascinating compounds containing iron in the unoxidized state, which are impossible to prepare by any other process. Skell's research has led to the preparation of sandwich compounds in which unoxidized tungsten or molybdenum atoms lie between benzene rings in a structure resembling a ball bearing placed between two hexagonal nuts. Neither of these molecules is likely to be a useful catalyst, but some similar new compounds will undoubtedly be superior to presently known catalysts.

Weird limestone. Carbon dioxide (CO_2) plays a vital role in the life processes of plants and animals, controls the acidity of the seas, and, when combined with water and calcium ions (Ca^{2+}), appears as useful deposits of limestone ($CaCO_3$). Toxic carbon monoxide (CO) and carbon suboxide (C_3O_2), a laboratory oddity, are other simple combinations of carbon and oxygen atoms.

Closely related to these oxides of carbon are the ringlike oxycarbon anions (negative ions) that contain an equal number of carbon and oxygen atoms. For example, red rhodizonate ($C_6O_6^{2-}$) forms a flat hexagon, and yellow croconate ($C_5O_5^{2-}$) anions have a five-sided structure. The square oxycarbon ion ($C_4O_4^{2-}$) was discovered accidentally in 1959, but the triangular $C_3O_3^{2-}$ ion remained beyond the reach of synthesis chemists until late in 1974.

Robert West and graduate student David Eggerding of the University of Wisconsin at Madison reported in January, 1975, that they had synthesized deltic acid, the acid containing $C_3O_3^{2-}$. The acid also has a delta structure. West suspected the molecule to be extremely stable because a double bond can form along a leg of the triangle between any two carbon atoms at two vertices. The double bond switches from leg to leg in a stabilizing process called resonance in such a way that each leg has an extra one-third bond.

West and Eggerding synthesized deltic acid in a two-step process that began with the acid of $C_4O_4^{2-}$ and involved the use of ultraviolet light. The deltic acid produced in this way is a white crystalline solid that must be heated to about $356°F.$ ($180°C$) before it decomposes. Preliminary evidence indicates that $C_3O_3^{2-}$ is quite stable in an aqueous solution.

The new synthesis closes the book on the oxycarbon anion series because $C_2O_2^{2-}$ and CO^{2-} cannot exist. It is interesting to note that the common carbonate ion (CO_3^{2-}) is also a flat triangle, containing one oxygen atom at each vertex. Consequently, mixing CO_3^{2-} and deltic acid would produce a precipitate that may be a weird limestone.   [M. Frederick Hawthorne]

Communications

Development of new communications technologies has greatly increased our ability to interconnect telephones. But along with development of call switching and transmission, the telephone itself is also changing.

A model of the Touch-a-Matic telephone for home use will be introduced in late 1975. The Bell System introduced this telephone for business offices in February, 1974. Like a small computer, it can electronically store up to 32 telephone numbers in 10 silicon large-scale integrated circuits, each of which has 1,500 transistors. The Touch-a-Matic for home use is the first application of silicon large-scale integrated circuits (circuits containing more than 500 transistors) in a home telephone.

A customer can automatically dial any of 31 numbers by pressing a button on the front panel beside a space containing the name of the party to be called. The 32nd button is used to redial the last number dialed manually.

Digital signals for local calls. For many years, some telephone calls have been transmitted over wires between switching offices in digital form, converting human speech into a stream of 0s and 1s. By transmitting speech as symbols, slight deviations caused by noise or distortion can be eliminated, and it is technically easier to send two or more calls simultaneously over the same wire.

In December, 1974, the American Telephone and Telegraph Company introduced the Subscriber Loop Carrier (SLC-40) system, which uses this technique to transmit calls from individual telephones to central offices. With this system, two pairs of wires handle up to 40 calls at the same time. The first installations of the system were made in rural areas where many phones are several miles from the central office. At these distances, it is more economical to mix digitally coded conversations on one wire than to have individual wires from each telephone to the central office.

High-frequency technology. High-frequency experiments were conducted during 1974 and 1975 with transmit-

A list of words most likely to be misspelled from a manuscript typed into a computer appears on a television screen. The list was generated by a new computer program as an aid to proofreaders.

Jul 10 15:10 Possible typo's in junk Pag

```
15 expersed        2 occupies
14 rreplacing      2 leaves
14 nd              2 en try
14 ingenuity       2 ended
14 ignored         2 ditto
14 encouraging     2 consisted
14 duplicates      2 com pression
13 nl              2 com pressing
13 investm ent     2 coded
13 constituent     2 chose
13 biggest         2 by te
13 aggreem ent     2 assigning
11 lookup          2 agreeing
 8 onw ard         1 codes
 7 didn t
 6 frewuencies
 4 pseudolam ellibranchi
 4 pseudolam ellibranchi
```

Communications

Continued

A television screen as thin as a picture frame uses about 36,000 tiny electronic parts to light phosphorescent material and produce an image. The screen is being developed by Westinghouse Research Laboratories.

ters and receivers aboard the Applications Technology Satellite (ATS-6) launched in April, 1974. The experiments tested new satellite broadcast frequencies of from 13 to 18 gigahertz. A gigahertz (GHz) is a billion cycles per second. These high frequencies are affected by rainfall and other atmospheric conditions. The studies will aid in the design of systems to overcome interference and make efficient use of the higher-frequency bands.

The ultimate number of satellite communication channels is determined by the number of radio-frequency bands available and the shortest possible distance between satellites using the same frequency band. An even higher-frequency ATS-6 experiment was designed to study microwave lengths above 30 GHz. Bell Telephone Laboratories is also developing better microwave antennas that can concentrate signals sent to a satellite into a narrower beam. Narrow-beam antennas are also needed to receive signals from a satellite without interference from its neighboring satellites. These improvements will permit the use of more satellites spaced closer together in synchronous orbit above the equator.

Higher frequencies are also being developed for terrestrial systems. Field-testing began in June, 1975, on an 18 GHz radio system for short-distance phone calls near New York City. Atmospheric conditions do not affect short-distance, high-frequency transmissions.

For longer-distance telephone calls, Bell Laboratories' field test of a high-frequency wave-guide system, which began in New Jersey in January, showed that the system is performing well. This system sends hundreds of waves at frequencies between 40 and 100 GHz through an underground pipe, or wave guide, with an inside diameter of 2.4 inches (6.09 centimeters). Because the pipe is underground, there is no atmospheric interference. The system can handle 230,000 phone calls.

To handle more overseas calls, a new transatlantic cable (TAT-6) was scheduled to be laid in 1975 between Rhode Island and France. It will handle 4,000 simultaneous conversations.

Electronic toll switching. Final development of a new Bell System Electronic Switching System (ESS) drew to a close in 1975. The first ESS to be used for toll-call switching, it was scheduled to begin service in Chicago in January, 1976. It will be the world's largest switching system, with a capacity of 350,000 toll calls per hour. This is more than three times the capacity of electromechanical toll-switching systems now in use. More than 30 of the huge new systems are to be installed before the end of 1978.

The new ESS employs time-division switching, in which the human voice is coded into 0s and 1s, known as data bits. These bits of information are placed in a time sequence with bits from many other phone calls so they can all be transmitted over the same wire inside the machine. The system separates individual conversations by recognizing only the bits that arrive at specific intervals of time.

This is an incredibly high-speed operation, since each call breaks down to about 8,000 bits per second. A special computerlike system controls the complex electronic switching circuits. Several of these machines are interconnected so if one fails, the switching office can continue to function.

Optical fibers. Development of thin glass fibers that can contain and guide light beams to transmit telephone signals over several miles continued. Transmission systems using fiber bundles would be especially useful in urban areas, where underground ducts are becoming choked with conventional wires. Pencil-thin cables containing about 100 fibers may be developed for use in the cities within the next 10 years. But as of mid-1975, bundles of six fibers cost approximately $4 per foot (0.305 meters), too expensive to be economically feasible.

However, a system that used a solid state crystal laser for a light source could carry about 4,000 one-way conversations on a single fiber. The cost of fibers per call would be greatly reduced by that volume of calls. One obstacle to development had been that solid state lasers lasted only a few hundred hours. However, in early 1975, Bell Laboratories developed techniques to eliminate strain in the crystal. Solid state lasers can now be expected to last more than 10 years.   [John A. Copeland]

Drugs

Scientists have turned increasingly to biologically active substances in marine organisms in their search for new drugs and new sources for old ones. At the Fourth Food-Drugs from the Sea Conference held in Puerto Rico in November, 1974, reports were presented on a wide variety of such compounds.

One is palytoxin, the most powerful constrictor of blood vessels known. It may be useful in the study of coronary artery disease, according to pharmacologist Pushkar N. Kaul of the University of Oklahoma in Norman. It is derived from marine coelenterates, a group of salt-water invertebrates with saclike bodies like the jellyfish. A number of chemicals extracted from marine tunicates, primitive relatives of the chordates, or animals with spines, have been found to inhibit tumor cell growth. The chemicals, whose structures are still unknown, also prolong the survival of mice implanted with leukemic cells and suppress the formation of certain antibodies. Further studies by microbiologist M. Michael Sigel of the University of Miami, Coral Gables, Fla., and his colleagues have demonstrated that these extracts inhibit the synthesis of deoxyribonucleic acid (DNA) and in this way slow the growth of cells in cultures of human lymphocytes, the white blood cells that make antibodies. Researchers believe that this may lead to greater understanding of the mechanisms of the immune system.

Other potentially important findings were also reported at the conference. For example, an antibiotic has been discovered in a Caribbean sponge and a prostaglandin with tranquilizerlike activity has been derived from gorgonian sea whips, a form of coral.

To give added impetus to undersea drug research, the Hoffmann-LaRoche Company of Switzerland, one of the largest international pharmaceutical firms, has established the Roche Research Institute of Marine Pharmacology in a suburb of Sydney, Australia. Director Joseph T. Baker has a group of more than 50 scientists and laboratory workers with the most up-to-date technical and computer facilities. They will conduct investigations of marine chemical compounds and their possible applications in human health care, veterinary medicine, and agriculture.

New codeine source. Dwindling opium imports and increasing demands for codeine, one of its derivatives, caused production problems for U.S. manufacturers in 1975. Codeine, a narcotic, is a popular ingredient of pain-relieving and cough-suppressing medicines. The opium shortage was estimated at 110 short tons (100 metric tons).

Traditionally, opium has been extracted from the poppy plant *Papaver somniferum,* and then further refined to yield codeine. However, it is illegal to grow the opium poppy in the United States because heroin, which is addictive and has no legal medical use, can also be obtained from it. To get around this problem, Mallinckrodt, Incorporated, a major U.S. producer of codeine, has been experimentally growing another poppy, *Papaver bracteatum,* which contains codeine but not heroin. Although the results thus far are encouraging, there are serious problems in obtaining sufficient seeds, which come chiefly from Iran, and in strengthening the plant against weeds, which easily choke it out. The United Nations is now funding studies of *P. bracteatum* in a number of countries, and other American codeine manufacturers are considering its cultivation. Within a decade, the United States may become self-sufficient in codeine.

Antacid labeling. The Food and Drug Administration (FDA) has ruled that the 800 antacids on the market must meet its new labeling standards by June 4, 1975. The antacids are the first of 27 over-the-counter (OTC) drug groups — sold without a doctor's prescription — to be evaluated by the FDA for safety, usefulness, and accuracy and completeness of labeling information. Antacids may now be labeled only for treatment of acid indigestion, sour stomach, and heartburn.

Only 13 chemical substances have been approved as active ingredients. These must be listed, along with the amounts of each, in every preparation. The label must now list appropriate warnings, such as "may cause constipation," if such an effect has been reported in 5 per cent or more of those taking the maximum recommended dosage.

In addition, substances in the compound which may be harmful to certain users, such as potassium to those with

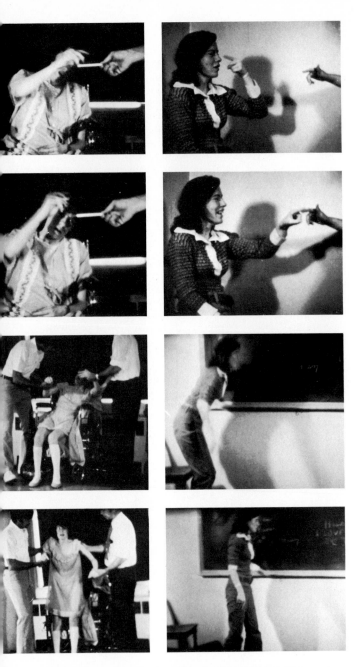

Victim of myoclonus, a nerve disorder that causes sudden uncontrollable muscle tremors, is not able even to grasp a pencil or get out of her wheel chair during an attack, *left column*. But after treatment with L-5-HTP, which controls the seizures, and carbidopa, which eliminates the unpleasant side effects of L-5-HTP, the patient can easily perform simple muscular tasks and walk short distances, *right column*.

kidney disease or lactose to persons allergic to milk, must be mentioned. An important new section of the label will be headed "Drug Interaction Precaution." The consumer can learn from this if a particular antacid can interact adversely with other medicines he may be taking. If, for example, the OTC drug contains aluminum, which is known to reduce the absorption of tetracycline antibiotics, the label must warn against taking the two medicines at the same time. During the next few years, new standards will be established and published for the estimated 100,000 to 500,000 OTC drugs in the remaining 26 classes.

Vitamins. Controversy over the effects of vitamin pills continued amid research that was inconclusive (see NUTRITION, Close-Up). In May, 1975, the FDA backed off from its 1973 proposal to classify all vitamin and mineral dietary supplements as drugs. Under proposed new rules, they would be considered foods in amounts generally recognized as safe. High potency vitamins A and D remain classed as drugs, because they have harmful effects in high doses.

New drugs introduced in the United States in 1975 included:

- Amoxil; Laroxin (Amoxicillin), a new oral penicillin that is effective against a large variety of bacteria, though not against so-called "penicillin-resistant" staphylococci. This compound is similar in action to the widely used ampicillin but is absorbed better from the gastrointestinal tract.
- Intropin (dopamine hydrochloride), a compound that increases the heart's ability to contract and may restore blood pressure when it is dangerously low during shock, such as after a heart attack or open-heart surgery. Although chemically related to adrenalin and other stimulant drugs, Intropin exhibits differences in activity, especially in increased blood flow to the kidneys and less abnormal heart rhythm.
- Adriamycin (doxorubicin hydrochloride), an antibiotic drug that shows promise in treating breast cancer, leukemia, lymph cancer, and certain malignant bone and muscle tumors. The compound is used alone and in combination with other cancer-fighting agents. In excess, it may adversely affect the heart. [Arthur H. Hayes, Jr.]

Ecology

Ecologists gathered new data in 1974 and 1975 on the roles that chemical compounds produced by plants play in defending the plants against insects. Knowing how such chemical barriers function may help scientists develop new techniques for the biological control of insect pests. It may also increase our understanding of what effects insects and plants had on each other during their evolution.

During the winter of 1974, ecologists James M. Erickson and Paul P. Feeny of Cornell University in Ithaca, N.Y., reported one of the most ingenious series of experiments with these compounds. They worked with the larvae (caterpillars) of the black swallowtail butterfly. The experiments showed how the larvae's feeding habits are restricted by a plant's chemical barriers and how a chemical compound (sinigrin) produced by the mustard family, *Cruciferae,* may protect those plants against insects. Earlier research showed sinigrin plays a key role in the plant's defense.

The larvae of the black swallowtail feed on species of the wild carrot family, which grow along roadsides and in fields and pastures. These habitats also contain plants from several other families, including the mustards. But the caterpillars never eat plants of the mustard family. So Erickson and Feeny placed these caterpillars in cultures containing food with various concentrations of the sinigrin compound derived from mustard plants and then measured their growth. The scientists hoped to simulate the consequences to the caterpillars if they mistakenly ate mustard plants.

They fed newly hatched caterpillars on celery leaves, the stems of which had been treated with solutions of sinigrin in distilled water. The concentrations of sinigrin ranged from 0.001 per cent through 0.01, 0.1, 1.0, and 2.5 per cent. **Groups of 16 larvae** were used in each of the experimental feedings on different concentrations of sinigrin. The scientists also kept a control group that did not receive sinigrin.

The scientists placed individual larva in Petri dishes lined on the bottom with filter paper so that larvae waste could

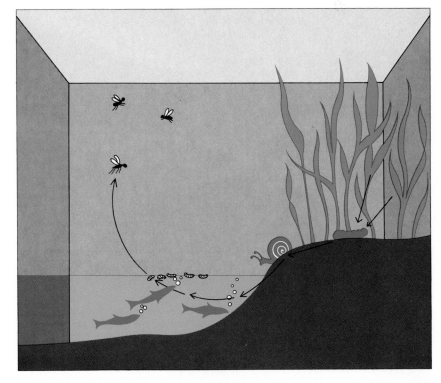

A mini ecosystem in a small aquarium is used to test persistence of radioisotope-tagged pesticides. Sorghum leaves sprayed with pesticide are eaten by caterpillars which contaminate the other species. The amount of pesticide each retains can then be measured.

be collected easily. They fed them 16-week-old celery plants, the stems of which had been soaked in the sinigrin solutions for 18 hours. The larvae were kept in a controlled-climate room.

The greater the concentration of sinigrin in the larvae's food, the longer it took the caterpillars to reach the pupa stage and then emerge as adults. The death rate also increased at higher sinigrin concentrations. Larvae on the 2.5 per cent and 1.0 per cent sinigrin treatments died before their first molting, and those on the 0.1 per cent treatment died soon after their fourth molting. The sinigrin either kills their appetites or poisons them or both.

When the surviving caterpillars—those fed on a 0.01 concentration or less—went into the pupa stage, the scientists placed each one in a large plastic box until it emerged as an adult butterfly. Then they fed the butterflies a honey-water solution that resembled the nectar they would ordinarily get from flowers. All females were mated and then placed in individual screened cages containing plants of the wild carrot family.

The scientists kept daily records of the number of eggs laid by each female and the percentage of these eggs that hatched. The females that had been fed sinigrin as larvae laid fewer viable eggs, with the number decreasing with increasing sinigrin concentration. For example, females that received no sinigrin produced a mean number of 142 eggs—half the females produced more and half produced less. Those that were fed a 0.001 per cent sinigrin solution produced a mean of 101 eggs; those fed 0.01 per cent, 73.

If this experiment adequately simulated the natural situation, it suggests that mustard-family plants contain enough sinigrin to prevent survival of black swallowtail larvae. It also suggests that chemical barriers created by plants partially restrict the feeding area of larvae in that ecosystem. Such plants have been likened to islands, defended in part by these chemical barriers. If insects could develop resistance to the chemicals, they could adapt more widely—feeding on a greater variety of plants—and suddenly expand their population. It seems increasingly clear that the sinigrin compound represents such a chemical barrier to the insect colonization of mustard-family plants.

New class of predators. Ecologists continually strive to understand the feeding relationships of organisms in order to unravel the mystery of natural food webs and their significance in the way ecosystems function. Terrestrial isopods, such as sow bugs or wood lice, are often found along foundations of buildings and among fallen leaves in forests and grasslands. Ecologists have assumed that they are scavengers that eat almost any kind of decaying material. But in the spring of 1974, ecologists Eric G. Edney and Warren Allen of the University of California at Los Angeles and Joan McFarlane of the University of California at Riverside reported on experiments which showed that isopods may also be predators.

The ecologists conducted laboratory and field experiments using fruit fly pupae to test for predatory feeding habits in isopods. In one set of laboratory experiments, 50 males and 50 females of two isopod species were each offered 16 fruit fly pupae. After 24 hours, the scientists found that males and females of both isopod species had consumed pupae.

To test this behavior over a longer time, they offered male and female isopods fruit fly pupae daily for seven days. The isopods consumed pupae daily but ate fewer and fewer each day. This may have been because they ate the larger pupae first, thus lessening their hunger. In another experiment, five male and six female isopods were offered 20 pupae of the eye gnat each day for seven days. Again, the daily consumption dropped. Some of the isopods were not feeding continuously.

The scientists then prepared an experiment that would test the isopods under more natural conditions. They constructed an environment in a metal tray measuring 45 x 35 x 15 centimeters (17.7 x 13.8 x 5.9 inches). They partly filled the tray with soil from a nearby citrus orchard and covered the soil with dead citrus leaves and other debris and three rotting citrus fruits. Then they fastened 80 fruit fly pupae that had been tagged with a radioisotopic tracer to leaves and placed them on the soil to mimic nature as closely as possible. Next, they added 20 isopods, which had

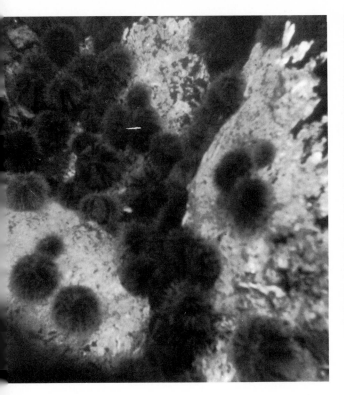

Ecology
Continued

Where sea urchins are plentiful, *above*, in shallow water off the Aleutian Islands, the area is almost bare of vegetation. Algae is abundant, *above right*, in similar areas where there are few urchins.

not been fed for 24 hours. A day later, the scientists collected the isopods and found that the bodies of nine contained radioactive material, clearly indicating that they had fed on pupae.

Finally, the scientists devised a field experiment to contain isopods and make them easy to collect. They built an enclosed circular arena, 55 centimeters (21.7 inches) in diameter and 15 centimeters (5.9 inches) high under a citrus tree in an orchard where isopods and fruit flies were abundant. Then they placed 120 pupae marked with radioisotope inside. After 68 hours, they collected 40 isopods and examined 25 of the largest for radioactivity. The amount of radioactive material in their bodies showed that some of them had eaten several tagged pupae, even though citrus leaves and other food were abundant. If, as these experiments indicate, isopods feed on small living invertebrates, our ideas about their significance in the food chain will need to be revised.

Ant colonies have long interested ecologists, not only because of ants'

complex social organization, but also because the large mound-building ants presumably have an effect on ecosystems. In December, 1974, ecologists Lee E. Rogers of the Battelle-Pacific Northwest Laboratory in Richland, Wash., and Robert J. Lavigne of the University of Wyoming in Laramie reported on their studies of the environmental effects of Western harvester ants on the short-grass plains ecosystem in Colorado's Pawnee National Grassland. The scientists wanted to determine how much area the ants clear and if this causes a significant drop in food resources for other plant-eating animals. They also investigated chemical modification of the soil, changes in soil moisture, and density of plant growth near ant colonies.

The scientists calculated the total area cleared of vegetation by ants by determining the average area each colony cleared and multiplying this by the estimated number of colonies in a given area. They estimated colony density by counting all colonies in 10 plots selected at random in each pasture under study.

They then selected 50 colonies at random for intensive study.

Their results showed that the colonies do not have a significant impact on the vegetation in the 15,000-acre (6,070-hectare) national grassland. The areas cleared for colonies by ants ranged from 1.9 square meters to 27.3 square meters (20 square feet to 294 square feet). The scientists found no detectable differences in plant production in the vicinity of the ant mounds, except for greater plant growth in the area immediately surrounding the colony. This was probably because there were no plants in the cleared area to compete with the bordering plants.

Lavigne and Rogers took soil cores from areas cleared by harvester ants and from areas of normal vegetation to determine if the elimination of plants significantly alters the soil moisture. The soil cores were 80 centimeters (31.5 inches) deep and 5 centimeters (2 inches) in diameter.

The soil measurements showed significantly higher moisture content in the cleared areas. The scientists attributed this to the lack of plants, which tend to draw off soil moisture.

The two researchers determined the amount of soil harvester ants move to the surface by calculating the volume of soil excavated by the ants in constructing tunnels and chambers in their mounds. They counted the number of tunnels and chambers in each mound, and measured the diameter and depth of the tunnels and the diameter and height of the chambers. The approximate mass of soil moved was calculated by multiplying the volume by the average soil density. Rogers and Lavigne estimated that each colony excavates about 3 kilograms (6.6 pounds) of soil.

They investigated the chemical modification of the soil by taking three soil samples from each of six colonies — one from beneath the mound, one from the surrounding cleared area, and the other from outside the cleared area. They found that the soil beneath the mounds has a higher content of phosphorus and nitrate, presumably because the ants store organic materials inside their chambers.   [Stanley I. Auerbach]

Electronics

Graphite button was implanted behind the ear of a deaf patient as a connecting plug between electrodes implanted in his inner ear and a computer. Coded information is translated into tones that he can "hear."

The electronics industry began to come to the rescue of the automobile industry in 1975, with the latter under the twin guns of improved emission control and energy conservation. The automakers are up against a series of demands and requirements that cry out for electronic solutions. The requirements include:
■ Emission control standards, which will go into effect in 1977, that can probably only be met practically by electronic fuel-management systems.
■ A call by President Gerald R. Ford to cut fuel consumption on cars by 40 per cent or possibly face stringent legislative penalties.
■ Continuing controversy over auto safety and how it can be improved.

To handle all these problems effectively, the auto industry is scrutinizing what is rapidly becoming the most important electronic development in a decade — the microprocessor. This is a tiny integrated circuit containing hundreds of transistors on a single silicon chip. The transistors are interconnected to form the central data-processing unit that is at the heart of a digital comput-er. With the addition of some external memory circuits and input and output devices for controlling the microprocessor, designers can build an inexpensive computer of up to 16-bit capacity. (A bit is a basic unit of information.)

Automotive engineers believe that they can put such a computer under the hood of an automobile to keep track of engine factors such as temperature, pressure, and speed. With linking electronic controls, the computer can automatically adjust the ignition timing and fuel-to-air ratio to produce the most efficient fuel consumption.

In late March, RCA Corporation chairman Robert W. Sarnoff announced that preliminary tests indicated that microprocessors could boost mileage by up to 40 per cent in standard-sized and large cars, and could be mass-produced for less than $100. Many firms, including National Semiconductors Corporation; Texas Instruments Incorporated; Intel Corporation; and Rockwell International, as well as RCA, have been working to develop microprocessors for up to two

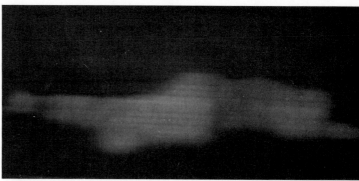

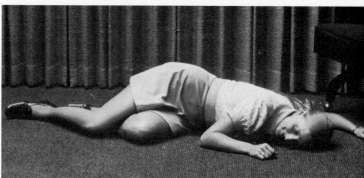

Electronics

Continued

Scanning a smoke-filled room with a Probeye detector, a fireman discovers a victim of the fire. The device converts infrared rays emitted from objects into television images. It is also used to find short circuits and other hot spots behind walls.

years. And most experts agree that it will be at least 1978 before these devices show up in cars. Even then, the fuel saving is expected to be considerably less than that predicted by RCA.

Theoretically, the microprocessor also has the capacity to handle guidance and safety functions. This would include automatic steering on specially designed highways, and automatic braking when, for example, the car comes too close to one ahead. In addition, the device may provide continuous diagnosis of engine performance.

Another electronic approach to fuel management, called the lean-burn system, has been developed by the Chrysler Corporation and is slated to be incorporated into its larger 1976 models. It will be used in conjunction with Chrysler's now-standard electronic ignition system to automatically advance or retard the spark timing applied to the engine's cylinders. The system monitors seven parameters: throttle position, rate of throttle change, temperature and pressure within the cylinders, timing, engine speed, and engine load.

Electronic circuits automatically adjust the spark timing to ignite the fuel-air mixture at the best time.

Semiconductor device development in 1975 reached new highs in the number of interconnected components that manufacturers could produce simultaneously on slices, or chips, of silicon. This improvement in large-scale integration (LSI) presages powerful computers in new and unconventional forms. The TRW Systems group in Redondo Beach, Calif., under the direction of Barry Dunbridge, produced a computer component about 0.3 inch (1.94 centimeters) square and bearing more than 17,000 individual transistor devices. The density of this device is 10 times greater than has been previously achieved.

Medical electronics. Scientists at Stanford University Electronics Laboratories, Palo Alto, Calif., have developed an experimental implantable stimulation system with the ultimate aim of producing a hearing aid for the totally deaf. The system, developed by Tushar Gheewala, Roger D. Melen, and Robert L. White, was first tested successfully on cats in 1974. In April, 1975, a University of Utah research group led by M. G. Mladjovsky announced that they had, with a similar system, succeeded in synthesizing sounds of varying pitch in the ear of a 62-year-old man who has been deaf since birth.

There are tens of thousands of people in the United States alone who are totally deaf because of the absence of, or a defect in, the tiny hair cells in their inner ear. These cells convert sound to electrical signals that the brain can process. The experiments indicate that it may be possible to build a device to provide a form of artificial hearing for these people.

The Stanford system, called Stimuliss IV (for Stanford Integrated Multi-electrode Implantable Stimulation System), directly injects electric pulses into the auditory nerve by means of four electrodes that are implanted in the cochlea (inner ear). These electrodes are connected to a small package consisting of integrated circuits and a pie-

zoelectric transducer, which converts sound waves to electric signals, implanted inside the skull behind the ear.

The circuits produce signals, or tones, of chosen intensity and frequency. Two signals are coupled simultaneously into the device through a second transducer housed in a package that is worn like a hearing aid. This arrangement makes it unnecessary to put wires through the skin. One radio-frequency channel provides about 5 milliwatts of power to the implant. A second channel carries ultrasonic signals, generated by a computer in digital form, which contain coded information on the frequency to be stimulated, and its intensity. The digital data is amplified, decoded, and reconverted to an analogue signal, and directed to the desired stimulating electrode by the circuitry.

In the University of Utah device, the electrodes were connected to a pluglike button protruding through the skin behind the ear. A cable connected this plug to a computer which generated the varying sounds that the patient could hear.

Drawing by Chon Day; © 1975 The New Yorker Magazine, Inc.

Electronics

Continued

Video record players. In 1975, the race to be first with a practical, low-cost video record player heated up on both sides of the Atlantic. A potentially lucrative consumer market awaits the successful producer of a unit that would work with an ordinary television set. But, just as in the early days of long-playing audio records, different and incompatible technologies are competing for the prize.

By the end of March, three different systems had been publicly demonstrated, and one was actually being offered for sale. This was a video disk player called TED, a mechanical system developed jointly by Telefunken in West Germany and Decca in England (Teldec). Of all proposed systems, the TED most closely resembles the conventional audio record player. A vibration-sensitive stylus rides on impressions in a rigid disk that spins at 1,800 revolutions per minute. TED is now being sold in West Germany for about $600.

In the United States, RCA has developed a video disk made of vinyl, with a metal coating protected by a styrene film and a lubricated surface to make the stylus glide easily. The picture information is encoded in a raised pattern pressed onto the disk, and is sensed by the stylus which detects minute changes in capacitance on the record surface.

The third contending video disk uses optical technology. A joint development of Philips Gloeilampfabrieken in the Netherlands and MCA Incorporated in the United States, it records video programs as a series of opaque and transparent patterns on a rigid plastic-coated disk. The disk is scanned by a laser and the resulting reflection variations are converted to appropriate electronic signals that are coupled into the TV receiver. The optical system has features that provide stop action, slow motion, and frame repetition. Other companies working on optical units include the Zenith Corporation and I/OMetrics in the United States and Thomson-Brandt in France. All of the optical units are compatible with each other, but are not compatible with mechanical ones.   [Samuel Weber]

Energy

The conservation of existing energy sources and the development of new sources were prime concerns throughout the world in 1974 and 1975. The United States formally established the Energy Research and Development Administration (ERDA) in October, 1974, to coordinate all its energy efforts. The new agency is responsible for research and development of fossil fuels and nuclear energy and for energy-related areas involving conservation, environment, and safety. However, responsibility for regulating the nuclear industry was given to the Nuclear Regulatory Commission, established in January, 1975. The Atomic Energy Commission was abolished. See ENVI-RONMENT; SCIENCE SUPPORT.

Fossil fuels. The U.S. government awarded the Foster Wheeler Energy Corporation of Livingston, N.J., an $8.7-million contract in early 1975 to build a pilot plant near Sioux Falls, S. Dak., for converting any type of coal to clean-burning gas.

Lawrence Livermore Laboratory, which is operated by the University of California for ERDA, announced in April that it would try to recover oil from shale with a minimum of waste and disturbance to the environment. The work will be done in western Colorado where shale deposits are 500 feet (152.4 meters) deep and up to 2,000 feet (609.6 meters) thick. The shale contains about 20 gallons (75.7 liters) of oil per ton.

The laboratory is developing an underground retort system for recovering the oil. After the shale is broken up by a series of underground explosions, engineers will set fire to the top of the shale layer and force air down through shafts to keep the fire burning evenly. At temperatures above 750°F. (399°C), the oil will be released from the rock, flow down to an underground collecting area, and then be pumped to the surface. The laboratory scientists estimate that about 300 billion barrels of oil could be recovered by this process.

Efforts were also underway to develop technology for conserving fossil fuels. Holifield (formerly Oak Ridge) National Laboratory in Oak Ridge,

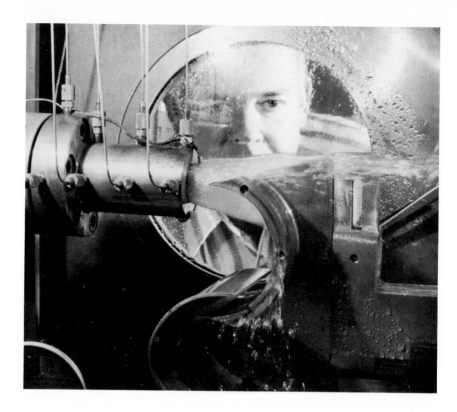

Special nozzle and
turbine blades,
designed to work
in geothermal brine
to convert its heat
to electricity, undergo
tests at the Lawrence
Livermore Laboratory.

Energy

Continued

Tenn., announced in April that its scientists and engineers had developed a new type of boiler for electric power plants that could cut fuel consumption by 25 per cent.

Most power plants use steam to drive the electricity-producing turbines and generators, but only about one-third of the heat energy used to produce steam is converted into electricity. This is true both of plants that use fossil fuels and those that use nuclear energy. The temperature of the steam drops as it is piped through the turbines. The remaining heat energy is wasted because the condensed hot water is dumped.

The hotter the steam, the more efficiently the heat energy can be used. But steam temperatures must not be raised above about 1000°F. (538°C). At this temperature, the tremendous pressure—about 4,000 pounds per square inch (281 kilograms per square centimeter)—is as much as the system's tubes and pipes can stand without weakening.

Holifield scientists developed a system for driving turbines that uses po-

tassium vapor and steam. Potassium, which melts at 146°F. (63°C), can be heated to 1500°F. (815°C) by nuclear or fossil fuels and vaporized in a special boiler. Even at such a high temperature, the potassium vapor pressure is only about 10 pounds per square inch (0.7 kilogram per square centimeter). Therefore, it will not strain pipes.

The hot potassium vapor is sent through a turbine to power a generator. When it emerges, it is still hot enough to heat water to steam at 1000°F. This steam is then used to drive turbines and generate electricity in the conventional way. By using potassium vapor, the engineers predict that power plants can convert more than half of their heat energy to electricity.

Nuclear power. General Atomic Power Company of San Diego in 1974 began studying the possibility of using nuclear reactors for steelmaking. The proposed process would use high-temperature, gas-cooled reactors to produce steam, which would then be mixed with methane gas to produce hydrogen and carbon monoxide. At

Energy

Continued

1650°F. (899°C), these gases can reduce iron ore to 95 per cent metal iron. The iron would then be made into steel in a conventional electric-arc furnace.

The National Aeronautics and Space Administration (NASA) announced in February that a computer study shows it is possible to build a nuclear-fission energy system capable of eliminating its own waste. Disposing of dangerous radioactive waste is a major problem in fission systems.

The study involved data for a theoretical system in which reactors would use a gaseous compound of uranium rather than a solid core for fuel. Gaseous fuel reactors create many neutrons, and radioactive wastes can be rendered harmless by bombardment with neutrons. The gaseous compound could be used first to create the energy-producing chain reaction and then, rich in neutrons, it could be recirculated in the reactor to treat the fuel's radioactive waste. According to NASA, such a system would reach equilibrium after three years of operation, creating and eliminating equal amounts of waste.

The gaseous reactor could also be designed to eliminate waste from conventional nuclear reactors.

Solar research. Colorado State University at Fort Collins began testing its experimental solar heated and cooled house in June, 1974. In August, the university reported that the solar collector delivered 16 gallons (60.6 liters) of water per minute heated to an average temperature of 173°F. (78°C) during daylight hours. This ran a 36,000 British thermal unit air conditioner, day and night. In such an air conditioner, the hot water sets off a chemical reaction in which water vapor acts as the refrigerant, taking heat from the surrounding air.

In February, Lawrence Livermore Laboratory began constructing a system to demonstrate an industrial use for solar energy. Six acres (2.4 hectares) of "solar ponds" will provide hot water for a uranium processing plant near Grants, N. Mex., owned by the Sohio Petroleum Company of Cleveland.

The laboratory describes a solar pond as a cross between a water bed

In an experimental solar pond, the sun heats water in plastic bags, and rigid plastic covers trap the heat.

and a greenhouse. The water is actually pumped through long, flat, plastic bags, which prevent evaporation. Each of these pond modules is 200 feet (61 meters) long, 12 feet (3.7 meters) wide, and about 4 inches (10 centimeters) deep. Black plastic under the bags absorbs the sun's heat, and an arch of transparent plastic placed over the bags creates a greenhouse effect, trapping the heat. The hot water from the ponds can either be pumped directly to the plant or to an underground reservoir.

Project scientists will first test three solar ponds. The full 6-acre system will provide 500 gallons (1,892 liters) of water per minute, heated to 140°F. (60°C). Excess hot water will be stored in the reservoirs for use at night. The ponds will provide almost all the plant's hot-water needs during the summer months. During the winter, when the sun is less intense, water will be heated as much as possible in the ponds and then brought up to the needed temperature by fuel burners.

Sandia Laboratories in Albuquerque, N. Mex., in 1974 began studies for a centralized solar-energy collecting system to serve an entire community. The new system would provide heat, cooling, electricity, and hot water for between 100 and 1,000 houses and commercial buildings by using solar energy to heat water. The designers of the proposed system claim a centralized system could operate more economically than individual collectors and storage tanks and would cut heat-energy waste. When the water is at its hottest, it would be used to generate electricity. Then it would be piped on to provide heat and hot water for homes and businesses.

Kelp power. The U.S. Navy established a 7-acre (2.8-hectare) kelp farm in the Pacific Ocean off San Clemente Island in late 1974. The seaweed, which grows about 2 feet (61 centimeters) a day, is harvested and placed in vats where bacterial action releases methane gas, which can be burned for power. Scientists claim the kelp can also be treated to produce a wide range of petroleumlike products, from fertilizer to plastics.   [Darlene R. Stille]

Environment

Two University of California chemists made an announcement on June 25, 1974, that overshadowed all other environmental news in 1974. F. Sherwood Rowland and Mario J. Molina said that laboratory experiments and calculations show that chlorofluoromethane gases used to pressurize many aerosol cans may be rising slowly to the stratosphere and destroying the layer of ozone that protects the earth's surface from damaging ultraviolet radiation.

Excessive ultraviolet light is harmful to humans and lethal to some life forms. Loss of the protective ozone layer in the earth's atmosphere might result in a greater number of human skin cancers and cataracts, as well as vast environmental changes with unpredictable effects. See Close-Up.

Rowland and Molina estimated that major effects of the aerosol-can gases would not show up for 50 years or more. However, other scientists believe that the effects might become apparent sooner and may already have begun to appear. Studies by other scientists in 1975 indicated a surprising amount of variation in the ozone layer in recent years. It will take at least three years to determine whether the theory that aerosol gases destroy the ozone layer is correct. In June, 1975, after a five-month study, a federal task force called for a ban on the use of the gases in aerosol sprays by January, 1978, unless conclusive evidence is found that they are not harmful.

Water quality. The U.S. Environmental Protection Agency (EPA) announced on Nov. 8, 1974, that a routine study had identified 38 toxic chlorine-containing chemicals in New Orleans' drinking water. However, the findings were not a complete surprise because a 1972 study had revealed the presence of this class of compounds in the Mississippi River, from which New Orleans draws its water. Some of the chemicals have caused cancer in test animals.

At about the same time that the EPA made its announcement, the Environmental Defense Fund released a report on research that seemed to show a slightly higher number of cases of cancer among people drinking the New

Orleans municipal water than among those drinking water from wells in the surrounding area. About 50 deaths per year were linked to the New Orleans water supply.

The publicity given these reports may have been a factor in the passage by Congress on November 19 of a Safe Drinking Water Act. This new law permits the EPA to set standards for the amounts of chemicals and bacteria to be allowed in water supplies. A few standards have been set under the act, but no action has been taken as yet to reduce the use of chlorine for water purification, although alternative water-treatment methods are available.

An earlier EPA report in July, 1974, announcing that water pollution from industrial sources and sewage had declined slightly over the previous 10 years attracted much less attention.

Atomic power. Nuclear power plants now account for roughly 7 per cent of U.S. power-generating capacity, but the industry is beset by a variety of problems. Higher prices and conservation of energy have cut the growth in demand for electricity. Electric power companies, already in financial difficulties, canceled or delayed about half of all planned nuclear power plants in 1974 and 1975, including several already under construction. In one instance, the Consolidated Edison Company ran out of money and could not complete the construction of a nearly finished nuclear plant, but was rescued when New York state purchased the facility.

To add to the industry's uncertainty, the Atomic Energy Commission (AEC), which had nurtured the growth of nuclear electric power, was dissolved on Jan. 19, 1975, and replaced by two new federal agencies. The Nuclear Regulatory Commission (NRC) took over the task of enforcing safety regulations, and the Energy Research and Development Administration (ERDA) took over the research and development of energy sources, including new types of nuclear plants. See ENERGY; SCIENCE SUPPORT.

The new agencies showed signs of following a different course than the

Shattering The Ozone Shield

For years, people have feared that some official finger might someday push the button that would trigger nuclear war, causing millions of deaths and drastically altering the environment. In 1974, these people learned that they themselves have innocently been pushing little buttons that may already be changing the environment for the worse. Every day, millions of aerosol cans of cleansers, deodorants, shaving cream, and other household items release chlorofluoromethanes, propellant gases that may be reducing the atmosphere's ozone layer, which protects the earth from the sun's ultraviolet (UV) radiation. See ENVIRONMENT.

Chemists F. Sherwood Rowland and Mario J. Molina of the University of California at Irvine showed, in June, 1974, that, once released, the gas molecules rise slowly to the top of the ozone layer in the stratosphere, about 20 miles (32 kilometers) above the earth, where one chlorine atom can be split off by UV radiation. In a chemical reaction, a single free chlorine atom may destroy thousands of ozone molecules.

Meteorologist Michael B. McElroy and his associates at Harvard University calculate that about 30 per cent of the ozone layer could be destroyed by 1994 if gas production continues to increase at its present pace. In that case, UV radiation would increase by about 60 per cent.

The energy that synthesized the earth's first organic molecules and led to the first living cells may have come from the sun's UV radiation. But this short-wave radiation can also break the chemical bonds that hold together the molecules that sustain life.

During the evolution of life, two events brought this danger under control. One was the development of enzymes that repair UV damage to such cellular materials as nucleic acids. The second was the generation of the atmosphere's protective ozone layer, which blocks the most dangerous wave lengths of solar UV radiation and reduces other wave lengths to tolerable levels.

The ozone screen arose from the oxygen produced by photosynthesis in plants. In a process still going on, certain wave lengths of UV radiation split oxygen molecules into two highly reactive oxygen atoms. One of these combines with another oxygen molecule to form ozone (O_3).

Biologists are trying to predict the consequences of greater UV radiation. The best-known effects are on human skin; it causes sunburn, skin aging, and skin cancer on the sparsely pigmented skin of Caucasians. If the ozone loss were 10 per cent and humans took no measures to protect themselves, one scientist estimates that there could eventually be as many as 1 million additional cases of skin cancer, causing 40,000 additional deaths, per year.

But to biochemist Kendric C. Smith of Stanford University and range scientist Martyn M. Caldwell of Utah State University, the possible effects of UV radiation on plants, animals, and microorganisms are even more disturbing. Elimination of key species could upset complex food chains, with serious consequences for all forms of life. The main areas of concern are the UV effects on insects and plants, particularly phytoplankton, the first step in the food chain of the sea.

Caldwell's group is studying the effects of additional UV radiation on plants. Any significant reduction in yields of rice, wheat, or corn, which supply about half of the world's food, would be most serious. Some plant species showed marked decreases in photosynthesis and growth and signs of cellular deterioration with increased UV.

Marine biologist Carl J. Lorenzen of the University of Washington, Seattle, has been investigating the impact of UV on phytoplankton. He has found that these single-celled plants grow much more rapidly when UV is filtered out of natural sunlight.

Nevertheless, any increased radiation should have little effect on overall marine productivity. Lorenzen found that UV does not penetrate deeply into ocean water, so only a small part of the planktonic growth zone would be adversely affected. UV penetrates the clearer waters of fresh-water bodies more deeply, however, and biologist John C. Calkins of the University of Kentucky has found that many fresh-water invertebrates may already be absorbing as much as they can tolerate.

Many aspects of insect behavior are known to be affected by UV light, including flower recognition, pollination, navigation, and mating behavior. Zoologists Ting H. and Catherine T. Hsiao of Utah State University found that additional UV does not interfere with pollination, but kills beetle larvae.

Although our knowledge of the ecological effects of solar UV radiation is clearly limited, a great deal of laboratory work has been done on the effects of artificial UV radiation on cells, particularly on bacteria. Solar UV radiation wave lengths range from under 200 nanometers (nm) to near-visible light at 400 nm. (A nanometer is a billionth of a meter.) The ozone layer permits no UV radiation under 286 nm to reach the earth's surface. But luckily, much work has been done in the range between 250 and 300 nm.

Researchers have found that much UV radiation is absorbed by the genetic material deoxyribonucleic acid (DNA). The greatest absorption occurs at 260 nm, and it drops off as the wave length approaches 300 nm. Mutation rates, cell damage, and cell death follow the same curve. So the greatest genetic damage to cells occurs at a wave length that is now completely blocked by the ozone layer.

Most cells have two enzyme systems that help repair UV radiation damage. The light-repair system is activated by visible light; the dark-repair system functions whether or not light is present. Because relatively little UV radiation presently reaches the earth's surface, some scientists had believed that these systems were evolutionary leftovers from a time when there was little or no ozone to block UV radiation.

However, no one knows if the enzyme repair systems, which have been found in the cells of all species studied so far, could handle the heavier UV radiation that would result if the ozone layer is depleted. Worse, no one knows if the systems can adequately handle new shorter wave lengths of solar UV radiation, particularly those near the critical 260 nm wave length.

Caldwell believes that the harmful effects so far established, though not dramatic, indicate that increased UV radiation would be detrimental to our environment. It will take years to resolve these questions. In the meantime, many people worry about the continued use of spray cans. [Joseph Eigner]

AEC. The NRC moved vigorously to shut down dozens of nuclear plants for inspection when cracks appeared in cooling systems. It announced that licenses for the import and export of nuclear fuels would be reviewed shortly after Congressman Les Aspin (D., Wis.) announced on March 23, 1975, that large quantities of extremely hazardous plutonium powder were being flown into New York City's Kennedy International Airport for shipment to a Pennsylvania processing plant.

The NRC announced on May 8 that it would not approve for at least three years the use of plutonium fuel in U.S. nuclear power plants. The agency backed off slightly from this position on May 27 and said the plutonium-recovery facilities would be considered on a case-by-case basis.

Delays may have very far-reaching effects because the nuclear power industry has been proceeding on the assumption that plutonium would be used as reactor fuel. Companies have accumulated large inventories of plutonium, and nuclear fuel-processing facilities have been designed for plutonium. Waste-extraction plants, necessary to process plutonium from spent nuclear fuel, cannot be operated, and spent fuel is piling up at nuclear power plants that have no way of disposing of it.

Much of the plutonium stockpiled thus far is intended for use in a new generation of nuclear plants, the liquid metal-cooled fast-breeder reactors (LMFBR) that theoretically can generate more fuel than they burn. But NRC reluctance to approve the use of plutonium fuel in existing plants—in part because of the difficulty of ensuring that it will not be stolen by terrorist groups to make nuclear bombs—casts a cloud over the prospects for plutonium-fueled breeders.

In one of its last official acts, the AEC issued an impact statement that tried to justify the breeder program in the face of strenuous objections raised by the EPA and citizens' groups. The statement was greeted without enthusiasm. The EPA said the report showed that the construction of LMFBRs could be delayed by as much as 12 years without serious effects on the U.S. energy program. The citizens' groups called the new statement unsatisfactory and the

A test tank, in which currents and waves can be generated, is used to study techniques to control oil spills and hazardous materials in rivers, canals, and along ocean shores.

Environment

Continued

program ill-advised. Nevertheless, President Gerald R. Ford's Administration devoted the lion's share of its proposed 1976 energy budget to breeder development. The program is now in the hands of ERDA, which will prepare its own evaluation of the $10-billion program before deciding whether to proceed. The new study will take at least 18 months.

Nuclear safety at existing power plants was questioned following the release on Aug. 20, 1974, of a massive AEC study made by a research group headed by Norman C. Rasmussen, professor of nuclear engineering at the Massachusetts Institute of Technology. The report was widely publicized as proving the safety of nuclear power plants. But the AEC's regulatory division, the Sierra Club, and a group called the Union of Concerned Scientists claimed that the report had misinterpreted research on radiation effects and so had greatly underestimated the number of deaths and injuries that would occur in a nuclear accident. These groups estimated as many as 23,000 persons would die and 100,000 would be injured in the worst nuclear power plant accident — figures that agreed with earlier studies.

The Rasmussen report purported to show that the probability of such an accident was less than one chance in a billion for any one power plant in any one year. The American Physical Society criticized the probability techniques used to develop this figure in an extensive report released in May, 1975.

Automobile pollution regulations were put off for another year amid widespread confusion. Automakers developed catalytic mufflers to break down harmful compounds in auto exhausts instead of developing cleaner engines as the government had hoped. On March 5, 1975, EPA administrator Russell E. Train said that he was granting automakers a one-year delay in the stricter regulations that were to have become effective in 1976. The delay was granted because EPA had discovered that catalytic mufflers create hazardous sulfur oxide compounds from auto exhausts.   [Sheldon Novick]

Genetics

With new methods, bacterial plasmids and animal chromosomes are taken from cells. After they are cut into pieces, they are mixed and ligase enzymes are added. This causes them to form plasmids with animal genes that can infect other bacteria, providing the bacterial cell with genetic data from the animal cell.

A conference of 86 U.S. scientists and 53 scientists from 15 other countries grappled with the new and potentially hazardous developments in genetic engineering — the artificial manipulation of genes, which carry the heredity of all living things. The scientists adopted a set of voluntary guidelines based on the degree of risk of specific experiments.

The conference was held at the Asilomar Conference Center in Pacific Grove, Calif., from February 24 to 27. It had been called for in an unprecedented letter, signed by 11 distinguished molecular biologists, in the July 26, 1974, issue of the U.S. journal *Science.* The letter suggested that a voluntary, worldwide ban on certain experiments be put into effect until their "potential hazards" could be determined and safeguards developed. The scientists, members of a committee of the National Academy of Sciences (NAS), included Paul Berg and Stanley N. Cohen of Stanford University, David Baltimore of the Massachusetts Institute of Technology, and James D. Watson of Harvard University.

Cause for alarm. The concern was prompted by recent advances in genetic engineering. The research began innocently enough in the early 1970s with the bacterium *Escherichia coli,* a normal and usually harmless resident of the intestinal tracts of human beings and many other animals. The genetic material of *E. coli* is a long strand of deoxyribonucleic acid (DNA), the ends of which are joined together to form a circle. *E. coli* cells can also carry other, smaller circles of DNA known as plasmids. Plasmids are infectious — they can be passed from one *E. coli* cell to another and to related species, sometimes even to more distantly related bacteria.

In a procedure that begins with the grinding up of plasmid-containing cells of *E. coli,* it is now possible to isolate chemically pure plasmid DNA by separating it from other cell substances in a centrifuge. This plasmid DNA can be reintroduced into bacterial cells.

The second research thread leading to the genetic-engineering controversy was provided in 1972 by the isolation of a class of enzymes called restriction

Shuffling a Deck of Genes

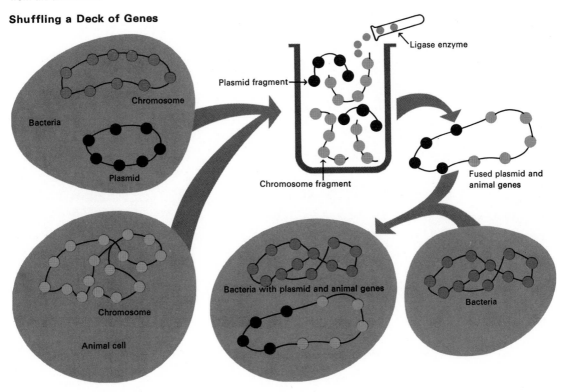

Bacteria

Chromosome

Plasmid

Plasmid fragment

Chromosome fragment

Ligase enzyme

Fused plasmid and animal genes

Chromosome

Animal cell

Bacteria with plasmid and animal genes

Bacteria

Leading bioscientists heard lectures
on genetic engineering, *left,* when
they met at Pacific Grove, Calif.,
in February to set guidelines for
gene-transplant research. Nobel
laureate Joshua Lederberg of Stanford
University, *below,* listens intently.
Stanford's Paul Berg, conference
chairman, makes a point, *below left.*
Nobelist James D. Watson and Sydney
Brenner of England argue the issues
during a coffee break, *bottom left,*
and others do so at meals, *bottom,*
using place mats as scratch pads.

Genetics

Continued

enzymes, which can recognize a particular tiny segment of a DNA molecule and cut the molecule at that site. Other enzymes, called ligase enzymes, can fuse together the free ends of two DNA molecules. Both kinds of enzymes can be extracted from some bacteria.

The third thread leading to the potentially dangerous research in genetic engineering grew out of the ease with which DNA from many organisms — plants, animals (including humans), bacteria, and viruses — can be purified.

Only recently, these threads were woven together to introduce a small segment of DNA from a completely different organism into *E. coli* cells.

The procedure for introducing foreign genes into bacterial cells was developed by Cohen and his colleagues at Stanford. They and a team led by Herbert Boyer at the University of California at San Francisco reported in the spring of 1974 that they had transferred a gene from a toad into *E. coli*. They cut the circular plasmid DNA molecule with a restriction enzyme, mixed the opened circles with fragments of purified DNA from the toad and then, with ligase enzymes, chemically linked the ends of the DNA fragments with the open ends of the plasmid DNA. It is similar to a jeweler enlarging a ring; he snips the ring open, inserts a small piece of material into the gap, and fuses the ends. The new DNA ring, too, was larger — by the length of the toad gene. See MICROBIOLOGY.

Tool for research. The ability to introduce foreign genes into *E. coli* provides a new tool for working on several scientific problems that could not be attacked otherwise. It enables researchers to obtain large quantities of an animal or plant gene by reproducing it in a bacterial culture. Only in this way will they be able to find out the complete chemical sequences of genes, because sequencing procedures require large amounts of genetic material. The action of genes transferred into easily studied bacteria like *E. coli* will also reveal important information about how the gene activity is regulated in plants and animals. The procedure will also be important in the study of the genetic character of cancer viruses, disease-causing bacteria, and drug resistance in bacteria.

The new ability to introduce foreign genes into bacteria may lead to bacterial "factories" containing functioning human genes that enable the bacteria to make such medically important molecules as insulin and antibodies. It may also some day be possible to introduce genes for nitrogen fixation into food crops such as corn (see THE NITROGEN FIX). The transfer of genes to bacteria may also provide a route for transplanting genes from one person to another to help cure hereditary diseases.

Potential for disaster. But certain kinds of gene-transfer experiments pose dangers as well. What would happen if an engineered strain of *E. coli* with genes, say, for resistance to many drugs escaped from the laboratory? Although the *E. coli* strains commonly used in laboratory studies do not seem to survive well under natural conditions, one might become widespread in human or animal populations. The great danger would be if such a strain transferred its plasmid carrying the resistance genes to other bacteria, some of which might be deadly disease organisms.

Such potential dangers led the NAS committee to urge the ban on further research in which genes for antibiotic resistance or toxin production, or from cancer-causing and certain other animal viruses were transferred.

The new guidelines were linked to the risk level of the experiments. To prevent the escape of bacterial strains, elaborate and expensive precautions, such as special work chambers with airlocks and special decontamination procedures for personnel, are now required for high-risk experiments. Only six such facilities exist in the United States, and there are few elsewhere in the world. Moderate-risk experiments will require expensive laboratory modifications to ensure containment of the laboratory organisms. Low-risk experiments will require only the use of accepted microbiological techniques and properly trained personnel.

Perhaps for the first time, scientists throughout the world have agreed, on moral grounds, to limit their research because of a potential public health hazard. The Asilomar agreement may have established a precedent of voluntary restraint in other potentially hazardous fields. [Daniel L. Hartl]

Geoscience

Geochemistry. Three University of Chicago geochemists in 1974 found what appears to be some of the oldest solid material in the solar system. Robert Clayton, Lawrence Grossman, and Toshiko Mayeda studied carbonaceous chondrites, a rare group of stone meteorites (as opposed to metal ones) that are rich in carbon and relatively volatile substances. There is some evidence that carbonaceous chondrites are samples of the solid grains that condensed directly out of the gaseous nebula that gave birth to the sun and the planets.

The scientists reported on this study and subsequent research on these and other chondrites at the Sixth Lunar Science Conference in Houston in March, 1975. They particularly studied Allende, the carbonaceous chondrite that fell in 1969 near the village of that name in northern Mexico. Allende is a rather strange mixture of white, coarse-grained silicates, surrounded by a black matrix rich in carbon and various types of volatile substances. The minerals and chemicals in the silicates indicate that they crystallized at about 2150° F. (1200° C). The matrix probably crystallized later at much lower temperatures, perhaps as low as a few hundred degrees Celsius. Because the Allende silicates differ in texture from silicates found on earth, the scientists believe they probably crystallized directly from a gas state rather than a liquid-rock state.

Analysis of the oxygen in both the silicates and the matrix indicated that it has an unusual composition that could probably be found only in the extremely hot gaseous nebula that existed before the solar system formed.

Oxygen exists in three isotopes of relative mass 16, 17, and 18. The isotopes of mass 16 account for 99.8 per cent of the oxygen atoms found in the solar system. However, the remaining isotopes are also important because they can be compared with oxygen 16 in a substance to determine how it was formed. Scientists know that chemical reactions or such physical processes as melting and evaporation produce small but readily measurable changes in the proportions of the three isotopes.

Raspberry-shaped aggregate of magnetite was found in Orgueil meteorite, one of the carbonaceous chondrites. New studies of this and other carbonaceous chondrites show that such meteorites are about the same age as metal meteorites, not many years older as previously believed.

A sample of Hebron gneiss from Saglek Bay in Labrador was dated at 3.65 billion years. Rocks this old found in Labrador reinforce the theory of Atlantic sea-floor spreading.

Geoscience

Continued

These changes are proportional to the differences in mass — that is, the percentage of change in the ratio of oxygen 18 to oxygen 16 for any process is twice that in the ratio of oxygen 17 to oxygen 16. The carbonaceous chondrites have an unusual isotopic composition. The oxygen isotopes vary from one sample to another in such a way that the percentage of change in the oxygen 17 to oxygen 16 ratio is the same as that in the oxygen 18 to oxygen 16 ratio. This must have been caused by the addition of pure oxygen 16. This is presumably because the carbonaceous chondrites were formed much earlier out of the gaseous nebula. The oxygen 16 probably came from nuclear reactions in stars, either in the early stages of our sun's activity or in another star beyond our solar system.

The scientists conducted the same type of experiments on noncarbonaceous chondrites and found, to their surprise, that these also showed the same kind of differences in the oxygen-isotope ratio as were found in the earth and moon rocks.

Age of Allende chondrite. Mitsunobu Tatsumoto of the U.S. Geological Survey in Denver and James Chen of the University of California at Santa Barbara independently calculated the age of the silicates in the Allende carbonaceous chondrite in 1974. They measured the rate of decay of its radioactive uranium into lead, finding an age of 4.57 billion years. This differs by no more than 0.01 billion years from the known ages of other chondrites.

The age determination shows that Allende is not much older than other meteorites. The measurements of the rubidium-strontium decay in Allende, made in 1973 by Christopher M. Gray, Dimitri Papanastassiou, and Gerald Wasserburg of the California Institute of Technology, indicated that the silicates might be 0.01 billion years older than the black matrix that forms the rest of the carbonaceous chondrite.

Labrador rocks. Richard Hurst of the University of California at Los Angeles, David Bridgewater of the Geological Survey of Greenland, and Kenneth Collerson of the Memorial

University of Newfoundland found ancient rocks, all gneisses, along the coast of northern Labrador in 1974. Using the rubidium-strontium decay method, they were able to set the age of the rocks at 3.65 billion years. The rocks came from the Saglek Bay area north of the town of Hebron on the Labrador Highland. The discovery lends support to the theory of Atlantic sea-floor spreading, which predicts that rocks of great age should be found in the Labrador region. See JOURNEY TO THE BIRTHPLACE OF CONTINENTS.

Jackson Barton of the University of Massachusetts in Amherst found rocks about the same age in 1974 just south of Saglek Bay. Rocks 3.7 billion years old, an age comparable to that of some of the older rocks on the moon, were found in Greenland. All these finds provide an opportunity for geochemists to compare the composition of rocks formed in the early stages of the earth's history with those being formed in volcanoes and elsewhere today.

Carbon dioxide in basalts. David Eggler of the Carnegie Institution Geophysical Laboratory in Washington, D.C., showed in laboratory experiments in 1974 that the amount of carbon dioxide present in volcanic lava can influence the type of basalt that is formed when the lava cools. His finding upset theories of basalt formation that relate type only to pressures and the presence of water deep in the earth.

Many types of basalts are found in the crystallized lava around volcanoes. One of the two main types, tholeiite, is rich in silica and poor in alkali elements such as sodium, potassium, and rubidium. The other, alkali basalt, is rich in the alkali elements and poor in silica. Until Eggler's discovery, scientists believed that the alkali basalts were always formed at greater depths than the tholeiites. Eggler found that the carbon dioxide content of the molten lava can also produce alkali basalts. Consequently, they may be formed at shallower depths than was previously thought. At a given depth, Eggler says, a series of lava types ranging from alkali basalt to andesite, a type of tholeiite, may be produced according to the proportions of carbon dioxide and water that may be found in the source material.   [George R. Tilton]

Geology. Geologists turned their attention to the Atlantic continental shelf in 1974 and 1975 in the hope of determining whether important new sources of oil can be discovered there. The largest unexplored area belonging to the United States, the Atlantic shelf is a shallow ocean area about 100 miles (160 kilometers) wide off the East Coast. It was not examined earlier because, until about 10 years ago, there was little reason to expect oil in this area. The few oil wells drilled many years ago on land along the coast were barren. In contrast, much oil was found along the coasts of Louisiana, California, and Texas, and drilling gradually extended from the coastland as much as $10^{1}/_{2}$ miles (17 kilometers) into the ocean.

During the past 10 years, geologists in other countries have discovered oil along continental shelves previously thought to contain none. For example, Australia has almost no oil underneath its continent, yet major oil discoveries were made in the shallow sea between Australia and Tasmania. Similarly, large oil fields were found in the North Sea off Norway. These discoveries led U.S. geologists to suspect that there was something special about the edge of the continent lying a few hundred feet beneath the sea surface, even though exploratory drilling off the coast of Oregon and Washington during the 1960s proved to be an expensive failure.

However, it is not enough to have a hunch that there could be oil beneath the Atlantic continental shelf because of the offshore oil discoveries of other countries. Possible environmental hazards and the immense costs of drilling make it imperative that there be a good chance of finding deposits of oil before offshore exploration is started.

Atlantic history. Two events during 1975 have helped geologists to better understand the origin, probable structure, and oil possibilities of the Atlantic Coast. The first was the publication of a detailed history of the geology of the Atlantic coastline, and the other was a special drilling project of the U.S. research vessel *Glomar Challenger*.

The history was published in April, 1975, by Princeton University geologist David Kinsman. He combined concepts of plate tectonics and sea-floor spreading with many observations to

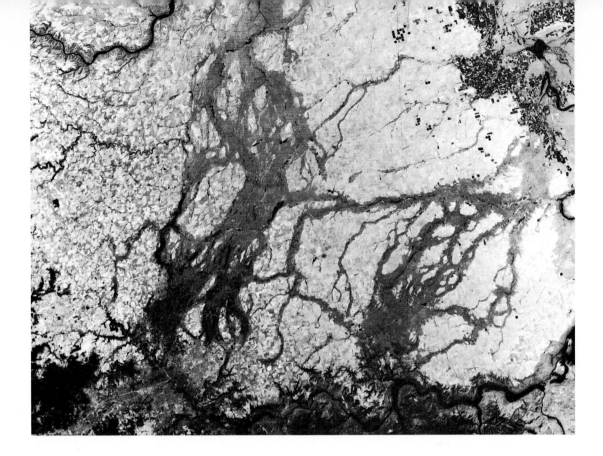

Geoscience

Continued

Satellite photo of the channeled scablands in eastern Washington confirmed a theory that the area's heavily scarred terrain was carved by the waters of an Ice Age flood of unprecedented size.

develop a detailed history that covered all coastlines formed by the splitting apart of a pre-existing continent, not just the Atlantic coastline. Kinsman recognized that any new ocean bed created by the splitting of a continent will at first be long and narrow.

Such long and narrow seas are usually poorly connected to the main ocean, Kinsman pointed out, with two probable results. First, normal evaporation would produce immense beds of salt if shallow water circulates only slowly into the narrow sea. Second, the oxygen in the bottom waters would be used up quickly if deep ocean water could not circulate into the newly formed sea. Dead plants and animals that sink into deep water that contains no oxygen will not decay. The result will be an organically rich accumulation of sediment on the bottom of the narrow sea. Burial and later heating of such organically rich sediment by radioactive forces deep in the earth then generate large amounts of oil into the sediments.

Kinsman's study shows that the possibility of finding oil off the East Coast of the United States is closely tied to events when the Atlantic was a narrow sea separating Africa from North America. The question of whether the bottom waters in that ancient sea ran out of oxygen can be answered only by exploratory drilling into the Atlantic continental shelf.

Challenger drilling. The *Glomar Challenger* drilled exploratory holes close to the coast of Africa in 1974, on the other side of the narrow sea that once separated northern Africa from North America. One drill hole off the coast of Angola came within a few feet of a suspected salt bed. More important, the drill cores contained an organically rich sediment. Therefore, at least part of the deep water in the Atlantic must have run out of oxygen.

The drilling took place off part of the African coast that was connected to Brazil before the continents split. So we must wait for further drilling off the part of the African coast that once fit against the United States East Coast before we know whether the northern part of the sea ran out of oxygen.

Geoscience

Continued

A 22-inch mirror reflects the sun's rays to passing satellite in technique developed to accurately map remote areas of world such as the Amazon jungle. The mirror flashes, noted on satellite photographs, can orient images to existing maps so that they can be corrected.

The *Glomar Challenger* is not equipped to look for oil directly. In fact, scientists must be careful not to drill into an oil or gas accumulation that might pollute the seas or be wasted. But these indirect tests can be used by geologists to evaluate the possibilities of finding oil.

Metal deposits. Geologists also gave increasing attention to the search for metals. Some predict a metals crisis corresponding to the present energy crisis, but others believe that the supply of metals in the earth's crust is much larger than the coal and oil supply. However, even if the metal supply does prove to be large, the environmental impact of extensive mining has to be carefully evaluated.

Toward this end, an important study, *Mineral Resources and the Environment,* was published in February, 1975, by the National Academy of Sciences. The 348-page report focused on three selected problems — the supply of copper, the impact of coal mining, and the supply and demand for energy. A large group of specialists under the guidance of geochemist Brian J. Skinner of Yale University prepared the report and interpreted the options for the future. Mining companies released a wide range of information to help the group evaluate future mineral supplies.

Origin of metal deposits. Yale University graduate student Neil Williams, working with Skinner, is completing a study of the origin of metals that promises to result in a major reinterpretation of how some of the world's most important ores were deposited. Present theories are based on work done by mining geologist Waldemar Lindgren about 1900. He showed that many deposits of metals such as lead, zinc, mercury, copper, and silver were formed by hot water circulating upward through cracks in rock. The hot water left the metal deposits in the cracks, and because of their hot-water origin, these were named hydrothermal ore deposits. Lindgren categorized these deposits as high-temperature, medium-temperature, and low-temperature hydrothermal ore deposits and showed geologists how to recognize each type. However, important ore deposits have since been found that looked like high-temperature hydrothermal ore deposits

but clearly had no evidence of having been formed by unusual temperatures.

Williams chose one of these, a large zinc deposit in Australia that has not yet been mined, for his study. He showed that all the ingredients of an ore deposit can come from ordinary sediment burial without any of the special hydrothermal deposits.

As in many deeply buried sediments, the enclosed water tends to become quite salty, and such strong salt solutions can transport metals, such as zinc, because they form metal-chloride complexes. The modest temperature increase with depth in the buried sediments surrounding the Australian deposit caused the zinc to resemble slightly the much-better-known high-temperature hydrothermal ore deposits. According to Williams, the zinc seems to have been swept up by the salty water from ordinary sediment.

Williams' and Skinner's studies are useful steps toward an understanding of the origins of mineral deposits. The new knowledge will greatly improve the chances of finding such deposits. Instead of searching blindly, geologists can use this new list of circumstances necessary for the formation of metal deposits to selectively explore areas where these circumstances are likely to have occurred. In his report, Skinner pointed out that, in the past, most ore discoveries have been made by surface prospecting or by indirect techniques, such as accidental discovery. He suggests that future mineral exploration will have to rely on greater understanding of the origin of such deposits.

Oldest rocks. The discovery of some of the oldest known rocks was reported in November, 1974, by Samuel S. Goodrich of Northern Illinois University in DeKalb and Carl E. Hedge of the U.S. Geological Survey in Denver. Analyses of the rocks, found in southwestern Minnesota, by two independent methods have confirmed that they are roughly 3.75 billion years old. The only rival, in terms of antiquity, is a formátion of similar age discovered in Greenland. Scientists believe that the earth was formed some 4.5 billion years ago. None of the earth's original rocks remain because they have long since been "plowed under" by the movement of the earth.   [Kenneth S. Deffeyes]

Geoscience

Continued

Rock formations that are on either side of known rift zone northeast of Reykjavík, Iceland, have moved apart 2 to 3 inches (5 to 8 centimeters) in recent years. This shows that the rift is indeed part of the zone where the earth's plates are separating, causing sea-floor spreading and an increase in the distance of Europe and Africa from the Americas.

Geophysics. New seismic reflection techniques were applied to study the structure of the earth in the spring of 1975 by the Consortium for Continental Reflection Profiling (COCORP). Headed by geophysicists Jack E. Oliver and Sidney Kaufman of Cornell University in Ithaca, N.Y., the consortium used the techniques to study the earth's deep crust and mantle beneath the continents. The study was part of the U.S. Geodynamics Program.

Similar seismic reflection techniques have enabled geologists and oceanographers to study oceans and to create detailed maps of all the ocean floors. They have also revealed the seabed structure along the continental margins to depths of 33,000 feet (10,000 meters). But, on land, they had previously been used only for oil exploration, never for studying the rocks that lie deep beneath land masses.

The Continental Oil Company developed the method used by COCORP. The technique, called VIBROSEIS, uses a series of computer-programmed shakers—sources that produce sound waves of 10 to 32 cycles. It directs these waves into the ground. The sound waves can be transmitted at gradually varying frequencies over a period of several seconds, and the seismic reflections are recorded and analyzed in a computer. The method of producing the waves is superior to explosives, which have been the traditional sound producers for such seismic exploration.

Preliminary reports on COCORP's studies in Hardeman County, Texas, are encouraging. Strong seismic reflections were observed at a number of depths within the earth's crust and below its sedimentary cover. The scientists received good indications of reflections from about 30 miles (45 kilometers) down, the approximate depth of the mantle in this area. These results suggest that it will be possible to study the crust and upper mantle beneath the continents in far more detail than ever before. Such studies may revolutionize our understanding of the continents in much the same way that the ocean studies revolutionized understanding of the geology of the oceans.

Geoscience
Continued

FAMOUS completed. The French-American Mid-Ocean Undersea Study (FAMOUS) was completed during the summer of 1974. French and American scientists in three submersibles explored the axis of the Mid-Atlantic Ridge about 200 miles (320 kilometers) southwest of the Azores. The scientists obtained thousands of photographs and rock samples, which are being analyzed. See JOURNEY TO THE BIRTHPLACE OF CONTINENTS.

Deep-sea drilling. While the submersibles were exploring the ocean floor, the deep-sea drilling U.S. research vessel *Glomar Challenger* drilled a series of holes into the basaltic rocks underlying the ocean sediments from near the ridge axis to a distance of about 112 miles (180 kilometers) from it. Analysis of the drill cores showed that the age of the basaltic rocks, predicted from magnetic measurements taken from the ocean surface, closely corresponded with that of the sediments lying just above the basaltic rocks.

However, the drill cores yielded some surprises. First, these basaltic rocks were found to be interbedded with sediments, and the amount of sediment decreased with depth until only basalts were found. Second, basalts were not magnetized uniformly in intensity or direction. The effects measured by the magnetometers at the ocean surface are apparently the sum of the magnetic effects of the individual basalts rather than the effect of a single, massive, uniformly magnetized sheet of basalt.

Sea-level variations. Peter Vail, a geophysicist who is employed at Exxon Production Research Company reported on detailed investigations of the deposition of sedimentary rocks in the ocean at the April meeting of the American Association of Petroleum Geologists in Dallas. The new data came from the seismic reflection records of the ocean floor taken over the last decade in many parts of the world. They found consistent patterns of rising and falling sea levels. Such worldwide changes in sea level, called eustatic changes, are caused by the melting of icecaps, variations in the rate of sea-floor spreading, or by other worldwide geological phenomena.

The cycles of sea-level rise and fall are irregular, with a gradual rise followed by an abrupt fall. Vail and his associates identified about 70 eustatic cycles and grouped them into 13 major cycles. The general trend they showed was a rise in sea level from the early Cambrian Period some 600 million years ago to a peak in the Cretaceous Period about 100 million years ago. The magnitude of this rise has not yet been determined, but there are suggestions that it could have been as much as 1,500 feet (457 meters).

A particularly interesting aspect of this study is the fact that the major rise in sea level during the Cretaceous Period correlates with the high rates of sea-floor spreading at that time. Such high rates have been observed by geophysicists Roger C. Larson and Walter C. Pitman of the Lamont-Doherty Geological Observatory at Columbia University.

Rapid sea-floor spreading would mean that a greater volume of the ocean basins would be occupied by the underwater mounds and basaltic ridges formed as molten lava wells up from the mantle below. This would cause the sea level to rise and the sea would cover more of the land. Furthermore, the production of more new crust at the ridges would be accompanied by a corresponding increase in the rate of crustal downthrust at the various ocean trenches. This movement of downthrusting crust would result in earthquakes and the eruption of volcanoes as lava from the geologic processes of downthrusting rose to the surface of the earth. Consequently, there should be more volcanic and earthquake activity when more new crust is produced.

Greater volcanic activity is known to have occurred during the middle Cretaceous Period. The greatest amount of lava to pour from volcanoes in the last 600 million years was produced then, particularly in the belt of extinct volcanoes extending from Alaska to Baja California. The great mass of granite and other igneous rocks found in and around the Cascade Range in Washington and Oregon, the Sierra Nevadas in California, and the Rocky Mountains in Idaho poured out of volcanoes at that time. The same pattern of Cretaceous rocks is found in Japan, Korea, eastern Asia, New Zealand, and in the Andes Mountains. [Charles L. Drake]

Geoscience

Continued

Paleontology. The remains of the largest creature known to have flown, an extinct winged reptile with an estimated wingspread of 51 feet (15.5 meters), was discovered in Texas by vertebrate paleontologist Douglas A. Lawson of the University of California at Berkeley. He announced the discovery on March 14, 1975.

The new species of reptile lived during the late Cretaceous Period, some 65 to 70 million years ago. It was bigger than the *Pteranodon,* the largest previously known flying reptile, and had a wingspread more than five times that of the condor, the largest flying creature now alive. Its 51-foot wingspread was 13 feet (4 meters) larger than that of the U.S. F-4 fighter plane.

The estimated size of the creature is based on various skeletal parts, including wings, neck, jaws, and vertebrae, found in Big Bend National Park in Brewster County, Texas, between 1972 and 1975. Other pterosaurs, to which this creature was related, have been discovered over the last century in many parts of the world. They domi-nated the skies between 160 and 65 million years ago.

Aerodynamic studies have shown that these reptiles could soar and glide efficiently on their thin, membranelike wings. However, scientific debate continues over whether such large creatures could take off from the ground simply by flapping their wings. Most authorities believe they launched themselves from high rocks and mountain peaks. At least they probably needed a long running start or a strong breeze to become airborne.

The Big Bend fossils were found in fresh-water sediments far from the oceans of that time. This suggests that the creatures did not feed on ocean fish, as scientists assume related species did. Lawson says that this species may have been a carrion eater. If this is true, it implies greater variation in this group of reptiles than scientists had suspected.

Lawson based his suggestion on the fact that the huge reptile had an unusually long neck, and each neck vertebra was attached to its neighbor by a complex interlocking process that must

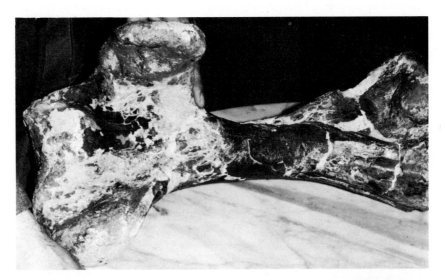

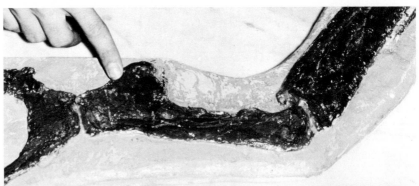

Bone from the wing of largest flying reptile now known was unearthed in Texas, *top.* It is much larger than that of another pterosaur, *bottom,* with wingspan of 20 feet (7 meters).

Geoscience

Continued

Standard X ray of an ancient trilobite, *top*, and an X ray taken while fossil was immersed in a xylene bath, *above*, provide different details that revise theories on the origin and descent of these arthropods.

have provided great flexibility and sturdy support. The creature's long jaws and flexible neck would have allowed it to reach deep into the carcasses of dinosaurs, much as vultures do when they consume dead animals.

Arthropods. John L. Cisne of Cornell University in Ithaca, N.Y., reported in October, 1974, that his studies indicate arthropods may have evolved as two separate groups more than 600-million years ago, rather than as a single group. There are 800,000 arthropod species living today, grouped in three classes—Crustacea (shrimp, lobsters, and their relatives), Chelicerata (spiders and horseshoe crabs), and Uniramia (centipedes and insects). All of these species have bodies divided into distinct segments and covered with a jointed external skeleton. Paleontologists have long considered the structural details of this complex outer skeleton as evidence of a common origin for all three classes. Trilobites, now extinct, had the same features and are therefore included in the arthropod group.

Cisne used 3-dimensional X-ray analysis techniques to study the details of various trilobite fossils. He found aspects in the internal anatomy of these creatures that had not been previously recognized. These included the structure of their food-gathering mechanisms, digestive systems, and musculature. All these findings pointed to a close relationship between trilobites and present-day Crustacea and Chelicerata. In contrast, there were many differences from present-day Uniramia.

Cisne proposed that the two separate lineages had evolved such jointed outer skeletons independently from Pre-Cambrian segmented wormlike ancestors more than 600 million years ago. One line became the insects and centipedes, while the other radiated into trilobites, spiders, and shrimps and their relatives. Cisne thinks that the evolutionary impetus behind the development of outer skeletal structures in both groups may have been a certain mechanical advantage or efficiency of motion, rather than protection from predators, as previously suggested.

Sauropod habitats. Paleontologist Walter P. Coombs, Jr., of the Pratt Museum at Amherst College presented evidence in February, 1974, against the

long-accepted theory that giant sauropod dinosaurs lived most of the time in water, crawling onto the land only occasionally. Sauropods were the largest of all land vertebrates, some almost 90 feet (27 meters) long and weighing up to 50 short tons (45 metric tons).

In the past, paleontologists argued that such large animals could support their bodies only with the buoyant action of water, and thus must have spent most of their lives swimming or standing in shallow lakes or swamps. Anatomical features such as a large nasal opening on the top of the skull (as in whales) and a long neck that would enable the animal to remain submerged in deep water with only its head above the surface, also were viewed as indicating that sauropods were aquatic.

Coombs pointed out that similar nasal structures are found in all land animals with a proboscis, including elephants. He further pointed out that most sauropods lacked a well-developed secondary palate in the mouth—a feature common in most aquatic vertebrates. Coombs noted that the peglike teeth, previously said to be adapted to feeding on soft aquatic vegetation, were features shared by land-dwelling herbivorous dinosaurs, so they are not clues to habitat.

Primates. Leonard Radinsky of the University of Chicago evaluated a series of fossil primate skulls in 1974 and found that those with larger brains had a smaller olfactory bulb—associated with smell—and a larger visual cortex—associated with sight. His studies included *Aegyptopithecus,* a primitive ape from the late Oligocene Period (26 to 28 million years ago) that was found in Egypt; *Apidium,* an early monkey from the same place and period; and *Dolichopithecus,* one of the oldest New World monkeys, from Patagonia.

Radinsky's findings imply that the development of sight and simultaneous decrease in the use of smell were important features in the emergence of higher primates from more primitive types. They also show that these features evolved early in the history of all anthropoid subgroups. Other changes, such as the physical separation of motor and sensory areas on the brain surface, were also noted, but their significance is not known. [Vincent J. Maglio]

Immunology

As complex details of the components of the immune system and the roles they play become unraveled, scientists are intensifying their search for broad mechanisms that must play a part in regulating various immune responses. In 1974 and 1975, several laboratories reported experiments that began to outline the general processes of the regulation of immunity.

From these studies, it is clear that a number of control pathways probably exist. Immunologists trying to trace these pathways see as their ultimate reward the ability to enhance immunity against disease agents and to depress immunity in cases where it might destroy a transplanted tissue or organ.

Alpha-fetoprotein (AFP) is the most plentiful substance found in the blood serum of the fetus during early and mid-embryonic life. Made in the fetal liver, AFP can also be found in the adult, but at concentrations of only 1/10,000 of that in the fetus. The substance's function remains a mystery.

However, Robert A. Murgita and Thomas B. Tomasi, Jr., of the Mayo Medical School in Rochester, Minn., reported some intriguing observations in February, 1975. They isolated AFP from mouse fetuses and found that it suppressed the immune responses of adult animals and of cells growing in cultures. The researchers do not know how AFP acts, but one possibility has been eliminated – it is not toxic to lymphocytes, the blood cells responsible for the immune reaction.

It is fascinating to speculate on what biological role AFP may play in such mysteries as these: Why does a mother's immune system not attack the fetus? Why can lymphocytes taken from newborn animals respond immunologically when placed in an adult, but lymphocytes from an adult cannot respond when placed in a newborn animal? What relationship exists between elevated levels of AFP seen in patients with liver cancer or hepatitis and the diminished immunity that is also characteristic of such patients?

Immunodeficiency. The most frequent known form of immunodeficiency in man is common variable im-

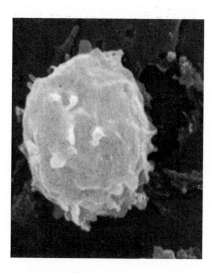
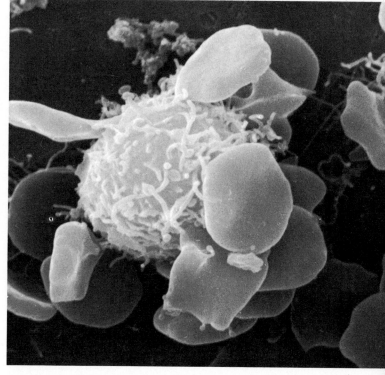

A T lymphocyte has a smooth outer surface, *above,* under most conditions. However, when mixed with red blood cells from sheep, such a cell produced many pronounced projections, *right.* This suggests that a common method of discerning B lymphocytes (many long projections) from T lymphocytes (few projections) may not always be reliable.

munodeficiency (CVI), a condition characterized by decreased numbers of antibodies, the blood proteins that help destroy bacteria and other intruders. People with CVI have normal numbers of B lymphocytes – the cells that make antibodies – but the cells fail either to synthesize or to secrete antibodies.

Most immunologists have believed for a long time that this failure was inherent in the B lymphocytes. However, Thomas A. Waldmann of the National Cancer Institute in Bethesda, Md., has demonstrated that this is not so. He described his experiments at the annual meeting of the Federated Societies for Experimental Biology and Medicine in Atlantic City, N.J., in April, 1975.

Waldmann eliminated a type of lymphocyte called a T lymphocyte from CVI blood cells. Once free of the T lymphocytes, the B lymphocytes produced normal amounts of antibodies. Furthermore, when the T lymphocytes from the CVI blood were added to normal blood, they suppressed antibody production by the B lymphocytes in that blood. Showing that T lymphocytes can sometimes block antibody formation represents an important beginning in assessing the mechanisms that underlie CVI and other kinds of immunodeficiency.

Transplantation. Most bioscientists have accepted the theory that transplanted tissues and organs are rejected because T lymphocytes recognize and attack foreign histocompatibility (H) antigens – protein fingerprints that project from the membrane of cells. However, Kevin J. Lafferty and his colleagues at the John Curtin School of Medical Research in Canberra, Australia, reported experiments in April, 1975, showing that this concept may be too simple. The researchers believe that a transplant's H antigens alone are not sufficient to trigger the attack by a host's T lymphocytes. Rather, some additional stimulus must be provided by lymphocytes or another type of blood cell called a macrophage that normally stow away within the donor's organ.

Lafferty's group kept mouse thyroid glands alive in cultures, then transplanted them into other mice. While the glands were in culture, their lym-phocytes and macrophages, both of which are mobile cells, gradually settled out into the surrounding medium. The "purified" glands, when transplanted into incompatible hosts, survived far longer than would be expected. Furthermore, the longer the glands had been kept in the culture, the longer they survived.

A new lymphocyte? Immunologists know that the body's immune system can kill foreign cells in two ways. First, B lymphocytes that have become sensitized to the invader by contact with it produce antibodies. These work with a series of serum proteins known collectively as complement to kill foreign cells. Second, T lymphocytes that have been sensitized to the invader can attack it directly and kill it. In September, 1974, Stig S. Frøland and his coworkers at the Institute of Immunology and Rheumatology in Oslo, Norway, reported a series of experiments which points to a third mechanism.

The experiments appear to confirm a concept introduced six years ago by Peter and Hedvig Pearlmann of the University of Stockholm in Sweden. They claimed that there was probably a form of lymphocyte action in which antibody-coated foreign cells could be destroyed – even in the absence of complement – by lymphocytes not previously sensitized to the invader.

Frøland and his colleagues reported that antibody-coated foreign cells were killed even though there were no complement, B lymphocytes, and T lymphocytes present. All macrophages, thought by some to be the Pearlmanns' elusive killers, were also eliminated. Evidently, the killer cells in Frøland's experiments recognized the part of the antibody molecules that remains exposed once the antibody has bound to a target cell. In this manner, antibodies apparently direct a lymphocyte of some kind to specific sites and it kills the invader.

Experiments supporting this mechanism have been performed by Frøland and others. They have found that antibody without the exposed part of the molecule cannot trigger this reaction. A paramount question now is just how important this pathway is in the immune response found in the normal person.   [Jacques M. Chiller]

Medicine

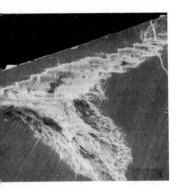

A tooth cross section, magnified 65 times by a scanning electron microscope, shows how decay progresses. Decay began on the surface, upper right, then dug in to make a cavity. For a 3-D view, see page 134.

Dentistry. Many researchers turned their attention in 1974 and 1975 to the immunological mechanisms in the mouth. They were searching for an immunological method to eliminate or reduce the number of bacteria — particularly *Streptococcus mutans* — that cause dental caries. An antibody, secretory immunoglobulin A (S-IgA), has been found in the saliva and may have potential for decay prevention.

Oral biologists Martin A. Taubman and Daniel J. Smith at the Forsyth Dental Center in Boston reported in June, 1974, that the level of salivary antibodies to *S. mutans* can be increased in rats by injecting killed bacteria near the salivary glands. This procedure seemed to decrease the number of *S. mutans* on the smooth surfaces of teeth, and lessened decay.

How S-IgA protects the teeth is not clear. In order to form dental plaque, a thin film of food particles and bacteria that leads to decay, *S. mutans* must form sticky substances which permit the bacteria to adhere to the teeth. S-IgA antibody makes it difficult for *S. mutans*

and other bacteria to attach to surfaces, presumably by inhibiting the enzymes that make the sticky material. On the other hand, it is conceivable that the S-IgA antibody might affect *S. mutans* in some other manner. Whatever the mechanism, this research has brought researchers one step closer to the development of a vaccine against decay.

In November, 1974, Stephan E. Mergenhagen of the National Institute of Dental Research, Bethesda, Md., implicated another part of the immune system in tooth loss among older persons. His research showed that toxic substances produced by bacteria stimulate the immune system to make macrophages, or scavenger cells, that fight the bacteria, but also produce collagenase, an enzyme that breaks down collagen, the fibers that anchor teeth to the jaw. Such elevated levels of collagenase have also been found in patients with rheumatoid arthritis, indicating that this disease and tooth loss from chronic gum inflammation may share a common immunological background. [Paul Goldhaber]

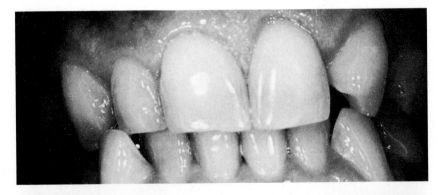

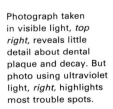

Photograph taken in visible light, *top right,* reveals little detail about dental plaque and decay. But photo using ultraviolet light, *right,* highlights most trouble spots.

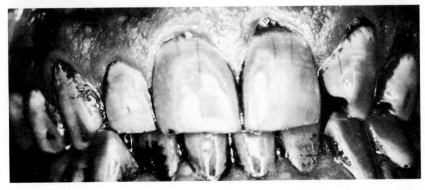

297

Medicine

Internal Medicine. Researchers are engaged in another tremendous effort toward preventing coronary artery disease, lessening its effect when it occurs, and improving survival chances after heart attacks. Clearly, prevention offers the most satisfactory approach.

Dr. K. M. Fox and his associates in Hull, England, reported in early 1975 that beta-adrenergic blocking drugs widely used to relieve angina – the pain associated with coronary artery disease – may also reduce the likelihood of a heart attack. Beta-adrenergic blocking drugs prevent the hormone adrenalin, which speeds up heart action, from attaching to beta sites, the receptor sites on cells in heart and blood vessel walls.

The researchers studied the case histories of 341 patients who were admitted over a one-year period to the coronary care unit of Kingston General Hospital in Hull with severe chest pain. The pain could have been either a heart attack or coronary insufficiency, a relatively mild heart ailment. From these patients, the researchers picked a group

of 90 with a previous cardiac history who had been treated with beta blockers for a least one month prior to admission and compared them with an untreated group of 90 matched for age, sex, and race. The most striking finding was that those treated with the drugs had less than half as many heart attacks as the control group. Heart attacks were diagnosed in 30 of the 90 who had been taking beta blockers, and 62 of the 90 who had not.

Researchers say these results indicate that, though beta blocking may not prevent a severe heart attack or change its outcome, it might prevent heart attack in marginal situations and improve the long-term outlook.

Beta blocking appears to produce its beneficial impact by slowing the heart rate and reducing blood pressure. These actions reduce the heart muscle's demand for oxygen, helping to relieve heart pain and improve the overall outlook for the patient's health. If further study substantiates this preliminary data, beta blocking may be used as a preventive treatment in all patients

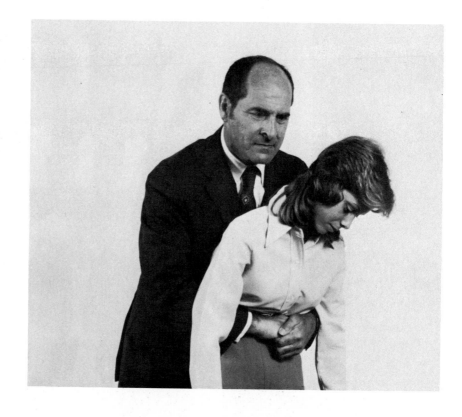

Dr. Henry Heimlich of Cincinnati shows his new first-aid technique for persons choking on food. The rescuer grasps victim from behind, locking his hands just below the ribcage. He then squeezes to force the diaphragm upward, compressing the lungs. This forces air back into the throat, which expels the food.

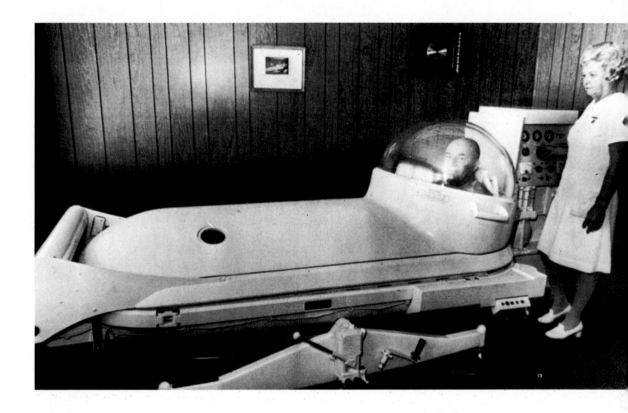

Medicine

Continued

An oxygen chamber designed for radiation therapy is now used in research on senility and memory loss. High-pressure oxygen forced into a patient's blood helps the brain function better.

who have been diagnosed as having coronary artery disease.

Lowering blood fats too late. It has been clear for some time that atherosclerosis – in which deposits of fat-rich materials on blood vessel walls narrow arteries and reduce blood flow – is more extensive in persons with elevated blood fats, notably cholesterol and triglyceride. These people are more likely to suffer from stroke, coronary artery disease, and heart attack. Several drugs are available that effectively and safely reduce blood fat levels. Their impact on the incidence of heart attack and subsequent survival has been the subject of a nine-year cooperative study supported by the National Heart and Lung Institute at 53 clinical centers.

More than 8,000 adult males volunteered for the study between 1966 and 1969. All had suffered a first heart attack and had recovered. They were treated with one of several drugs.

Niacin and clofibrate, two widely used drugs, were evaluated during the entire period. Although patient tolerance of the drugs, absence of side ef-

fects, and the reduction of blood fats proved satisfactory, it was not possible to show that treated patients were less likely to experience a second heart attack or that they enjoyed a more favorable survival rate. The study suggests that fat reduction in patients with well-established cases of atherosclerosis is not likely to help. Long-term studies are now underway to determine if fat-lowering drugs may help persons with slight cases of atherosclerosis who have not yet had a heart attack.

Finding curable hypertension. Another important risk factor associated with increased incidence of stroke, heart attack, and kidney disease is high blood pressure, or hypertension. More than 20 million adult Americans have hypertension. It is perhaps the most important risk factor affecting cardiovascular deaths. Most people with high blood pressure have no definable causes for the disease and are, therefore, classified as essential hypertensives. For them, lifelong drug therapy is necessary. But there is a small group, probably less than 10 per cent, who have a

299

Tracing A Human Cancer Virus

In January, 1975, Robert E. Gallagher and I, working at the National Cancer Institute's Laboratory of Tumor Cell Biology in Bethesda, Md., reported finding a virus in the blood cells of a woman suffering from a form of leukemia. We believe that the virus is of human origin and probably caused the leukemia – a type of cancer in which the white blood cells multiply in an uncontrolled manner. If so, this is probably the first time that scientists have actually isolated and grown a leukemia virus from humans.

The virus is a type C ribonucleic acid (RNA) tumor virus. These viruses are associated with specific kinds of cancer in animals. In some cases, such as leukemia in cats, scientists have found that the viruses are the natural cause of the disease. In cancers in other species, the exact role of the virus is uncertain.

In the early 1970s, scientists isolated type C viruses from primates, such as gibbons, suffering from leukemias and malignant tumors. They named this class of virus the gibbon ape leukemia virus (GALV). At about the same time, they isolated another type C virus, the simian sarcoma virus (SSV), from a malignant tumor of the woolly monkey. Both viruses caused cancerous disease when injected into healthy primates.

Ordinarily, a virus consists of a protein coat and nucleic acids, which contain genetic material. Although scientists may find the genetic material of a type C virus in a cancer cell, often they cannot find evidence of the mature whole virus. In these cases, the genetic information for forming the virus is either dormant or defective. In most primates, including man, visible evidence of whole type C virus is extremely rare. Because of this, investigators have had to use biochemical and immunological techniques to detect virus "footprints" in the form of virus proteins and nucleic acids.

In 1970, we detected one of these footprints, reverse transcriptase, an enzyme related to one in GALV and SSV, in the blood cells of some patients with acute leukemia. Other researchers later found a structural protein closely related to one from GALV and SSV in human leukemia cells.

These findings strongly suggested that some human leukemia cells, and possibly other human cells, contain a type C virus that is in a dormant state or is defective – that is, with incomplete genetic information. Also, the virus is related to the tumor viruses isolated from some primates. But we wanted to get the whole virus out of the cell and grow it in the laboratory so that we could study its nucleic acid and proteins much more thoroughly. We would also be able to look for antibodies against the virus, determine how it is transmitted, and study the interaction of the virus with its host cells.

Investigators working with animal cells have developed various techniques to stimulate production of detectable whole viruses from dormant or defective viruses. One of the tricks they use for dormant viruses is to add certain chemicals to the host cells to coax out all the virus's genetic information. Scientists are not certain how these chemicals work, but they are using them to look for viral causes of human cancers.

My colleagues and I used similar tricks to isolate the type C virus. We discovered a factor, probably a protein, in a strain of human embryo cells that stimulated the leukemic cells to grow in tissue culture. After the cells were in culture for several weeks, whole type C virus began to bud off from them. As we expected, the virus seemed to be closely related to GALV and SSV.

In this case, it appears that the complete genetic information for the virus was present in at least some of the patient's cells. So the virus was dormant, rather than defective. Yet, even though researchers found footprints related to GALV and SSV in the cells of other leukemia patients, the stimulation of cell growth did not release detectable whole virus. In these cases, the virus could be defective or there might not have been enough virus produced to be detected. We believe that the cells of the woman from whom we obtained whole virus contained the right genetic information, but that it was dormant until her cells were stimulated.

Cancer researchers must develop increasingly sensitive methods for detecting small amounts of virus and improve methods for isolating dormant and defective viruses. Such viruses from human tissues may shed light on leukemia in man.   [Robert C. Gallo]

surgically curable form of the disease. It has been difficult to identify these people in the past.

The most common form of curable hypertension develops when an arterial obstruction interferes with blood flow to a kidney. This causes increased production of the kidney hormone renin, which, by activating the blood hormone angiotensin, constricts all blood vessels and contributes to hypertension. Many such cases can be cured by surgery. But since the operation involves risk, it would be useful to be able to predict which patients will respond favorably to the surgery.

In March, 1975, Dr. David H. P. Streeten at State University Hospital in Syracuse, N.Y., provided physicians with a technique for discriminating between patients likely or unlikely to be helped by surgery. The drug Saralasin inhibits angiotensin by competing with it for binding sites on the arterial wall. It lowers blood pressure in patients whose high blood pressure is caused by kidney secretion of renin. When Saralasin produces a prompt and significant decline in blood pressure, surgical relief of arterial obstruction will probably reduce pressure. If it does not, surgery ·will probably not help.

Useful as it is for identifying curable hypertensives, Saralasin may be of even greater value in differentiating the more common essential hypertensives, who can be grouped into high-renin and low-renin types, according to the amount of the hormone in their blood. The use of Saralasin to identify the high-renin, or renin-dependent, essential hypertensives may allow doctors to prescribe better treatment.

Friendly fever. Fever, common during bacterial infection, has long been described as a protective device used by the body. Indeed, for years before specific antimicrobial drugs were available, raising body temperature artificially was a favored therapeutic technique. Although this often proved beneficial, its artificial nature and the fact that the induced temperatures were higher than natural fever temperatures, left uncertain what role fever plays in fighting infection.

Dr. Matthew J. Kluger and his associates at the University of Michigan reported in April, 1975, on experiments which suggest that even a modest increase in body heat may have important protective effects. Aware of the fact that reptiles alter their temperature by as much as 8°C (14.4°F.) to match the surrounding environment, the Michigan group carried out their studies on 141 lizards (*Dipsosaurus dorsalis*).

The researchers divided the animals into groups and kept them at temperatures ranging from 32° to 42°C (90° to 108°F.). They inoculated some animals at each temperature with disease-causing bacteria and control groups were given sterile salt water. The reptiles maintained at 38° and 40°C (100° and 104°F.) were protected from the bacteria, which had lethal effects at lower temperatures. Kluger and his associates also grew the bacteria in cultures at the same temperatures. The differences in heat had little or no effect on bacterial growth. Because the bacteria themselves were not affected by temperature, the high temperatures must have improved the host's defenses against the bacteria.

It is reasonable to assume that fever mechanisms have a common origin in mammals and reptiles, so the experiments may be relevant to clinical medicine. The experiments suggest that aspirin and other drugs that are used to suppress fever may not always be the best course of action in patients with moderate fevers.

Hematology. Two Rockefeller University scientists, Anthony Cerami and Charles M. Peterson, reported on a series of experiments in April, 1975. Their work brings physicians one step closer to a treatment that can prevent the occurrence of "sickling" and perhaps extend the lives of its victims. Sickle cell disease is a blood disorder that produces anemia, fever, dehydration, leg ulcers, and pains in the joints or in the abdomin.

The disease is caused by abnormal hemoglobin genes. Those who inherit a single abnormal gene have the sickle "trait," rather than the disease. They have no symptoms, but can transmit the disease to their children if their marital partner also has the sickle trait. Offspring who inherit the sickle gene from each parent have the disease. Sickle cell disease is found almost exclusively in blacks. Roughly 10 per cent of

Medicine

Continued

A victim of psoriasis undergoes a new type of treatment by bathing in ultraviolet light after being coated with salve made from coal tar.

all black Americans carry the trait and about 3 in 1,000 are born with the hereditary disease.

The sickle cell, a distorted red blood cell, has the amino acid valine in the hemoglobin molecule instead of the normal glutamic acid. This appears to make the molecule react adversely to water, and transforms the normally spherical red cell to a sickle shape. The sickle-shaped red cells clump and plug the smallest blood vessels, blocking the flow of oxygen to the tissues.

Cyanate, a chemical formed when urea is dissolved in water, is known to combine with hemoglobin. This union, called carbamylation, is irreversible. The researchers demonstrated that cyanate could combine with hemoglobin in blood cells without harming the cell itself; in fact, it prevented the cell from sickling. Next they showed that sickle cells carbamylated in a test tube actually lived longer when reintroduced into patients than the patient's other blood cells and seemed to reduce the frequency of sickle crises. Future study will be required to determine the value of this treatment for the world's 2 million sickle cell victims.

Sickle cell disease may someday be prevented, but painful and often life-threatening sickle crises are still frequent realities for the disease's victims, and a sickle crisis sometimes causes death in spite of the best medical treatment. In March, 1975, Dr. Malcolm Green and his associates at St. Thomas Hospital in London reported dramatic results in treating two patients who had not responded to conventional treatment and were very near death.

The physicians took more than 12 pints (6 liters) of blood from each patient and transfused them, replacing more than 90 per cent of each patient's red cells. The sudden removal of the cells with sickle hemoglobin brought prompt improvement to the patients. Although this drastic draining technique has been used in the past to prevent any complications in pregnant women or in sickle cell patients awaiting surgery, this is the first time it was used on patients in actual sickle cell crises.   [Michael H. Alderman]

Medicine

Continued

Surgery. A dramatic new milestone in heart transplantation occurred in November, 1974, when Christiaan N. Barnard of Cape Town, South Africa, implanted a donor heart next to and joined to that of a patient with heart failure. In all previous cardiac transplants, the recipient's own heart was removed, requiring a heart-lung machine to maintain circulation during surgery. There is little that can be done to save such a patient if his system should reject the donor heart.

Parallel cardiac transplantation can be carried out without a heart-lung machine. And if the patient's system rejects the transplant, he can be restored to his pretransplantation state and may still have a chance for life with his own heart.

Dr. Barnard's operation was actually a partial parallel heart transplant. Only the left side of the patient's heart was attached to the donor heart. He sutured the left atrium of the donor heart to the corresponding chamber of the patient's heart. All of the patient's blood returned to his own heart, and his right ventricle had to pump it through the lungs. The blood leaving the lungs then went to the left atrium, and at that time it could flow into either heart's left ventricle. Because the left ventricle of the recipient's heart was failing, the abnormally high pressure in that chamber would force most of the blood into the donor heart. The blood flowed from the donor heart to the descending aorta and from there to the rest of the body.

It was not necessary to synchronize the two hearts; each continued to beat at its own pace. The different rhythms did not cause any circulation problem. The continued presence of the patient's own heart did not change the rejection threat associated with all organ transplants, and this patient had some rejection problems which were treated with immunosuppressive drugs. The patient survived for four months with the parallel heart. The cause of death was unclear because of many complications.

Artificial organs. Researchers also reported the first long-term survival of animals with implanted artificial hearts. Surgeon Willem J. Kolff, who

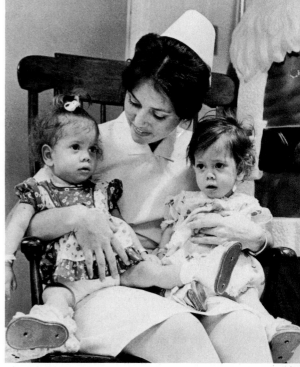

Clara and Alta Rodriguez, 13-month-old Siamese twins, *below,* were joined in the abdominal area and shared a common liver, colon, and part of the lower intestine, as well as an interconnected urinary system. They were separated, *right,* by a surgical team in an eight-hour operation at Philadelphia's Children's Hospital.

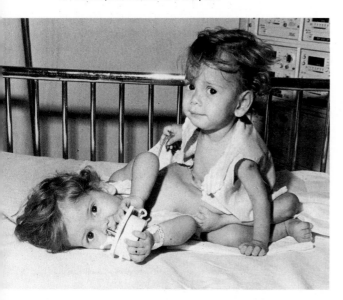

heads the Division of Artificial Organs at the University of Utah in Salt Lake City, described in April, 1975, his work with pneumatically powered implants in calves. The first long-term survivor lived 36 days with a silicone rubber heart while a second calf lived for 94 days. Dr. Kolff and his colleagues were encouraged because the animals remained in excellent condition and continued to grow.

A major problem with artificial hearts is that they require an external power source. Dr. Kolff inserted the pneumatic-drive lines into the calves' flanks and the air pump that ran the heart was connected to the lines outside the animals. The lines, unfortunately, provide potential entry points for bacteria and infection occurs easily. But these developments indicate that it is possible to construct an artificial heart for long-term use. A totally implantable heart driven by an internal power source would solve the infection problem.

Yet another artificial organ, the membrane lung, came into wide use in 1974. This device oxygenates the blood of patients with pulmonary lesions while the lungs are healing. The commonly used heart-lung machines are not well adapted to be used for more than 24 hours, because they injure red blood cells. But when a thin membrane is interposed between the blood and the oxygen, the blood can be oxygenated with minimal cell damage.

The National Institutes of Health, Bethesda, Md., underwrote the first large-scale test of this device on patients with acute respiratory failure in 1974. Dr. J. Donald Hill of the University of California at San Francisco believes that this procedure saved at least 15 patients during the test. The membrane lung has been used for as long as 31 days on a patient.

Artificial blood. Dr. Gerald S. Moss of the Department of Surgery, Cook County Hospital, Chicago, and his co-workers reported significant results in 1974 in the continuing effort to find the ideal substitute for blood. Dr. Moss prepared a hemoglobin solution from broken blood cells by carefully removing all other cell material. This eliminated the need to match blood types.

The researchers entirely replaced the blood of baboons with the hemoglobin solution. Even though the animals no longer had any red cells circulating in the blood stream, their tissues were able to take up normal amounts of oxygen.

Surgeons fight diabetes. In recent years, surgeons have joined the battle against diabetes mellitus. Most diabetics can control the disease by following a prescribed diet or by taking oral hypoglycemic drugs, but many require daily insulin injections.

Surgeons first attacked this disease by transplanting the entire pancreas, the organ that makes insulin. Although one such transplant continues to function after more than two years, the technical problems and limited number of donors compared to the large number of diabetic patients have forced surgeons to look for a new approach.

Insulin is produced in microscopic cell islets scattered throughout the pancreas. Recently developed techniques make it possible to separate these islets from the remainder of the pancreas. In February, 1975, Dr. Walter F. Ballinger of Washington University in St. Louis reported that he had successfully transplanted pure islets without the difficulties associated with pancreas transplants, and completely reversed the diabetic state in experimental animals. Although pancreatic islets can be injected under the skin or into the abdominal cavity, Dr. Ballinger showed that they were most effective in replacing pancreatic hormone function when they were injected into the large vein leading to the liver. The islets then become lodged in the liver, where they have an excellent blood supply.

Dr. John S. Najarian of the University of Minnesota and Dr. Richard Weil III of the University of Colorado at Denver have shown that, in addition to controlling the abnormal blood sugar levels seen in diabetes, islet transplants prevent the more serious complications of diabetes, such as kidney failure.

The main barrier to islet-cell transplantation in human beings is the low yield of islets from a single human pancreas. Dr. Russell K. Lawson of the University of Oregon Medical School in Portland reported in 1974 that he had increased the number of insulin-producing cells by growing islets in tissue culture. Dr. Lawson was able to keep these cells alive for up to five

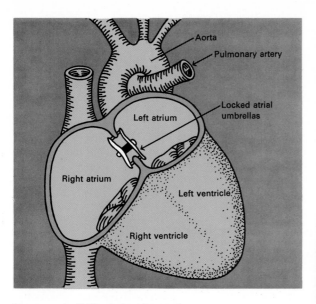

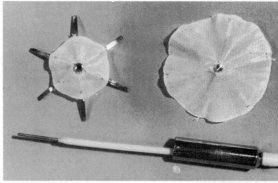

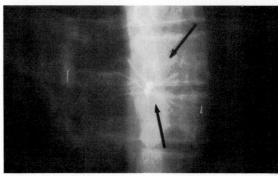

Dacron umbrellalike devices, inserted into the heart by catheter, *top right,* are used to correct heart defects. To seal a hole between the left and right atria of the heart, the umbrellas are placed on both sides of the hole, *above,* and locked in place. An implant can be checked by X ray, *right.*

Medicine

Continued

weeks using a culture technique in which he alternated high and low sugar concentrations in the culture medium. His data confirmed the replication of the cells and the production of insulin in response to increased glucose.

Parting twin sisters. In an extremely complex eight-hour operation carried out on Sept. 18, 1974, a team of 23 doctors, nurses, and technicians separated Siamese twins. The team was headed by Dr. C. Everett Koop of the Children's Hospital in Philadelphia.

Twins Clara and Alta Rodriguez, 13 months old, were brought to Philadelphia from the Dominican Republic. An extremely rare variety of Siamese twins, they were joined along the abdomen and perineum – the genital and rectal area. This type occurs in only 6 per cent of Siamese twins and poses a multitude of surgical challenges. The twins had normal upper bodies and each girl had her own pair of legs. However, they shared a liver, large intestine, and anus, and they had abnormal connections between their kidneys and bladders. Because of the common liver, doctors in

the Dominican Republic were unwilling to attempt a separation.

Dr. Koop and his colleagues studied the twins carefully and corrected a number of medical problems, such as anemia and kidney infection, before the surgery. After careful preoperative rehearsals, the surgeons began the major surgery. First, they divided the common liver so that each twin was left with half of the organ. Because of the remarkable regenerative power of the liver, this reduction in liver size was no problem for the infants. The intestine was divided so that one twin kept the anus, while muscles in the pelvis were used to create an anal opening for the other girl. The surgeons then sorted out and corrected the malconnections in the urinary system. Finally, the pelvic bones were sawed apart and the twins were separated. The instability of the remaining pelvic bones, plus the fact that the twins had never walked, were major problems, but time and more surgery should correct them. Dr. Koop believes that the girls should be able to live normal lives.   [Frank E. Gump]

Meteorology

Several developments in 1974 and 1975 highlighted progress in understanding climatic change and searching for ways to blunt or avoid its possible disastrous consequences. Evidence of the changing climate includes: the continuing severe drought on the southern edge of the Sahara; clear-cut increases in the average temperature of cities and decreases in urban rainfall, combined with heavier precipitation downwind of cities due to pollution; and the steady and significant decrease in average temperatures in the Northern Hemisphere since 1940. Other aspects of climate change are the increase in atmospheric carbon dioxide and other pollutants and the implications of potential ozone losses in the upper atmosphere (see ENVIRONMENT, Close-Up).

Rainfall and plant cover. Jule G. Charney and Peter Stone of the Massachusetts Institute of Technology (M.I.T.) in early 1975 presented the details of a new model of general global circulation, the first to relate changes in rainfall and plant cover to one another. They noted that changes in plant cover

cause changes in the ground albedo — the amount of solar radiation reflected by the surface. The more radiation that is reflected, the less there is absorbed to provide heat to generate showers.

Ground covered by plants may have an albedo of from 10 to 25 per cent and normal precipitation. If the vegetation is significantly reduced or eliminated, the albedo may rise to from 30 to 45 per cent. Using conditions found in the area south of the Sahara as a base, the scientists computed the difference in rainfall caused by changes in albedo. Assuming full grass cover with an albedo of 14 per cent, they computed that rainfall during a seven-week test period would average about 0.17 inch (4.5 millimeters) per day. With a 35 per cent albedo — bare ground — average rainfall would drop to about 0.1 inch (2.5 millimeters) per day. The results clearly showed that reducing the plant cover by overgrazing, for example, could cause weather changes which would speed later plant losses.

Atmosphere and the ocean. Just as winds create ocean waves and major air

Inside hallways that open to the north or east provide the best tornado shelter in single-story schools.

School Hall Haven from Whirlwinds

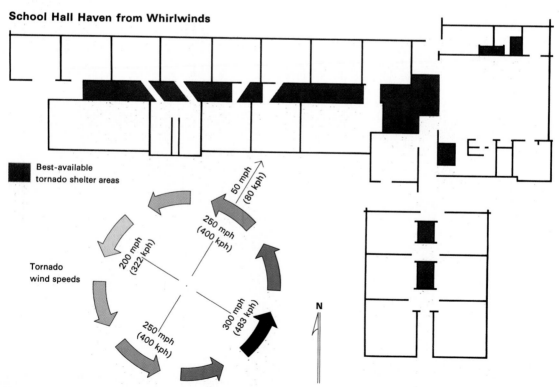

Best-available tornado shelter areas

Tornado wind speeds

50 mph (80 kph)

250 mph (400 kph)

200 mph (322 kph)

300 mph (483 kph)

250 mph (400 kph)

N

CONTRAIL →

← William B. Bankhead
National Forest

GUIN O

← TORNADO TRACK

Meteorology

Continued

Path of destruction 57 miles (90 kilometers) long left by a 1974 tornado that swept through parts of Alabama is the longest one ever identified on a satellite photograph.

currents influence ocean currents, so do the oceans affect the atmosphere. Storm waves drag the air with them and set up surface winds; warm water adds heat to cold air masses; and the oceans provide limitless reservoirs of moisture for rainfall. Nevertheless, these interrelationships are so complex that meteorologists have been able to create only very limited computer models to explain how the atmosphere interacts with the ocean on a global scale.

In January, 1975, Syukuro Manabe and his colleagues at the National Oceanic and Atmospheric Administration's (NOAA) Geophysical Fluid Dynamics Laboratory at Princeton University announced the first significant breakthrough. Their model postulates 9 levels in the atmosphere and 12 in the ocean. It includes crude topographies of the land and ocean floor and allows for moisture and heat to move freely across the surface of the ocean. The distributions of carbon dioxide, ozone, and clouds are kept constant over time—an admitted oversimplification—but the basic characteristics of

the model agree with long-term observations of the real world. The first results showed that ocean currents significantly affect distributions of surface temperature and precipitation.

Satellite contribution. Harry G. Stumpf of NOAA's National Environmental Satellite Service announced in April, 1975, that the new high-resolution infrared radiometer aboard the NOAA-3 satellite can detect the upwelling of cool ocean water caused by strong northerly winds in the Gulf of Tehuantepec, Mexico. This water, which remains near the surface only a few days, is rich in nutrients and would be an obvious feeding area for fish. The ability to detect how the ocean responds to weather adds a new way to manage tuna resources in the eastern tropical Pacific Ocean.

Divergent evolution. Analyses based on the Venus cloud photos sent back from Mariner 10 in February, 1974, and data from other spacecraft probes inspired an international conference on the Venusian atmosphere in October, 1974, in New York City. Perhaps the

Winding Up The Winds

Can driving on the right side of the road cause tornadoes? It seems absurd but, surprisingly, the more we look into the idea, the more plausible it becomes. Calculations and data analyses by John D. Isaacs of the Scripps Institution of Oceanography in San Diego, graduate students James W. Stork and David B. Goldstein, and myself have been increasingly convincing.

One of the necessary ingredients for tornadoes is rotation of the air. Convection—moist, warm, rising air—is another requirement. Tornadoes always occur with thunderstorms, which are convective storms, but not every thunderstorm produces tornadoes. This indicates that the heat involved in a convective storm provides sufficient energy to produce tornadoes, but that the missing ingredient is rotation.

It is clear that moving automobiles and trucks can cause rotation in the atmosphere. Opposing streams of traffic passing on the right side of the road create eddies of air that whirl counterclockwise. Tornadoes in the United States also spin in this direction, due to the influence of the earth's rotation on air currents. But the 2.5 million cars and 600,000 trucks on the road in the United States at any moment can create enough atmospheric rotation each year for up to a million tornadoes.

The annual number of tornadoes reported in the United States has increased tenfold in the past 40 years. The number of cars and trucks on the road has increased by a similar amount over the same period. However, this type of correlation, although significant, is not particularly meaningful. As one of our detractors commented, we might as well correlate tornadoes with X-rated movies because tornadoes depend on hot, steamy air to form.

To test our hypothesis, we obtained all records of some 15,000 tornadoes that occurred in the United States from 1950 to 1973. When we tabulated the data, we found a weekly cycle, with an extreme low on Saturdays.

We could find only one published reference to a weekly cycle: "On the seventh day, God ended His work which He had made and He rested on the seventh day from all His work which He had made" (Gen. 2:2). Thus, there are two possible sources of a seven-day cycle in natural events—the Biblical Creator and people. No other natural cycle—solar, lunar, or terrestrial—has a weekly component. The only scientific explanation is the human one. And, of the possible human sources, we credit the weekly cycle of tornado frequency to automobile traffic.

No other explanation appears to fit the data. Particles and heat emitted by industry cannot explain the observations. Exhaust fans on buildings do not all rotate in the same direction, so they cannot be responsible.

Interestingly, the incidence of tornadoes returns to normal on Sundays. Thus, whatever is missing on Saturday afternoons seems to be added on Sunday afternoons. On Friday evenings and Saturdays, a large amount of traffic flows out from major metropolitan areas. This one-way traffic does not provide the coupled forces necessary to create rotation in the atmosphere. But when the traffic returns on Sundays, it may provide the other half of the couple. This theory needs more testing since other explanations are possible.

Other tests seem to confirm our conclusions. The percentage of clockwise tornadoes reported—an extremely small number to begin with—decreased over the period of our records. This tends to support the idea that increasing amounts of counterclockwise rotation are being added to the atmosphere. As air flows across the United States from west to east, it acquires counterclockwise rotation from more and more traffic. Thus, we might expect the Saturday lows to be more pronounced on the West Coast. We have observed this. For example, the state of Washington has never had a tornado on a Saturday or Sunday.

More work must be done before we unabashedly promote the obvious corrective measure—driving on the left side of the road. Also, automobile races such as the Indianapolis 500, which circle counterclockwise, appear to go in the wrong direction. The racing cars introduce enough rotation into the air for 30 tornadoes. We have not yet checked the Memorial Day incidence of tornadoes around Indianapolis, but if they peak at the end of May, race drivers may have to learn to lean to the other side. [Gerald L. Wick]

Meteorology

biggest puzzle discussed is the reason for the widely separated evolutionary paths taken by the atmospheres of the "twin" planets, Earth and Venus. John S. Lewis of M.I.T. and James Walker of Arecibo Observatory in Puerto Rico pointed out that because the two planets are so similar in mass, density, and solar-system position, they both started with the same initial composition and similar amounts of the same atmospheric gases – carbon dioxide and water vapor.

Apparently, three factors influenced the divergence: Venus was just close enough to the Sun to prevent its water from existing as a liquid; life developed on Earth; and the Earth possesses a massive satellite – the Moon. It appears that Venus lost its water vapor over a period of about 30 million years as ultraviolet light from the Sun broke down the water into hydrogen and oxygen. The lightweight hydrogen rapidly escaped and the oxygen slowly combined with the surface materials. Venus still has most of its initial carbon dioxide, the major component of its atmosphere.

On Earth, the water vapor condensed and helped to dissolve the carbon dioxide, most of which then became locked in the surface rocks. Plant life not only replaced oxygen lost by surface weathering, but also increased the supply, making possible the protective ozone layer and the oxygen environment needed for animal life. Finally, the Earth's Moon may have helped stabilize the Earth's orbital motions and thereby stabilize its climate.

Other Venus reports covered atmospheric structure and dynamics, and cloud composition and motions.

Venus' unseen lower atmosphere – where surface pressure is at least 90 times that on the Earth and the average temperature is 900°F. (480°C) – acts as a huge convection cell. Air rises from that part of the planet directly facing the Sun, and sinks on the opposite side of the planet. There is a slow, even flow of from 3 to 6 feet (1 to 2 meters) per second from one side to the other.

Although Venusian clouds are featureless in visible light, the upper clouds can be observed with ultraviolet photography. The upper atmosphere, from about 12 to 45 miles (20 to 70 kilometers) above the surface, is an area of broad turbulence with many streaks and spirals. High-speed winds of 330 feet (100 meters) per second circle the planet from east to west in four days. There are strong variations during this four-day period, with "morning" winds exceeding "evening" winds by some 135 feet (40 meters) per second. The temperature at the top of the clouds is about −20°F. (−30°C).

The atmosphere on Venus probably is almost totally carbon dioxide, possibly with some nitrogen or neon. The clouds consist of minute droplets of concentrated sulfuric acid, something like a moderate smog. The smog is heaviest about 25 or 30 miles (40 or 50 kilometers) above the surface. The only water vapor exists in extremely limited quantities high above the smog.

One of the most intriguing problems is identifying the material that produces the features in the ultraviolet photos. Suggestions include variations in cloud droplet sizes, the presence of hydrobromic acid, or the possibility that elemental sulfur particles may be involved.

If the last suggestion proves true, Venus may have sulfur in liquid, solid, and gaseous form, as is the case for water in the Earth's atmosphere. Bruce Hapke and Robert Nelson of the University of Pittsburgh noted that sulfur is an important energy source for many Earth bacteria. Such bacteria may have been present on the Russian Venera spacecraft that have landed on Venus, and they might be thriving and multiplying there.

Program for action. The National Academy of Sciences in early 1975 issued a major report on understanding climatic changes. The report, written by W. Lawrence Gates of the Rand Corporation and Yale Mintz of the University of California at Los Angeles, stresses the importance of devising models of global circulation to study how the atmosphere interacts with water, land masses, and ice and snow deposits. It recommends summarizing all available sources of climatic data, and proposes monitoring changes in such things as surface vegetation, soil moisture, ground water, river flow, lake levels, total precipitation, chemical pollution, ice cover, radiation, and cloud cover.   [John R. Gerhardt]

Scientific advances predicted by science fiction writers have often become reality. In 1974 and 1975, genetic engineering joined rocket ships, organ transplants, moon flights, and the countless other science fiction themes come true.

In genuine genetic engineering, a specific gene is transferred from one kind of organism into another, with the second organism acquiring the trait or function controlled by the gene and then passing the newly acquired gene on to its offspring.

Several research groups succeeded in manipulating specific genes into bacteria and demonstrated that the foreign genes conferred new properties upon the bacteria and were passed on to future generations. Further, they showed that the bacteria could be used as a "gene factory," each bacterium producing countless identical copies of a selected gene.

Making the transfer. In the spring of 1974, two teams of scientists, one led by microbiologist Stanley N. Cohen at Stanford University, the other by microbiologist Herbert Boyer at the University of California at San Francisco, reported that they had transferred a gene from a South African toad into the common intestinal bacterium *Escherichia coli.* The gene was one involved in the toad cells' synthesis of ribosomal ribonucleic acid (RNA), a necessary part of the protein-producing machinery employed by all cells.

The scientists first isolated ribosomal RNA from the toad cells. They also isolated deoxyribonucleic acid (DNA) from the cells. DNA is the chainlike material that contains a cell's genes. The scientists then mixed the ribosomal RNA with the DNA. The ribosomal RNA and the gene involved in its synthesis stuck together because of a similarity in their chemical structures.

This created hybrid DNA-RNA molecules which were easily separated from the remaining unhybridized DNA by spinning the mixture in a centrifuge. The hybrid was then separated chemically into the DNA carrying the genetic information for ribosomal RNA synthesis and the ribosomal RNA itself. Next, the scientists added an enzyme to the DNA to cleave it into gene-sized fragments that included the gene that was to be transferred.

To get these toad ribosomal RNA genes into *E. coli,* the scientists took advantage of a strain of the bacterium that is resistant to the antibiotic tetracycline because of a gene within a plasmid. A plasmid is a small piece of DNA, often composed of only a few genes, that floats free in the cytoplasm of some bacterial cells and replicates, or makes copies of itself, independently of the bulk of the cells' DNA. Scientists know that bacteria easily transmit these plasmids to one another.

The scientists broke the bacterial cells and collected the plasmid DNA. Then, using the cleaving enzyme they had used before, they broke this DNA into smaller pieces. To these pieces, which included copies of the gene for tetracycline resistance, they added the fragmented toad DNA containing the ribosomal RNA genes. Then they added ligase, an enzyme that joins DNA pieces.

They added the resulting mixture, which now contained DNA chains made up of plasmid DNA and toad DNA, to a strain of *E. coli* that had no plasmid DNA and was, therefore, susceptible to tetracycline. After giving the bacteria sufficient time to take up the plasmid-toad DNA, the scientists added tetracycline. This killed all the bacteria which had not taken up the resistance gene carried by the plasmid. In this way, the scientists had also isolated the *E. coli* cells that now possessed the toad ribosomal RNA genes.

The critical step of proving that the bacteria now possessed the toad ribosomal RNA genes remained. To do this, the scientists grew a large population of the bacteria, broke them open, and isolated their plasmid DNA. Then they mixed the plasmid DNA with toad DNA, which carried the gene for synthesis of toad ribosomal RNA. If the plasmid DNA also contained toad genes, it would combine with the toad DNA because of their chemical similarity. The two DNAs did combine with each other.

Not only did the bacteria contain the toad gene, but they also had replicated the plasmid DNA during their growth, replicating the toad gene just as if it were a bacterial gene.

Next, the scientists wanted to determine if the function of the gene, the

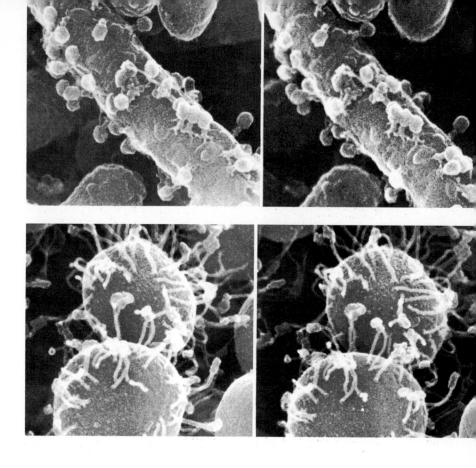

Paired views of cells of bacteria being attacked by spidery, clinging viruses produce a 3-D picture with the aid of a paper divider about the size and thickness of a business envelope. Hold the divider perpendicular to the page with a short edge against and along the line separating the pair of photos. Put your head against the other edge of the divider so that each eye sees only one photo. Focus beyond the pictures, and they will gradually merge to produce the three-dimensional view.

Microbiology

Continued

synthesis of toad ribosomal RNA, was expressed in the bacteria. They mixed cytoplasm from the bacteria with toad DNA and found that RNA in the bacterial cytoplasm combined with toad DNA. This, then, was the toad ribosomal RNA that the bacteria had produced with the toad genes they had harbored.

Gene amplification. Prospects for using plasmid-borne genes in genetic engineering got another boost in January, 1975, when microbiologists Teruo Tanaka and Bernard Weisblum of the University of Wisconsin at Madison reported they had developed a procedure for amplifying these specific genes. Gene amplification entails greatly increasing the number of plasmid-borne genes in a bacterial cell without increasing the number of the rest of its genes. To do this, the scientists used techniques that are similar to those already described.

They attached plasmid genes conferring resistance to the antibiotics streptomycin and sulfonamide from one strain of *E. coli* to plasmids from anoth-

er *E. coli* strain that had no genes for resistance to the antibiotics. Then they transferred the hybrid plasmids into another *E. coli* strain that previously had no genes for resistance to the two antibiotics, thus making the bacteria resistant. The Wisconsin scientists allowed these genetically engineered organisms to reproduce until they had a large population. Then the antibiotic chloramphenicol was added. This substance causes bacteria and the main portion of their DNA to stop reproducing. However, the mechanism for plasmid replication is not affected by chloramphenicol. As a result, each bacterial cell continued to produce copies of the plasmid DNA and soon contained hundreds of them.

These experiments demonstrate that potentially unlimited amounts of plasmid DNA containing specific genes can be stockpiled through relatively simple manipulation. The day may come when a scientist reaches for a vial of a specific gene for an experiment as readily as he now reaches for a particular chemical reagent.

Microbiology

Benefits and risks. The potential benefits of such genetic engineering were immediately apparent to scientists. For instance, genes for nitrogen fixation, for resistance of plants to insect pests, or for almost anything imaginable could be linked to plasmids, amplified in a bacterium, and transferred into some new organism (see THE NITROGEN FIX). The potential for exchanging genes between species, and thereby for creating completely new ground rules for evolution, staggers the imaginations of most scientists.

The possible dangers of genetic engineering are also apparent. Genetic monstrosities – for instance, bacteria resistant to all known antibiotics or with the potential to produce new diseases – could be created and inadvertently escape the laboratory and cause epidemics. Experiments in which viral genes that cause cancer in hamsters were transferred into bacteria and amplified have already been performed. What if bacteria carrying human cancer viruses should be created and then escape to infect masses of humans?

These and other considerations brought leading scientists from every nation capable of doing genetic-engineering experiments to a meeting at Pacific Grove, Calif., in April, 1975. The meeting, which was without precedent in scientific history, was called to establish limitations on genetic-engineering experiments.

After several days of debate, the scientists reached agreement. Some kinds of experiments – for instance, those involving DNA from organisms that cause dangerous diseases in human beings or DNA that produces virulent toxins – are not to be done at all. Also, the bacteria used to receive plasmid hybrid genes must have so many special nutritional requirements for growth that they can survive only in carefully regulated laboratory environments.

The message from the scientists at the meeting is clear. Potentials for genetic engineering are so great that the work must go on. But scientists must take the responsibility for doing the work in ways that do not threaten mankind. See GENETICS. [Jerald C. Ensign]

Neurology

Neuroscientists Robert Galambos, Kurt Hecox, Terence Picton, and Steven A. Hillyard of the University of California at San Diego in 1974 continued work on a new method to discover hearing loss in a newborn child. The method is likely to give neurologists and pediatricians a head start in remedying defects that often are not discovered until the child is older.

The fact that the very young cannot communicate is a major problem in testing hearing in newborn children. By the time a child is old enough to respond meaningfully to tests, the loss of hearing may have permanently affected speech and even understanding.

The diagnostic method developed by the San Diego scientists is based on a phenomenon called the evoked auditory potential. When a person is exposed to a distinct sound, such as a click, an electric response is evoked from the brain. The response can be recorded through electrodes attached to the scalp. When the responses to a series of clicks are measured very precisely and averaged by computer, the evoked response can be seen to consist of 15 separate, identifiable wave forms. The first of these occurs as early as 1.5 milliseconds (thousandths of a second) after the click, the latest from 300 to 500 milliseconds after. By comparing wave forms recorded from normal people with those from someone with a suspected hearing impairment, the examining audiologist can detect differences that suggest hearing problems.

For example, the early wave forms have been identified as reflecting the progressive activation of the cochlea, a cavity in the inner ear that contains the sensory ends of the auditory nerve, and auditory tracts in the brain stem. If close analysis of the data from a test on a child shows these waves to be abnormal, not only has hearing loss been determined, but also its probable cause – conduction loss to the cochlea or damaged hair cells in the cochlea.

Pain research. The relief of pain has always been a central goal of medicine. Many theories of pain have been proposed, but none so far has effectively explained it. Is pain a separate sense, or

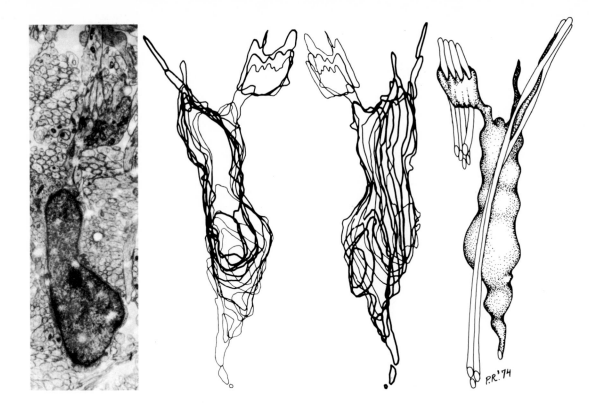

Neurology

Continued

By turning a brain cell, above, in an electron microscope, a series of cell contours can be seen and sketched at various angles, center left and center right. A computer scans the sketches and produces a 3-D view, far right, that is interpreted to give new information about the cell's volume and surface area.

is it a dimension of other senses? How and where does the brain regulate and control pain? How do drugs and other therapies relieve pain?

Perhaps the research of psychologist John C. Liebeskind of the University of California at Los Angeles and his co-workers David J. Mayer, now at the Medical College of Virginia, and Huda Akil, now at Stanford University Medical College, may provide answers to these questions and produce an understanding of pain that will allow physicians to control it completely. Their studies, reported in October, 1974, involved the electric stimulation of the brains of rats and cats with tiny electrodes placed in the brain.

They have demonstrated conclusively that pain relief can be produced by stimulating specific places in the brain. When stimulated in these areas, animals are completely undisturbed by stimuli that would ordinarily cause them great pain. At the same time, the animals respond normally to other sensory stimuli, which indicates that the electric stimulation is not just knocking

out all general response systems but is affecting a specific pain system.

From these and other findings, Liebeskind and his co-workers propose that the electric stimulation directly activates a pain-suppression system that blocks a chemical pathway in the brain that transmits pain information.

Mental retardation. New studies linking brain abnormalities with mental retardation in children may give scientists a new basis for understanding the mechanisms of the normal brain as well as defective development of the brain. Pathologist Miguel Marin-Padilla of Dartmouth Medical School in Hanover, N.H., has been searching for such links. He reported late in 1975 on the results of his study of the brain of an infant who died suffering from a genetic defect similar to Mongolism.

Marin-Padilla concentrated his studies on the defective brain's motor cortex, the part of the outer portion of the brain that controls movement. However, he also examined sensory areas, particularly those concerned with vision, hearing, and touch. Under microscopic examination, the sensory areas seemed to be normal, but the motor area revealed great abnormality.

For one thing, it contained considerably fewer cells than a normal brain. In addition, dendritic spines, tiny outgrowths of the dendrites, were sparse and seemed distorted and underdeveloped. The dendrites are the fibers on the brain nerve cells that make contact with and receive chemical and electric information from other brain nerve cells. If the dendritic spines are defective, incoming signals will not be properly received.

Neuroscientist Dominick P. Purpura of the Albert Einstein College of Medicine in New York City is also studying the relationship of nerve structures to mental retardation. He compared dendritic spines from the brains of six severely retarded children with normal child brain material. The dendritic spine distortions he found in the cerebral cortex of the retarded children emphatically demonstrated that defects in these tiny nerve parts are related to mental and behavioral deficiency.

These clinical studies also furnish valuable data for basic brain research.

Nerve growth factor. Drs. Michael Young, Barry G. W. Arnason, and their co-workers at Harvard Medical School and Massachusetts General Hospital reported in May, 1975, that a biologically active substance very much like the nerve growth factor (NGF) found in cancer cell fibroblasts is also produced by normal fibroblasts, which are cells that form connective tissues.

More than 20 years ago, Rita Levi-Montalcini and her colleagues at Washington University School of Medicine in St. Louis discovered NGF in the salivary glands of male mice. When applied to embryonic rat or chicken tissue, NGF caused nerve fibers to grow in profusion.

Because fibroblasts are involved in healing wounds, the new findings suggest that NGF may play other roles besides its known stimulation of sensory and sympathetic nerve growth. The substance may take part in the healing of many other kinds of brain and body tissue as well.   [George Adelman]

Nutrition

Nutritionists in 1974 questioned the use of Recommended Dietary Allowances (RDAs) as the sole criteria of a sound diet. RDAs are guidelines set by the Food and Nutrition Board of the National Research Council on the amount of 17 nutrients required daily, based on an individual's age, sex, height, weight, and other factors. Research findings and their interpretations led to a controversy over the RDAs of various vitamins. See Close-Up.

D. Mark Hegsted of the Harvard University Department of Nutrition in January, 1975, criticized the widespread reliance on RDAs as a measure of dietary quality. He noted that people in special categories, such as pregnant or nursing women, have difficulty in following the recommendations.

Hegsted also pointed up the difficulty in correlating daily intake with long-term nutrient balance. For example, a person may eat a food one day that is quite rich in a vitamin that the body can store, then go several days with only minimal intake of that vitamin. An analysis of such a diet would

indicate a deficiency for most of the days, even though the overall supply of the vitamin may be adequate. Hegsted proposed that RDAs be amended so that they are expressed as so much of a vitamin or mineral per 1,000 calories of food energy.

Not by cereal alone. A study of 44 popular brands of processed breakfast cereals by Consumers Union of the United States, Incorporated (CU), of Mount Vernon, N.Y., exemplifies the difficulty in applying RDAs to a dietary staple. CU reported in February, 1975, that the cereals lacked total nutrient content. Some of the cereals contained up to 100 per cent of vitamin RDAs.

In the study, rats fed a diet made up only of breakfast cereal did not gain weight and grow as well as those on a balanced, fortified, scientifically designed laboratory ration. With an arbitrary score of 100 per cent assigned to the laboratory diet, eggs alone rated 82 per cent and milk rated 55 per cent. CU did not give numerical ratings for each of the cereals, but they ranged from 70 per cent down to 20 per cent.

Vitamins: Too Much Of a Good Thing?

Controversy flared up during 1974 and 1975 surrounding research on vitamins, the organic nutrients that usually combine with proteins in human cells to produce enzymes necessary for normal metabolism. Some researchers reported possible benefits from large doses of certain vitamins, while others reported possibly harmful side effects. Much of the research raised anew the basic question: How much of the various vitamins should a person take?

Vitamin C is a well-publicized case. The recommended dietary allowance (RDA) set by the National Research Council's Food and Nutrition Board ranges from 35 to 80 milligrams daily, enough to prevent scurvy and provide a margin of safety in most humans. In 1970, Nobel prizewinning chemist Linus C. Pauling suggested 5 to 100 times that amount might prevent the common cold. Sales of vitamin C supplements soared, and researchers began to test Pauling's suggestions through clinical studies of volunteers.

After reviewing most of the studies, Michael H. M. Dykes of the American Medical Association's Department of Drugs and pharmacologist Paul Meier of the University of Chicago reported in March, 1975, that vitamin C does not prevent colds; at best, it may slightly reduce cold symptoms.

Reports began appearing in 1974 that vitamin C could effectively treat or prevent arteriosclerosis, the progressive thickening and hardening of artery walls that is a major cause of heart attacks and strokes. These claims were based on experiments done by pathologist Constance R. Spittle in Wakefield, England, and Emil Ginter of the Institute of Human Nutrition in Bratislava, Czechoslovakia.

Spittle measured the blood cholesterol levels of some 60 healthy volunteers who took 1 gram of vitamin C daily for 6 weeks in 1971. Cholesterol is a fatty substance found in many foods and synthesized in the body to perform important functions. High levels of blood cholesterol, however, are considered a leading risk factor linked to arteriosclerosis. In persons under age 25, the cholesterol levels dropped 8 per cent on the average. For those between 25 and 45, the levels rose 4 per cent, and for those over 45, the levels remained the same. Among the older test subjects and in a group of 25 arteriosclerosis patients, individual results varied widely. Most other tests have produced similar inconsistent results. Spittle suggested that the increases in the older test groups and the patients resulted from movement of cholesterol from the arterial walls into the blood stream, but based on such variable results, her suggestion is controversial.

Ginter's work, reported in 1973, showed that vitamin C lowered the blood cholesterol level in guinea pigs that originally lacked a normal amount of the vitamin, but had enough to avoid scurvy. At the New York Academy of Science's Second Conference on Vitamin C in October, 1974, Ginter reported on tests with humans. Persons over 50 reduced their cholesterol levels by an average of 17 per cent by taking 1 gram of vitamin C daily for six months.

In June, 1975, however, Virginia E. Peterson and her co-workers at the Stanford University Medical Center in Palo Alto, Calif., reported no significant changes in cholesterol levels in nine patients who took 4 grams of vitamin C daily for two months.

There have also been claims that large doses of vitamin C can be harmful. Here, too, research is controversial.

Suddenly stopping large doses may produce scurvy even though a normal quantity of vitamin C is consumed. This is because the body's machinery for destroying excess vitamin C has become so effective that even normal amounts of it are quickly destroyed. This withdrawal effect and the related threat of rebound scurvy may last for several weeks.

My colleague Elizabeth Jacob and I reported in October on the effect that vitamin C has on vitamin B_{12}. We blended foods with various amounts of vitamin C and incubated the mixtures for 30 minutes at 98.6°F. (37°C) — stomach temperature. We found that 500 milligrams of vitamin C destroyed 50 to 95 per cent of the vitamin B_{12} in these test meals.

The toxic effects of large doses of some vitamins, especially vitamins A and D, are well established. This, along with unproved claims of benefits, suggests that taking large vitamin supplements can be risky.   [Victor Herbert]

The researchers found no match between the growth test results and such measurable factors as the grain or mixture of grains for each cereal, the amount of sugar in each, whether cooking was required, or how much the cereal was fortified with nutrients. "A cereal can be stuffed with just about every nutrient for which daily requirements have been established and still not be able to support the lives of test animals," the organization reported.

Some nutritionists criticized the tests because the missing nutrient responsible for limiting growth was not identified. Eaten with other foods such as milk, which might make up the deficiency, the cereals could be an adequate part of a total diet. Most nutritionists are more concerned with the large amounts of fats and sugars often put into cereals while parts of the grains rich in fiber, such as bran, are removed.

Benefits of bulk. Fiber is the material surrounding the cell walls of plants that both supports and protects the cells. Because nonnutritive fiber, such as cellulose or pectins, cannot be digest-ed by the intestinal tract, nutritionists have generally assumed that fiber plays a minor role in determining a person's health. But medical research scientist Denis P. Burkitt and his associates at the Medical Research Council in London reported in August, 1974, epidemiological findings that showed a relationship between dietary fiber and many of the diseases that are characteristic of industrialized societies.

Burkitt first examined differences in the workings of the intestinal tract between diets that are low in fiber, as in Western European nations, and diets high in fiber, as in African nations. The speed with which food residue passes through the intestinal tract increases remarkably as fiber content in the diet increases, and so does the volume of feces and its water content. These factors make for a more easily passed stool.

The investigators related the lowered straining required for bowel action when fiber is abundant to fewer cases of appendicitis, diverticular diseases, hiatus hernia, and hemorrhoids. They also suggested that the high-speed move-

Drawing by Lorenz; © 1975 The New Yorker Magazine, Inc.

"I'll have the Hamburger Helper with soybean extender
and a side order of hydrolyzed vegetable protein."

Nutrition
Continued

ment through the intestines associated with high-fiber diets might decrease the incidence of cancer of the colon and rectum. Any cancer-causing compounds found in the foodstuffs would be rapidly eliminated, they concluded.

The most startling correlation between fiber and disease was between low levels of fiber and the high incidence of ischemic heart disease (IHD), in which the flow of blood to the heart is obstructed. IHD is responsible for over 30 per cent of all deaths in the United States. In Africa, IHD is virtually unknown, although it is increasing in large cities.

Burkitt's colleague, Hubert C. Trowell, formerly a physician at Makerere University in Uganda, analyzed heart-disease statistics and fiber intake. He concluded that fiber plays an important role in the development of heart disease. For example, Bantus living in rural South Africa eat four times as much fiber as Bantus living in urban areas and have less heart disease.

Trowell suggests that the link between fiber and heart disease is the blood cholesterol level. About a decade ago, David Kritchevsky of the Wistar Institute in Philadelphia demonstrated that dietary fiber binds to bile salts by various mechanisms. He reported in 1975 on studies of baboons, rabbits, and rats and concluded that the liver produces as much cholesterol in animals fed a high-fiber diet as a low-fiber diet. For test animals on a high-fiber diet, however, cholesterol synthesized by the liver is converted more rapidly to bile salts. In other words, fiber acts like a blotter; it allows more liver cholesterol to be excreted instead of entering the blood stream.

Not all investigators accept the epidemiological findings on the importance of fiber in the diet. In these studies, the effect of reduced fiber cannot be isolated from the effects of such accompanying dietary changes as increased intake of animal fats and refined sugar. But, like them, fiber will probably have to be monitored to ensure the healthful aspects of the prudent diet. [Paul E. Araujo]

See also PUBLIC HEALTH.

Oceanography

The Deep Sea Drilling Project (DSDP) continued to explore the composition of the ocean floor in 1974. During the summer, DSDP research ship *Glomar Challenger* drilled into the volcanic layer of the sea floor near the Mid-Atlantic Ridge, an underwater mountain range that bisects the ocean basin. The drill site was near the dive area of the French-American Mid-Ocean Undersea Study (FAMOUS). See JOURNEY TO THE BIRTHPLACE OF CONTINENTS.

The DSDP researchers drilled a record 583 meters (1,913 feet) into the volcanic layer to take core samples. Analysis of the cores showed that the upper part of the volcanic layer is basalt rock with alternating layers of deep-sea sediments. Scientists believe the sediment was deposited between periods of volcanic activity.

Oceans and ice ages. By dating and analyzing rock and sediment cored from the floor of the Norwegian-Greenland sea, the DSDP scientists discovered that Greenland had separated from Norway about 55 million years ago. Further analysis showed the underwater Jan Mayen Ridge off the Norwegian coast was torn off Greenland 25 million years ago. They also found there was a land bridge, formed about 50 million years ago by volcanic action, that had connected Iceland with the British Isles. The bridge, which sank 20 million years ago, is part of a submerged formation that controls the beginning and end of ice ages.

Oceanographers and geologists theorize that as the sea level rises, warm Atlantic water flows over the formation into the Norwegian and Arctic seas and melts the polar icecaps. This causes continuous fog and heavy snowfalls. The snowfall on land gradually builds up huge glaciers. However, this process of taking moisture from the ocean lowers the sea level until warm Atlantic waters can no longer cross the underwater formation in sufficient quantities to keep the Arctic Sea from freezing. Then the tremendous precipitation stops, and the glaciers melt. As a result, the sea level gradually rises until the Atlantic waters can again move into the Arctic, and the cycle is repeated.

In October, the *Glomar Challenger* investigated the western part of the South Atlantic Ocean from the mouth of the Amazon River to Argentina. The scientists discovered that the São Paulo Plateau and the Rio Grande Rise, submerged formations off Rio de Janeiro, were once land masses up to 2,000 meters (6,560 feet) high that sank beneath the surface about 80 million years ago. This caused major changes in prehistoric ocean currents. Drilling 900 meters (2,950 feet) into ocean-floor sediments near the mouth of the Amazon, the researchers found that the massive river system began flowing about 10-million years ago.

Ocean-mining impact research. A group of scientists, led by geologist Oswald A. Roels of Columbia University's Lamont-Doherty Geological Observatory, studied the ecology of the Pacific Ocean floor within a 52,000-square-kilometer (20,000-square-mile) area southeast of Hawaii to determine the probable environmental impact of deep-sea manganese-nodule mining. The researchers sampled salt content,

water temperature, dissolved oxygen, sediment, phytoplankton, or tiny marine plants, nutrients, and trace metals. They gathered more than a ton of manganese nodules, measured deep-sea currents with special submerged meters, and took more than 2,000 photographs with an underwater camera.

Deep-sea mining activities will bring up large quantities of bottom water and sediments. To determine the effects of this on organisms that dwell near the surface, the scientists ran tests in aquariums. They found that phytoplankton developed better when exposed to deepwater nutrients and sediment.

The scientists also discovered phytoplankton spores that they knew must be very old because of the depth at which they were buried in the cores of sediment. When exposed to sunlight, these spores bloomed to life. The scientists kept cultures of them for later identification and analysis. They also planned to investigate the possible environmental impact of resurrecting the spores.

Photographs showed few bottom-dwelling animals. Large animal life,

A diver drills a core from coral off the coast of Florida for a U.S. government study of coral rings. X rays of the cores reveal that coral rings, like tree rings, are records of changes in climate. The wide, dark rings mark the years when the water was unusually cold.

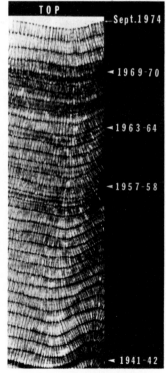

TOP
Sept. 1974
1969-70
1963-64
1957-58
1941-42

Oceanography

such as sea cucumbers, anemones, sponges, and corals, was as sparse as one animal per 100 square meters (1,076 square feet).

Ocean engineering. In April, 1974, the National Oceanic and Atmospheric Administration's National Sea Grant Program announced the development of an inexpensive, submerged breakwater system. Developed at Scripps Institution of Oceanography in San Diego, the system can calm waves even in mid-ocean. It can be used to create inexpensive harbors along coastlines and help prevent beach erosion.

The breakwater consists of submerged spherical floats anchored to the sea floor. Swinging like upside-down pendulums, they absorb the energy of passing ocean waves. Construction began in early 1975 on a test model for California's Channel Islands. Twenty-five rows, each containing about 50 hollow steel floats 1.5 meters (5 feet) in diameter, will be anchored to the sea floor. By increasing the number of rows, the size of waves can be reduced proportionally.

The cost of constructing a rock or cement breakwater in depths of more than 15.2 meters (50 feet) is prohibitive. With the pendulum system, the only added expense is more anchoring cable. Also, the anchored breakwater does not interfere with cleansing of a bay or harbor as do solid types, which reduce normal tides and currents.

Superwaves. Oceanographer Walter O. Duing of the University of Miami's Rosenstiel School of Marine and Atmospheric Science in the summer of 1974 led an expedition in which several research ships monitored the meandering of the equatorial undercurrent. The study was part of the Atlantic Tropical Experiment of the Global Atmospheric Research Program.

The equatorial undercurrent is an underwater river that begins somewhere off Brazil and flows rapidly toward Africa's Gulf of Guinea. It lies from 50 to 150 meters (164 to 492 feet) below the surface and is only about 100 meters (328 feet) deep and 150 kilometers (93 miles) wide. Above it, the equatorial surface current flows in the opposite direction. The researchers discovered that the current meanders back and forth between 1° south latitude and 1° north latitude during a one-week period. The meandering motion is apparently caused by enormous waves, 2,400 kilometers (1,500 miles) long, flowing westward.

Sea-grass transplants. Sea grasses are a vital part of shallow-water marine ecosystems. Many human activities in coastal zones have killed or displaced marine grass beds.

Marine botanist Anitra Thorhaug of the Rosenstiel School reported in May that she successfully transplanted one variety, turtle grass (*Thalassia testudinum*), to the bottom of Miami's Biscayne Bay, where effluent from a power plant had killed the original grasses. She obtained seeds from Bahamian turtle grass, treated them with root-growth hormone, and suspended them in a culture of running seawater until they sprouted. Then divers planted the seedlings, which took root and grew quickly.

Turtle grass is important in stabilizing sediments and providing a habitat for marine organisms. Transplantation of turtle grass can lead to rapid recovery of areas blighted by silt, chemicals, sewage, or channel construction.

Marine pollution. The National Academy of Sciences in January, 1975, published a comprehensive analysis of the environmental effects of petroleum in marine environments. The report indicates that about 6.6 million short tons (6 million metric tons) of petroleum enters the oceans each year, most of it from oil tankers, loading terminals, and river and urban run-off.

Most spilled petroleum follows a set pattern of changes. Some is oxidized by chemical and biological action to carbon dioxide, some evaporates, and the rest forms tar lumps. Microorganisms consume the least poisonous parts in a few days or months. Highly toxic parts break down more slowly.

Fish and other marine animals take up petroleum products. Scientists have found petroleum hydrocarbons in the bodies of marine organisms, from clean as well as polluted areas, in amounts ranging from 1 to 400 parts per million. But fish and lobsters can rid their bodies of most petroleum hydrocarbons within two weeks, and petroleum hydrocarbons do not become concentrated in marine life high up in the food chain.

Oceanography

Continued

A variety of undersea vessels were used by diver-scientists of the Scientific Cooperative Operational Research Expedition to study a coral reef near the Bahamas. The vehicles provide undersea labs, house oceanographic equipment, and permit exploration of the reef's vertical walls.

The most damaging effects of petroleum spillage are the oiling and tarring of beaches, the death of sea birds, and the disruption of shoreline marine animal communities.

Fish are generally less susceptible to petroleum poisoning than sea birds or bottom-dwelling animals. If continuously exposed to petroleum, however, the fish flesh may acquire an unpleasant oily flavor. But this apparently poses no imminent threat to human health. The report states that concentrations of cancer-producing substances in sea foods are no greater than in foods grown or raised on land.

Using detergents to disperse oil at the surface is better than using sand to sink the oil to the sea floor. According to the report, experiments in the North Sea showed that sinking oil with sand tainted fish and shellfish for several weeks.

Man in the sea. Duke University researchers in January reported overcoming two serious deep-diving problems by developing new breathing-gas mixtures and improved decompression timetables. Previously, dives to 300 meters (1,000 feet) or deeper were impractical because it took 24 hours to get a diver down and 14 days to retrieve him. The new method can get a diver to 1,000 feet in only 20 minutes and decompress him in only 4 days.

Instead of pulling a diver up rapidly about halfway to the surface and then slowly pulling him up to allow gas bubbles in his blood to disperse, the new method slowly raises him to the surface at a uniform rate. This prevents gas bubbles from forming as the pressure nearer the surface decreases.

Tests at Duke University Medical Center also showed that by adding 18 per cent nitrogen to the helium-oxygen gas mixture normally breathed by divers, scientists could eliminate the dangerous high pressure nervous syndrome. Nitrogen counteracts the symptoms of the syndrome, which occur during rapid descent to 150 meters (500 feet) or more—nausea, dizziness, trembling, and lapses of consciousness.

The new mixtures and timetables were successfully tested in December, 1974. [Richard H. Chesher]

Physics

Atomic and Molecular Physics. Various research groups reported breakthroughs in 1974 and 1975 in the use of lasers to separate isotopes of different elements, including uranium. The new laser processing methods promise to lower costs and make isotopes of many important elements much more widely available. See CHEMISTRY (Chemical Dynamics), Close-Up.

X-ray lasers. French researchers reported a significant first step in the development of X-ray lasers in October, 1974. Physicist Pierre Jaeglé and his co-workers at the University of Paris Physical Chemistry Laboratory located in Orsay, France, found evidence that triply ionized aluminum (Al^{3+}) – atoms from which the three outer electrons have been stripped – store energy that can be released in a burst of X rays.

Their study is based on the fact that electrons can occupy only certain energy states in atoms or ions. An electron can move from a low- to a high-energy state by absorbing the fixed difference in energy from radiation hitting the atom. Normally, the electron spontaneously returns to a lower energy level by emitting a photon having precisely the energy difference between its new state and the excited one.

Lasers work by exciting more electrons than normal to certain high-energy states and orchestrating their transitions to certain lower ones with a stimulating wave of radiation having just the right energy. As the wave moves from atom to atom, other electrons change states, adding their energy to the intensifying laser beam.

The Orsay physicists generated a hot aluminum plasma – a mixture of Al^{3+} ions and stripped electrons – by focusing a neodymium laser with 20 joules of radiation output on an aluminum target. Many of the Al^{3+} ions emitted X-ray radiation as electrons changed states. Other ions absorbed the X-ray radiation by boosting their electrons to higher energy states.

To observe the absorption process in greater detail, the researchers split the neodymium laser beam and focused the two beams on adjacent spots, creating two plasmas. They used one plasma as the source of X-ray radiation that the second plasma absorbed. By comparing the intensity of radiation emerging from the second plasma to that entering it, the group measured the ease with which Al^{3+} ions absorbed and radiated X-ray energy during each of the various energy-state transitions.

In most cases, the X-ray radiation given off was less intense than that absorbed. For radiation with a wave length of 117.41 angstroms (A), however, the X-ray beam came out of the plasma nearly 17 per cent stronger than it went in, indicating stimulated emission from a highly filled high-energy state. Although much remains to be done before an X-ray laser is constructed, the Orsay group showed the potential for its development.

Electron-positron waltz. Brandeis University researchers reported in January, 1975, that positronium, a short-lived matter-antimatter atom, radiates light in a manner similar to ordinary atoms. Just as the simplest atom, hydrogen, consists of a single electron bound to a proton, physicists believe that positronium forms when an ordinary electron combines with its positive antiparticle, the positron. In less than about a millionth of a second, the matter-antimatter waltz ends in mutual annihilation in which the mass of the two particles is completely converted into energy in the form of a gamma ray.

Theoreticians believe that before this happens, positronium exists in several different energy states. An electron changing from a high- to a low-energy state in positronium should radiate light. The theorists calculated the wave length for the simplest transition radiation to be 2430 A, which is ultraviolet.

Physicists Karl F. Canter, Allen P. Mills, Jr., and Stephan Berko detected such radiation. They slowed down positrons produced by radioactive cobalt-58 by passing them through a film of gold and magnesium oxide. Guided by a magnetic field, the slow positrons struck a germanium target 4.9 feet (1.5 meters) away. Some of them captured electrons and formed positronium in the target.

The Brandeis scientists looked for the predicted preannihilation radiation at 2430 A with a photomultiplier tube that viewed the target successively through three filters. One filter passed only ultraviolet radiation at the predicted wave length while the others

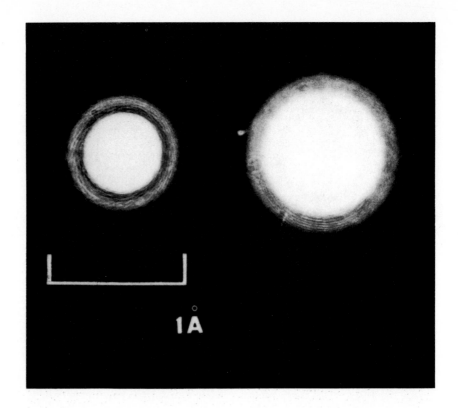

Photographs of the electron clouds that surround argon atoms were made with a holographic microscope that can resolve detail 100 times better than a conventional electron microscope can. Magnification is 200 million times. Dark rings are due to diffraction effects within the instrument.

1 A

Physics

Continued

passed light of slightly longer and shorter wave lengths. The physicists measured a significantly larger signal at 2430 A — convincing proof that the predicted transition was occurring.

Avogadro's number is up. Many of the laws of physics are expressed as equations containing such fundamental constants as the speed of light. Physicists continually strive to measure these constants with greater precision in order to test the validity of the currently accepted laws.

Richard D. Deslattes and eight other researchers at the National Bureau of Standards in Washington, D.C., determined a new and more precise value for another fundamental constant, Avogadro's number. This dimensionless number represents the number of molecules in a mole of any substance — an amount of that substance, measured in grams, numerically equal to its molecular weight. For example, the molecular weight of oxygen is 32. Hence a mole of oxygen has a mass of 32 grams and contains Avogadro's number of oxygen molecules — about 602 billion trillion.

To determine the new value for Avogadro's number, Deslattes and his colleagues counted the number of molecules in a portion of a nearly perfect silicon crystal they had sliced in two. They moved one piece parallel to the other very slowly and accurately while passing X rays through both. As rows of atoms moved past each other, the atoms in one piece shadowed those in the other.

By counting the passing shadows, the physicists tallied the number of atomic rows passed. At the same time, they measured the total relative distance the pieces moved with a laser interferometer. Knowing the number of crystal rows in a given distance, they could calculate the distance between rows and the number of atoms per unit volume. They also measured the mass density of the crystal, enabling them to calculate a new value for Avogadro's number — 602.20943 billion trillion molecules per mole. Good to one part per million, the measurement has 30 times less experimental error than earlier direct measurements.   [Karl G. Kessler]

Elementary Particles. The big news in 1974 and 1975 was the discovery of three new unpredicted particles (see A SUBATOMIC SURPRISE). Otherwise, the year might be aptly named the "year of the quark." New experimental results, accumulated over the preceding three years, grew to a flood and completed the dethroning of the proton and neutron as elementary particles, along with their 300 or so relatives in the hadron family. Hadrons respond to the strong force – the force that holds protons and neutrons together in atomic nuclei. Electrons, however, are still regarded as elementary, pointlike objects.

The most striking evidence of more fundamental particles comes from studies of collisions of electrons with protons and neutrons at the Stanford Linear Accelerator (SLAC) in Palo Alto, Calif. The new data leaves little doubt that hadrons are composed of much smaller objects, usually called quarks, and their antiparticles, antiquarks.

Until recently, physicists were wary of taking quarks too literally. Despite intensive searches for them, no quarks have yet been found in a free state, uncombined with other quarks. Moreover, most probing experiments were too crude to reveal small-scale structure inside the proton and other hadrons. Physicists use the collisions of particles as their principal tool for determining particle structure.

In a collision between subatomic particles, the most significant clue is the amount of momentum transferred from one particle to the other. The more momentum transferred, the finer the detail that can be observed.

The SLAC experiments that use electrons as projectiles are easy to interpret because the electron always survives the collision. Its precollision direction and momentum is known and the momentum transfer can be measured by observing its direction and momentum after the collision. Spectrometers for this purpose, built by a team headed by physicist Charles Sinclair, have been showing high-momentum transfers in collisions since 1971.

By 1974, enough data was in hand to make it clear that all of the electric charge inside a proton or neutron is concentrated in three tiny centers that, in addition to charge, have other prop-

erties such as spin, exactly as predicted by the quark model.

The most recent SLAC result was reported in April, 1975, by a team from American University in Washington, D.C., headed by Benson T. Chertok. Using Sinclair's spectrometer, they studied what happens when an electron hits a deuteron, a nucleus consisting of a proton and neutron loosely bound together. Their results showed that, when probed by a sufficiently energetic electron, a deuteron appears to consist of six small, independent, electrically charged objects.

Further clues come from the study of collisions of neutrinos with nuclei. Neutrinos are related to electrons, but they have no mass when at rest. Because they are electrically neutral, neutrinos do not interact with other forms of matter via the electromagnetic force. They interact only through the weak force, which is 10 billion billion times weaker than electromagnetism.

Experiments that use neutrino probes are nearly as simple as are the SLAC electron experiments. When the neutrino hits a proton or neutron, it is converted to a muon, another variety of heavy electron. A part of the neutrino's energy goes into creating hadrons. If both the muon and hadrons are observed, the momentum of the incoming neutrino can be estimated, and comparison with the muon gives the momentum transfer.

In 1974, a group at the European Center for Nuclear Research (CERN) in Geneva, Switzerland, and two other groups at the Fermi National Accelerator Laboratory (Fermilab) in Batavia, Ill., conducted neutrino probe experiments. The multinational CERN team of nearly 100 researchers observed the collision products by the tracks they left in a large bubble chamber named *Gargamelle*, built under the direction of the late André D. Lagarrigue.

The Fermilab groups used spark chambers to track the neutrino collision particles. These detectors reveal less detail in each collision but collisions occur more frequently. One team, headed by Barry C. Barish, was a joint Fermilab project with researchers from the California Institute of Technology in Pasadena. The other was a team with members from the universities of Wisconsin, Pennsylvania, and Harvard and headed by David B. Cline, Alfred K. Mann, and Carlo C. Rubbia.

High-momentum transfers in collisions between hadrons were also observed in more than a dozen other Fermilab experiments with a variety of projectiles and targets. These confirmed results from proton-proton head-on collisions observed earlier at CERN. In hadron-hadron collisions, however, the evidence is indirect because a high-momentum transfer collision usually severely disrupts both particles. Eventually, some portion of the momentum may be transmitted to a single new hadron or group of hadrons emerging from the collision, but it is difficult to trace back to the original encounter that produced the transfer. Nevertheless, the very existence of high-momentum transfers in hadron-hadron collisions indicates that something smaller than a hadron is present.

Quark spoken here. Taken as a whole, evidence from CERN, Fermilab, and SLAC indicates that hadrons are built up from constituents at least

100 times smaller than a proton, and this has changed the language of particle physics. For several years, groups of physicists known as phenomenologists have analyzed data about hadrons without referring directly to any particular picture of their internal structure. Their calculations are now taken seriously as measures of such properties as the number of quarks present or their distribution inside a hadron. The new experimental results have encouraged vigorous efforts by theorists to clear up the inconsistencies in the quark theory.

Foremost among these loose ends is an understanding of the force that holds quarks together in hadrons. The only hadrons the force produces are those containing three quarks, known as baryons, and those containing a quark and an antiquark, known as mesons.

The fact that a free quark has yet to be found is a puzzle that is all the more baffling because it is rather easy to push quarks around inside a hadron. Hundreds of billion electron volts (GeV) have been poured into hadrons without knocking a quark free. But it takes only

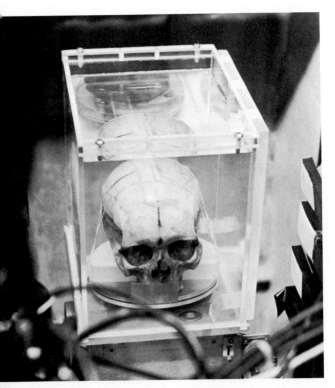

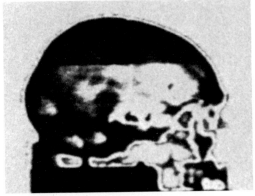

Charley, *left,* a skull housing a human brain with a blood clot in its right temporal lobe, sits in a lucite tank of water at Argonne National Laboratory. Charley's radiograph, *above,* was generated by a computer with data from firing beams of energetic protons through the skull. This process may enable physicians to spot clots at lower exposures than X rays.

Physics

Continued

a fraction of a GeV to dislodge a quark from its normal motion and start it moving differently in the hadron.

This tenacious grip may be contrasted with that of the electromagnetic force that holds electrons in their orbits around the nuclei of atoms. It takes only 33 per cent more energy to remove an electron from a hydrogen atom than it takes to lift it out of its normal orbit to the next largest one. Theorists are trying to find a simple, natural picture of a force that does not permit this for a quark in a hadron.

No new facilities. American particle physicists will have to use weapons already in their arsenal to attack these problems. They have proposed several new research facilities, but funds for most new research tools were not approved during 1975.

Physicists are particularly interested in such colliding-beam machines as the Stanford Positron Electron Annihilation Ring (SPEAR) at SLAC. When a speeding energetic particle hits a stationary target, it can transfer only a small fraction of its momentum. This is because the moving particle becomes heavier as it gains energy and a heavy object cannot transfer much of its momentum to a light one. For example, a hammer cannot bend a light piece of metal without the aid of an anvil.

Colliding-beam machines such as SPEAR, which store particles racing in opposite directions in a closed course, solve this problem by producing head-on collisions in which both particles have increased mass and energy.

The Energy Research and Development Administration (ERDA), which funds nearly all of the major high-energy physics laboratories in the United States, has deferred all plans for new facilities except for one to be built as an adjunct to SPEAR at SLAC. The proposed proton-electron-positron (PEP) machine will store electrons moving in one direction and protons or positrons moving in the other.

Funds for the final design of PEP were cut from the fiscal 1976 budget of ERDA. Because the design and construction of such a project takes at least five years, this ensures that no new particle research facilities will be available in the United States until at least the 1980s. [Robert H. March]

Nuclear Physics. Georgii N. Flerov of the Joint Institutes for Nuclear Research in Dubna, Russia, announced in June, 1974, at the International Conference on Heavy Ion Science in Nashville, Tenn., that his research group had synthesized element 106, an atom that has 106 protons in its nucleus. In September, Albert Ghiorso and his co-workers of the Lawrence Berkeley Laboratory (LBL) in California reported their synthesis of element 106.

The Russian experimenters bombarded lead targets with energetic beams of chromium and molybdenum nuclei and studied the spontaneous fission of the new element. The LBL team used very heavy, radioactive targets such as californium 249 and lighter projectiles such as oxygen. The LBL procedures led to the formation of different isotopes from those studied at Dubna, that is, the nuclei contained different numbers of neutrons.

Disagreements arose concerning the experimental results. The Russians and an American group from the Holifield National Laboratory in Oak Ridge, Tenn., subsequently negotiated a joint experiment, to be conducted in 1976, to settle the differences. Curtis Bemis and his team at Holifield will equip a trailer truck with detectors, computers, and other instruments, then take it to Dubna. Coupled to projectile beams from the Russian accelerators, the equipment will permit a collaborative study of element 106 and possibly heavier new elements. See CHEMISTRY (Chemical Synthesis).

New nuclear reaction. Using the LBL SuperHILAC accelerator, John R. Huizenga and his colleagues at Rochester University in New York reported evidence of a new kind of nuclear reaction in 1974. Independently and somewhat earlier, a Russian team led by Vadim V. Volkov used the Dubna cyclotron to study similar reactions.

When heavy nuclear projectiles and targets collide, they usually either graze one another and exchange a few neutrons or protons, or they hit almost head-on and fuse to form a compound nucleus. But Huizenga and Volkov discovered heavy-ion collisions in which the two nuclei rub together and form a viscous neck that holds them together while they spin many times like a tum-

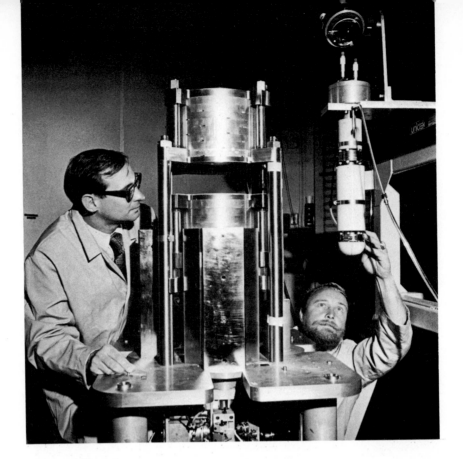

A carbon monoxide laser, right, was excited by atoms from the fission of uranium 235 hit by neutrons from a reactor, center. The process may allow large amounts of energy to be pumped into large lasers.

Physics

Continued

bling dumbbell. The nuclei exchange many neutrons and protons, then fly apart because of their electrostatic repulsion, just as in normal fission. But there is no complete fusion.

Although interesting, this process is disappointing to nuclear physicists who hope to synthesize supertransuranic nuclei—those containing 110 to 114 protons—that could last for tens to thousands of years instead of for a few thousandths of a second like element 106. If the new reaction took place in all the collisions, no supertransuranic compound nuclei would form in collisions attempting to form them.

A more clean-cut example of two heavy nuclei colliding to form a molecular complex was reported in March, 1975, by Eric R. Cosman of the Massachusetts Institute of Technology (M.I.T.) and his colleagues at Brookhaven National Laboratory in Upton, N.Y., and at the Argonne National Laboratory near Chicago. They studied energetic magnesium nuclei produced from collisions of carbon nuclei. In its lowest energy state, a magnesium

nucleus has a stable football shape and spins end over end when given additional energy. Given more than 6-million electron volts (MeV) of energy, the nucleus normally decays spontaneously by emitting neutrons, protons, or alpha particles.

By giving magnesium nuclei more than 20 MeV in collisions, Cosman discovered what appears to be an entirely new family of nuclear shapes—dumbbells in which the magnesium nuclei had separated into two connected carbon nuclei. The dumbbell could be spun end over end to generate a series of new excited states.

Robert G. Stokstad and his collaborators at Yale University reported evidence in January, 1975, of even more complex high-energy nuclear molecular states in silicon nuclei. In them, two carbon nuclei bind together by exchanging an alpha particle. The work of Cosman and Stokstad opens a new field of both experimental and theoretical nuclear spectroscopy.

Meanwhile, Karl Erb and Robert Ascuitto, also of Yale, demonstrated in

1975 for the first time that quantitative results of complex nuclear interactions involving heavy nuclear projectiles can be predicted. They obtained new experimental evidence and developed new, more complete theoretical models for the reaction mechanisms that bridge the gap between models of nuclear reactions and dynamics and models of microscopic nuclear structure.

Bevalac born. In August, 1974, the LBL SuperHILAC first injected nuclei into the Bevatron, the LBL proton synchrotron. The Bevatron was coupled to the SuperHILAC in a technological *tour de force* by bringing the SuperHILAC beam out of its laboratory, across a parking lot, down a 100-foot (30-meter) hillside, over a superhighway, and into the Bevatron laboratory. After acceleration in the system — now called the Bevalac — carbon, nitrogen, oxygen, and neon nuclei had energies of over 2 billion eV per nucleon.

In preliminary measurements with the Bevalac, Erwin Schopper and his colleagues from the University of Frankfurt in West Germany found evidence of shock waves within silver nuclei when Bevalac's oxygen projectiles plow through them like the shock waves that aircraft create in air when exceeding the speed of sound. Walter Greiner at the Institute for Theoretical Physics in Frankfurt analyzed the data and found the speed of sound in nuclear matter to be about one-tenth the speed of light. The measure of the compressibility of nuclear matter that follows from this is of fundamental importance not only to nuclear and subnuclear scientists but also to astrophysicists studying the extremely dense stars believed to form following the gravitational collapse of normal stars.

Preliminary nuclear experiments were conducted in 1974 at four new intermediate-energy nuclear facilities — the Los Alamos Scientific Laboratory's Meson Physics Facility in New Mexico; the TRIUMF negative hydrogen ion cyclotron in Vancouver, Canada; the similar SIN cyclotron in Zurich, Switzerland; and the Bates Laboratory Electron Linear Accelerator at M.I.T. These machines use secondary beams of mesons as well as protons and electrons to probe the detailed internal structures of nuclei.   [D. Allan Bromley]

Plasma Physics. The study of plasma for controlled thermonuclear fusion reactors continued in 1974 and 1975 along three parallel paths: magnetic confinement of plasma, laser pellet fusion, and electron beam pellet fusion.

Magnetic confinement. Researchers in the United States, in Russia, and in Europe are gearing up to achieve break-even in their quest for fusion power. They are seeking to build a laboratory magnet device that confines and heats its fuel — a tritium-deuterium plasma — to produce more power from nuclear fusion of the fuel than it uses to generate the magnetic field and operate other equipment. Tritium and deuterium are heavy isotopes of hydrogen.

To achieve break-even, the device must meet the Lawson criterion, which states that the product of the plasma density (expressed in ions per cubic centimeter) times the length of time the magnetic field can compress the plasma (in seconds) must exceed 10^{14}. Existing doughnut-shaped devices, called tokamaks, have reached only 3×10^{12}.

Four large new tokamaks are under construction, at least two of which are expected to achieve break-even. Two are planned for Princeton University's Plasma Physics Laboratory. They are the Princeton Large Torus (PLT), which will be completed in 1976 or 1977, and the Tokamak Fusion Test Reactor (TFTR), scheduled to be in operation by 1980 or 1981. PLT is a larger version of various existing tokamaks, designed to reach within a factor of two of break-even.

The TFTR is expected to achieve break-even. It will test the principle of the two-component torus in which a high-energy deuterium ion beam striking a heated tritium plasma produces more energy through nuclear fusion before it slows down than was needed to generate the beam. In principle, such a device can circumvent the Lawson criterion and achieve break-even about 10 times easier. It would also have the advantage of making possible a far smaller fusion reactor that would generate hundreds instead of thousands of megawatts (Mw) of electric power.

The Joint European Torus (JET), a large tokamak expected to reach break-even, is scheduled to start operating by 1980 or 1981 at an as-yet-

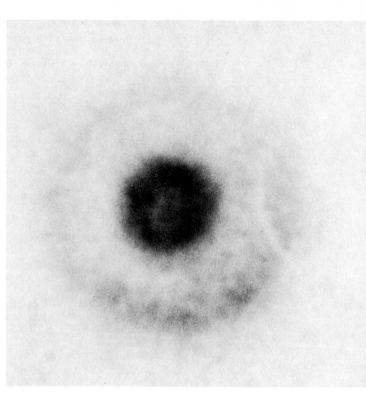

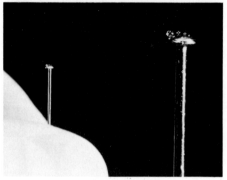

Physics

Continued

Microballoons, hollow glass spheres smaller than a pinhead, *above,* are filled with gaseous thermonuclear fuel and mounted under a microscope, *top left.* An X-ray image, *above right,* shows a shell (outer ring) as laser light hits it. The heated shell then implodes, compressing the fuel (inner dark region) and starting a fusion reaction.

undetermined European location. Unlike conventional tokamaks, it has a very fat plasma torus with a D-shaped cross section instead of a circular one. These features reduce the magnetic-field strength needed to confine a given amount of plasma and thereby greatly trim costs. Plasma currents caused by the confining field in JET will reach about 6 million amperes – one-third the amount needed for a reactor. To boost plasma temperature to the fusion ignition point, about 100 million°C, a beam of neutral atoms will be injected at a beam power of 1 or 2 Mw to begin with, eventually reaching 10 to 20 Mw.

A large Russian tokamak, T-10, being built at the Kurchatov Institute in Moscow is expected to be in operation in 1976. It is similar to PLT, but is not expected to attain break-even.

Laser pellet fusion. Researchers reported important advances in this approach to controlled thermonuclear fusion. In laser pellet fusion, powerful laser beams from many directions are focused on a tiny pellet of deuterium and tritium. The laser pulse heats and

vaporizes the pellet, ionizing the fuel to create a plasma. The plasma implodes violently, ideally compressing to a density 1,000 times that of solid hydrogen, and heats to the fusion ignition point.

Details of the pellet target were made public for the first time in October, 1974. The pellets are microballoons – tiny hollow glass spheres filled with a gaseous fuel mixture at a pressure 50 to 100 times atmospheric pressure. The thin glass shell is coated with a plastic material to absorb the laser light better.

The microballoon targets were developed at the Los Alamos Scientific Laboratory (LASL) in New Mexico, the Lawrence Livermore Laboratory (LLL) in California, and by KMS Fusion, Incorporated, in Ann Arbor, Mich. When fired on, much of the laser energy is absorbed and converted into kinetic energy of the imploding shell, which becomes an efficient piston for compressing the gaseous fuel inside. Initial experiments by Roy R. Johnson at KMS, Gene H. McCall at LASL, and John F. Holzrichter at LLL have

Physics

Continued

shown this compression and detected neutrons that they believe came from a fusion reaction.

More powerful lasers are needed now to reach break-even. Estimates as to precisely how powerful vary, but most researchers agree that the pulse must last less than 1 billionth of a second – 1 nanosecond (nsec) – and have an energy of from 10 kilojoules (10 kJ) to as much as 1,000 kJ.

Currently, the closest approaches to this goal have been attained by Moshe J. Lubin at the University of Rochester's Laboratory for Laser Energetics (1 kJ lasting 0.1 nsec); by Philip J. Mallozzi at the Battelle Memorial Institute in Columbus, Ohio (1.5 kJ lasting 3.5 nsec); and by Sidney Singer at LASL (1.25 kJ lasting 1 nsec). Very large sub-nanosecond laser systems are under construction at LASL (10 kJ), LLL (10 kJ), and the Kurchatov Institute (40 kJ). They will use glass rods and disks, which cannot endure the stresses when the power is eventually boosted.

Gas lasers appear to be the only feasible alternative. The CO_2 laser is the only gas laser powerful enough, but its wave length is too long. Thus, one more breakthrough is needed – a powerful, short-pulse, short-wave-length, gaseous laser.

Electron beams. The third approach to fusion power has two great advantages over lasers. Beam generators with enough power already exist, and the targets absorb up to 75 per cent of the beam energy. They also have two disadvantages – the difficulty of focusing such a beam onto the small pellet, and of shortening the pulse duration.

A team of 50 researchers led by Gerold Yonas at Sandia Laboratories in Albuquerque, N. Mex., began testing Proto 1 in May, 1975. This is the first high-power, short-pulse electron beam accelerator expressly designed to irradiate fuel pellets. The machine will enable them to test the concept of self-pinching, whereby electrons from two facing hollow cathodes 12 inches (30.5 centimeters) in diameter are pinched down to the BB size of the pellet by their own magnetic field. Operated at a peak output of 3 million volts, the beams will have a maximum combined power of about 2.5 million Mw and last about 25 nsec.   [Ernest P. Gray]

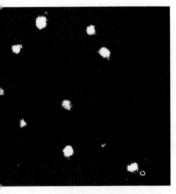

Whirlpools, viewed end-on in a small, rotating container of superfluid liquid helium, offer direct visual evidence that motion in superfluids is quantized similar to energy states in atoms and molecules.

Solid State Physics. Experimenting at temperatures near absolute zero, solid state physicists in 1974 and 1975 visualized the effects of quantum mechanics by photographing for the first time whirlpools in superfluid helium and electron-hole drops in germanium crystals. Also during the year, theorists used a computer to accurately model water, and an experimental team discovered superconductivity in a polymer.

Quantum whirlpools. Gary A. Williams and Richard E. Packard of the University of California at Berkeley reported in July, 1974, that they had photographed arrays of vortex lines (whirlpools) in a thimble-sized rotating container of liquid helium chilled to within a fraction of a degree of absolute zero. At this temperature the liquid helium becomes a superfluid, able to flow without resistance through small openings. The photographs provide direct visual evidence that the flow of fluid around the center of a vortex line in a superfluid is quantized in much the same way that the motions of electrons in atoms are limited to certain energy and momentum states.

This quantum effect was first suggested independently in the 1950s by two Nobel prizewinning physicists, Lars Onsager of Yale University and Richard P. Feynman of the California Institute of Technology in Pasadena. Onsager and Feynman predicted that the number of vortexes in the macroscopic rotating superfluid would be proportional to the speed of rotation and inversely proportional to Planck's constant, an important fundamental constant, like the speed of light, in the world of atoms.

In order to see the whirlpools, the researchers inject electrons into the superfluid to create negatively charged bubbles that are about 32 angstroms (A) in diameter. The bubbles are sucked into the vortex cores by the Bernoulli effect, the same force that sucks dust into a tornado. After about 10 seconds, the scientists turn on an electric field parallel to the vortex axes, drawing the electrons to and through the top surface of the rotating liquid and accelerating them into a phosphorescent plate. A fiber-optics light pipe carries the weak light signals from the plate to a room-temperature image intensifier, an electronic device that amplifies the images 100,000 times so that they can be photographed on high-speed film.

The first attempts to photograph the whirlpools yielded only a blur. Reasoning that the vortex lines might be moving around faster than their image was collected, Williams and Packard added less than 1 per cent of the rare isotope helium 3, a normal fluid, to the superfluid helium 4. The helium 3 damped the vortex fluctuations but required the researchers to build a refrigerator capable of chilling the mixture to 0.18°F. (0.1°C) above absolute zero.

Theorists had predicted that the vortexes might form a triangular array, but the Berkeley researchers observed only random patterns. This may be caused by vibrations in the rotating refrigerator. The researchers hope to produce a moving picture of the vortexes. But photographing them is already a striking confirmation of the Onsager-Feynman theory and a dramatic example of quantum effects on a macroscopic scale.

Electron-hole drops. Berkeley physicists James P. Wolfe, Robert S. Markiewicz, Carson D. Jeffries, and Charles Kittel, and their Lawrence Berkeley Laboratory co-workers William L. Hansen and Eugene E. Haller reported in May, 1975, that they had photographed electron-hole drops in a pure crystal of germanium that they had cooled to −456°F. (−271°C). The metallike drops measure about 0.03 inch (0.08 centimeter) in diameter and each contains about 10 thousand billion mobile pairs of electrons and *holes* – the term solid state physicists use for vacant spots in a semiconductor crystal, such as germanium, where electrons are missing. Holes behave in a fashion opposite to electrons, as if they were actually positively charged particles.

Even though they are embedded in the lattice of a rigid, solid material, the overall uncharged electron-hole drops behave like a liquid. They have a surface tension like raindrops, and they can flow slowly throughout the rest of the crystal lattice. At higher temperatures, the drops boil away into dis-

Drawing by Porges; © 1974 The New Yorker Magazine, Inc.

"But I digress."

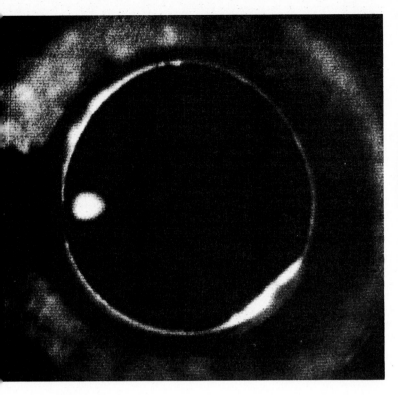

A bright spot, *left,* is the image of an electron-hole drop magnified 20X. The drop is a collection of trillions of electric charges that acts like a liquid in a crystal of germanium. A halo of infrared light, scattered from the crystal, surrounds the image, viewed on a television screen, *above.*

Physics

Continued

persed electrons, holes, and bound pairs known as excitons.

About seven years ago, Russian physicists inferred that excitons can form droplets in low-temperature germanium. They measured a sudden increase in electric conductivity while shining light on the crystal, but did not attempt to directly observe an image of the droplets.

To create and photograph a large electron-hole drop, the Berkeley-LBL physicists focus an intense green laser light on the surface of a germanium wafer about 0.16 inch (0.4 centimeter) in diameter that is held inside a plastic ring by a tiny plastic screw. With the assembly inside a container of liquid helium, they monitor the opposite side of the wafer with a television camera sensitive to infrared light.

The laser photons knock electrons free from their normal lattice sites, creating both free electrons and holes that combine to form neutral excitons and a fog of small electron-hole droplets. When the electrons rejoin the holes, they emit infrared photons.

Pressure applied by the screw causes enough stress at a point in the crystal for the droplets to coalesce and form a glowing drop large enough to be seen by the infrared camera. Amplified and converted to a visible image, the photograph reveals a large bright dot – the image of the drop – surrounded by a bright halo of infrared light from the drop that has been scattered from the circular crystal edge.

Computer water. Calculating the motion of an individual molecule in a liquid is difficult because the molecule is strongly influenced by the behavior of its neighbors. While this situation also exists in a solid, the molecules in a crystal form regular arrangements that simplify the calculations. However, Aneesur Rahman of Argonne National Laboratory near Chicago and Frank H. Stillinger, Jr., of Bell Telephone Laboratories in Murray Hill, N.J., recently used a technique known as molecular dynamics to calculate some properties of water.

The basic idea behind molecular dynamics is to use a large digital comput-

Physics
Continued

er to solve Newton's second law of motion for the individual motions of a small number of molecules that interact inside a cubical region. The number of molecules must be small, otherwise the cost of the computations becomes prohibitively large. For example, one might study 216 molecules, which would correspond to a cube 18.6 A on a side for the density of water. This is, of course, a very small number relative to the more than 30 billion trillion molecules in a gram of water.

Stillinger and Rahman had to develop a model law describing forces between water molecules that accurately mirrored the true forces and was also simple enough to allow detailed calculations. In addition to the motion of the center of mass of a water molecule, they had to allow for the rotational or tumbling motion of a molecule about its center of mass and for the relative orientation of two molecules.

By solving the combined equations of motion for the rotations and translations of the water molecules interacting through the model force law, the two scientists computed for the first time the manner in which individual water molecules move under the influence of their near neighbors. They were then able to calculate accurately the density and thermal expansion of water, and the velocity of sound in it. The scientists plan to study computer salt water next.

Superconducting polymer. Richard L. Greene and G. Brian Street of the International Business Machines Corporation Research Laboratory in San Jose, Calif., and Laurance J. Suter of Stanford University discovered superconductivity at about 0.5°F. (0.3°C) above absolute zero in the crystalline polymer polysulfur nitride. Below this temperature, the chainlike organic structure consisting of nonmetallic elements sulfur and nitrogen loses all resistance to the flow of electric current.

Reported in March, 1975, their findings represent the first discovery of superconductivity in a polymer. It may be the first of an entirely new class of metallike organic materials that could help scientists understand superconductivity better.   [John B. Ketterson]

Psychology

Psychological research published in January, 1975, shed new light on the relationship of family size and birth order to intelligence. Studies going back to the 1940s have shown that the brightest children tend to come from small families. Further, in any given family, those born early tend to be more intelligent than younger brothers and sisters. But researchers have been at a loss to explain these findings.

Robert B. Zajonc of the University of Michigan believes he has found an answer. With Greg Markus, he constructed a mathematical model that shows how the intellectual climate of a family changes with the birth of each child. They began by arbitrarily assigning each adult an intellectual level of 100, and giving a newborn infant a level of near-zero. (They were careful to warn against confusing these figures with IQ scores.) A couple without children would have an average intellectual environment of 100 (100 plus 100, divided by 2). Add a new baby, however, and the intellectual environment of the family drops to about 67, because the baby contributes nothing. If a second child is born in a year or two, the level drops still further. And, in most cases, it continues to drop with each successive child.

Zajonc explains that a child's intellectual growth is partly controlled by the overall intellectual climate of his home. Thus, children from large families, who spend more time surrounded by child-sized minds, develop slower intellectually than children from small families, who have more contact with adult minds.

One apparent flaw in the theory is that only children are not the highest scorers on intelligence tests. Zajonc gets around this by explaining that the only child shares a common disadvantage with the youngest child in any family – he has no younger brothers or sisters to teach. The chance to teach, Zajonc maintains, is an important boost to intellectual development. The best situation, according to Zajonc, is to be the older of two children.

The path of learning. James Olds of the California Institute of Technology

After learning to place features on a blank face, *above,* a chimpanzee was allowed to try on hats in front of a mirror, *top.* She then tried to change the face model to reproduce what she saw in the mirror. First, she turned over the mouth piece and used it as a hat, *above right.* Then, becoming more inventive, she used pieces of banana peel to make a hat without disturbing basic face features, *right.*

reported in May that he had tracked the earliest stages of learning in a mammal's brain. In his experiments with rats, Olds attached a skullcap with electrodes to the head of each rat. The electrodes penetrated the brain and recorded electrical activity. A plug on the outside was attached to equipment that recorded changes in electrical impulses from different parts of the brain during the rat's training.

Olds recorded the changes that occurred as a hungry rat learned to associate the sound of a tone with the arrival of food. The wire electrodes allowed him to trace the effects of the tone signal on different parts of the brain. Olds recorded the messages for the first 10 milliseconds (thousandths of a second), and drew successive maps that showed the changed responses along the pathway as the animal learned to associate the tone with food. After 10 milliseconds, the changes became so complex that he could not keep track of them.

However, during that brief period, Olds found that different parts of the

brain appear to learn at different times, and the reactions in different parts of the brain were correlated with changes in behavior. The responses that Olds measured are called unit responses—increases or decreases in the firing rate of nerve cells, as shown on an electronic readout. He found that training left the beginning of each response unchanged, but changed the peak and length of time at each point in the brain.

Changes appeared first in the hypothalamus—the part of the brain that controls temperature, hunger, and thirst—before the training had any effect on the rat's behavior. The reticular formation—a network of small nerve cells that controls sensory and motor activities—responded as the rat began to look toward the tone. The extrapyramidal motor system—which directs movement—showed changes while the rat learned to move toward the food. The sensory cortex—which perceives sensory impulses—responded as the rat's movements became smooth and fast. Only after the rat's movements were perfected did changes appear in

"The thing to bear in mind, gentlemen, is not just that Daisy has mastered a rudimentary sign language, but that she can link these signs together to express meaningful abstract concepts."

Psychology
Continued

the frontal cortex, which may be important in higher intellectual activities.

The song that distracts. People who hum, sing, or talk while doing work that requires delicate hand movements may be interfering with their own efficiency. Robert E. Hicks of the State University of New York at Albany reported in March that talking reduced a right-handed person's ability to balance a rod with the right hand. Some left-handed subjects had trouble balancing with the left hand while talking, while others had difficulty balancing with either hand.

The results corroborate studies by other psychologists, such as Roger Sperry of the California Institute of Technology and Brenda Milner of the Montreal Neurological Institute, that indicate that verbal abilities are controlled by the left side of the brain. The left side of the brain also controls the right hand. In right-handers, talking while using the right hand requires the left side of the brain to handle two competing tasks at the same time, and the balancing ability of the hand suf-

fers. However, talking had no effect on balancing with the left hand, which is controlled by the brain's right side.

The situation is more complicated for left-handers. Hicks found that in left-handers with a family history of left-handedness, talking interfered only with left-handed performance, indicating that in such persons speech control may be located in the right side of the brain. In other left-handers, speaking interfered with the balancing performance of both hands.

Shy company. Shyness may afflict more people than generally thought. Three Stanford University psychologists reported in May that 40 per cent of the 800 persons they interviewed considered themselves shy. Three-fourths of the shy ones did not like their predicament, and more than half expressed interest in going to a shyness clinic for therapy, if one existed. The researchers suggest that the prevalence of shyness may be a consequence of an overemphasis on competition, individual success, and personal responsibility for failure.   [Kathryn Sederberg]

Public Health

A report issued in February, 1975, by public health specialist Alan R. Dyer and his associates at Northwestern University Medical School in Chicago cast doubt on the belief that excessive body weight reduces life expectancy. This belief is so firmly rooted that extremely overweight persons have had to pay higher rates for life insurance. Scientists believed that obesity increases stress on the cardiovascular system, and broad population studies generally supported the thesis that any excess weight increases cardiovascular risk.

Dyer's group observed 1,233 white male employees of People's Gas Company in Chicago for 14 years. Initial weights of all the healthy men from 40 to 59 years old employed by the company in 1958 were available, and researchers tabulated mortality for the entire group through 1972. If the influence of other factors that increase the risk of cardiovascular disease were statistically eliminated, the study showed that the death rate was linked to weight, but in an unexpected way. Surprisingly, the men at the so-called

ideal weight level had the highest mortality rate. The next highest was among the heaviest men—more than 50 per cent heavier than recommended for their age and height. But those with the best survival rate were the moderately overweight, weighing 25 to 35 per cent more than the ideal suggested by the Metropolitan Life Insurance Company.

The researchers suggest that this evidence makes it necessary to re-evaluate the traditional view that there is a straight-line correlation between weight and health, with health deteriorating as weight rises. They also point out that these data apply only to healthy adults. Overweight persons who smoke, have high blood pressure, or have high fat levels in the blood might still run greater risk than slimmer persons, and weight reduction might help them significantly.

Diphtheria epidemic. Diphtheria cases have increased noticeably during the past six years, according to reports received by the National Center for Disease Control (CDC) in Atlanta, Ga. Until recently, however, these minor epidemics have been confined to non-immunized children living in Southern States. But public health officials in Seattle have uncovered hundreds of cases and carriers in the past 30 months. In fact, 155 of the 184 cases reported to CDC in the first five months of 1975 were from the Seattle area.

Diphtheria was a common, and often fatal, disease until immunization became widespread more than 50 years ago. This contagious disease, which occurs predominantly in young children, is caused by the bacillus *Corynebacterium diphtheriae* and is spread most commonly by coughing and sneezing or by contact with contaminated sheets or clothing.

The microorganisms develop in the nose and throat and produce a toxin that destroys cells and forms a yellowish-gray membrane which can obstruct the throat and choke victims to death. The toxin may also cause serious heart and nerve damage.

However, the Seattle epidemic shows a different clinical pattern. Although some children have developed throat infections, this new trend seems to be toward diphtherial skin infections among adults, of whom from 80 to 90 per cent are alcoholics. Most of the victims are migrant workers, derelicts, and indigent American Indians living in Seattle's skid row area. Thus the new disease pattern seems to be based on life styles. Poverty, crowding, and the poor personal hygiene that probably accounts for the disease spreading by skin infection are all factors in the epidemic. Poor diet, another poverty-associated problem, lowers immunity and also encourages transmission of the disease. Unlike the superficial sores noted in previous cases of diphtherial skin infection, these infections have sometimes spread to the throat, heart, and nervous system, and have caused several deaths.

Having identified the unusual nature of the epidemic and the vagrant character of the victims, the city of Seattle has organized an imaginative protection program that includes immunization clinics in neighborhood bars.

Industrial hazard. An unusually large number of cases of hepatic angio-

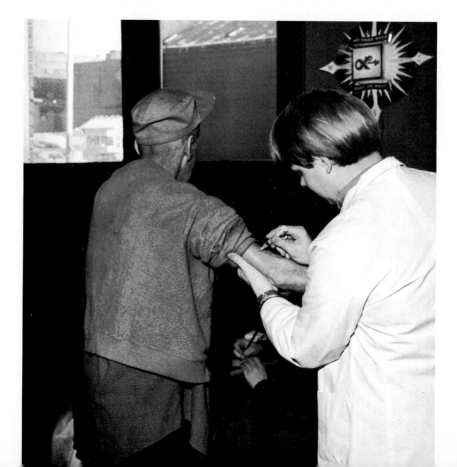

In a Seattle tavern, a public health worker inoculates a patron against diphtheria. The rare disease has broken out among the city's indigent alcoholics.

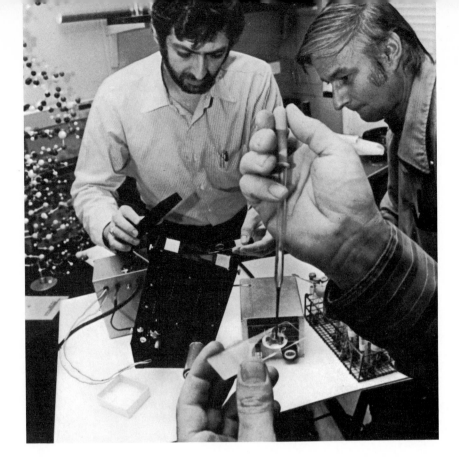

A new instrument to detect lead poisoning requires only a single drop of blood to measure lead levels in the body. Portable, the equipment can be used to test children in schools and urban neighborhood centers.

Public Health

Continued

sarcoma, an exceedingly rare form of liver cancer, have been found since 1974 in persons who had been exposed to the vinyl chloride gas used in making polyvinyl chloride, a common plastic for the past 40 years. The first clue that this might be harmful to the liver came in a 1965 report from Russia suggesting that plastics workers were likely to have scarring of the liver. Animal studies revealed in 1971 that tumors could occur after exposure to vinyl chloride, and the first report of human hepatic angiosarcoma after vinyl chloride exposure came in 1974. Reports on about 10 cases confirm that this hazard is associated with industrial exposure.

Further study in the United States has now revealed that workers exposed to the gas are more likely to develop scarring of the liver, which can lead to liver failure and even death. Liver scarring has also been shown to occur following exposure to another chemical called Fowler's solution, which has been widely used in the treatment of psoriasis, an inflammatory skin disease. In addition, there is evidence that

this kind of chemically induced liver scarring occurs much more commonly in certain geographic areas.

All of this supports the idea that environmental factors may cause liver scarring. Relating liver scarring to the development of cancer is still speculative. However, scarring by itself is very dangerous. Therefore, epidemiologists are trying to identify other possible liver-damaging chemicals.

Refugees' health. The collapse of South Vietnam and the flight of more than 120,000 refugees to Guam and Wake Island posed potentially serious public health problems. Officials feared that many of the Vietnamese might be carrying tropical diseases such as malaria and that crowding in the reception camps might add to the danger.

Medical examination and immunization centers were set up on both islands and at receiving camps in the United States. The refugees proved remarkably healthy. By June 1, 1975, only 90 cases of malaria had been found. There were nine cases of typhoid fever. [Michael H. Alderman]

Even though the United States economy was in a severe recession, overall support for science and technology maintained a steep upward trend in 1974 and 1975 as the government poured resources into programs for developing energy technologies.

However, relations between the scientific community and the federal government continued to be strained. Disagreements that had erupted during the Administration of President Richard M. Nixon remained unresolved under President Gerald R. Ford, who took office in August, 1974.

Prominent scientists argued that science policy was drifting aimlessly, and they urged Ford to reinstate a science policy office in the White House. Researchers in the nation's universities complained of declining support for basic research, and there was dissatisfaction with the Administration's strategy of targeting specific diseases, such as cancer, rather than allocating more funds for basic medical research.

A new and extremely important factor in biomedical research policies emerged during the year — ethical concerns over research involving human subjects. This led to some severe restrictions on the freedom of scientists to plan and conduct their experiments.

White House and science. Responding to agitation by scientific organizations, President Ford had Vice-President Nelson A. Rockefeller study the role of science advice in presidential decision making. As a result of Rockefeller's study, the Administration introduced a bill in Congress in June, 1975, to establish a small office of science and technology policy in the White House, headed by a science adviser to the President. A similar agency had been abolished by Nixon in 1973.

Energy agency reorganization. The event that had the greatest impact on support for science and technology in 1974 and 1975 was the Arab oil embargo of late 1973 and early 1974. The embargo led former President Nixon to launch Project Independence, a massive effort toward making the nation self-sufficient in energy. This project brought about phenomenal growth in budgets for energy research.

But even before planning for Project Independence got underway, it became evident that federal energy programs had to be reorganized. Their activities were scattered among various departments and agencies, making coordination and priority-setting a nightmare.

In October, 1974, Congress approved a sweeping reorganization of the government's energy bureaucracy. The reorganization abolished the Atomic Energy Commission (AEC) as of Jan. 19, 1975, and set up the Nuclear Regulatory Commission (NRC) to regulate and license the nuclear power industry, and the Energy Research and Development Administration (ERDA), which became responsible for all energy development programs and related conservation and environmental concerns. See ENERGY.

Not only did the reorganization group most federal energy programs into one agency, but it also removed the long-standing complaint that the AEC suffered from a conflict of interests because it both regulated and promoted nuclear energy.

William A. Anders was appointed chairman of the NRC. Robert C. Seamans, Jr., president of the National Academy of Engineering, became the first administrator of ERDA.

President Ford's budget, which was sent to Congress in February, proposed that $2.1 billion be spent on energy research and development and related environmental research during fiscal 1976 (July 1, 1975, to June 30, 1976). This was a significant increase over the estimated $1.6 billion spent in fiscal 1975 and $1.1 billion in fiscal 1974.

The budget proposed for ERDA included $279 million for research on coal technology, $57 million for solar energy, $28.3 million for geothermal energy, $32.1 million for conservation, and $120 million for nuclear fusion.

In spite of the energy-budget increase, President Ford's energy proposals ran into considerable criticism from some members of Congress, scientists, and the public for relying too heavily on nuclear power to meet the nation's energy needs by the 1990s. In his State of the Union message in January, 1975, Ford set a goal of bringing 200 more nuclear power stations into operation before 1985.

The Administration requested a total of $1.6 billion for nuclear energy dur-

Deceivers In The Laboratory

Scientists learn early in their training that plagiarism and fraud are absolutely forbidden. The penalty, they are told, is expulsion from science or exile to its outer margins. Yet, stories about fraud in science occasionally make headlines. In 1974, they made several.

In April, William T. Summerlin, an immunologist at the Sloan-Kettering Institute for Cancer Research in New York City, was suspended from his post for painting mice to make it appear that they had accepted skin grafts from genetically different tissue grown in cultures. In August, Walter Levy, Jr., director of the Institute for Parapsychology in Durham, N.C., was forced to resign after allegedly tampering with an experiment on the ability of rats to mentally control physical objects.

In November, *New Scientist* magazine reported that Indian nutritionist M. Swaminathan had published excessive — some said false — claims about irradiated wheat strains, without mentioning that other scientists had found his claims to be groundless. It was reported in December that Steven Rosenfeld, a student assistant at Harvard University, may have falsified research results on the transfer of immune reactions from one animal to another.

These cases raise a number of questions: Why the sudden rash of cases of scientific fraud? Is it something new? What drives presumably dedicated scientists to commit fraud?

One of the most famous hoaxes was probably the work of amateur paleontologist Charles Dawson, who claimed about 1911 that he had discovered bones of a 250,000-year-old man called Piltdown man in England. But dating techniques developed in the 1950s indicated that a much younger human skull and a modern ape's jawbone apparently had been planted at the site.

Another famous scandal involved Austrian biologist Paul Kammerer, who tried to prove that species pass on characteristics acquired in response to environmental changes. He forced midwife toads to breed in water rather than on land and in 1909 claimed that they developed nuptial pads — rough black areas on their forelimbs — typical of water-breeding toads. However, another scientist discovered in 1926 that black ink had been injected into the forelimbs of the only existing toad specimen from Kammerer's experiments. It is still a mystery whether it was Kammerer or one of his assistants that committed this fraud.

Deliberate falsification is one thing; unconscious deception, another. The history of science abounds with examples of such scientists as French physicist René-Prosper Blondlot, who in 1903 believed he had found a new type of radiation similar to X rays, which he called N rays. Other reputable scientists then claimed they also had observed the radiation. It took two years for the N-ray claims to be discredited.

There is no reliable way of finding out how often deception occurs. Some potential frauds never get into print because persons who monitor research for scientific publications recognize the invalid claims, usually assuming them to be error. Scientists are also reluctant to accuse colleagues of fraud. Findings that cannot be duplicated come from the best laboratories occasionally, resulting from contamination, lack of controls, or just plain error.

The line dividing acceptable from unacceptable practice is not sharp, and scientists disagree about such things as the grounds for separating relevant from irrelevant data. This involves individual judgment, and one scientist's selectivity is another's "fudge."

Fraudulent claims have been made for a variety of motives, including the quest for fame and fortune and religious or political zeal. But these are minor compared to the recognition and the research funds a scientist can win by making an original contribution.

Nevertheless, most scientists believe that dishonesty among them is rare. So, if 1974 was an unusual year and fraud is relatively infrequent, and if the chances of its being discovered increase with the significance of the claim, does it really matter that it occurs occasionally? It does. Even if all deceptions were uncovered, they would still burden scientists with false leads and waste their time and energy on detecting false claims. Equally important, fraud undermines public confidence in science and scientists' confidence in their colleagues' work. Honesty is not only a moral value in science, it is also a pragmatic one.   [Harriet Zuckerman]

Federal Spending for R & D

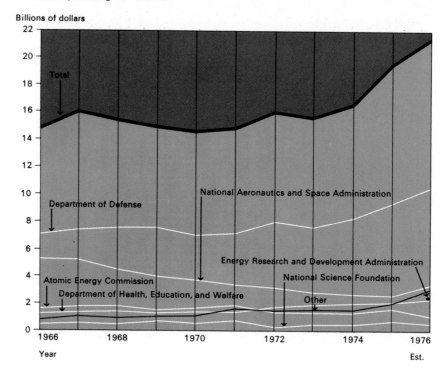

Billions of dollars

Science Support

Continued

ing fiscal 1976, with the Liquid Metal Fast Breeder Reactor (LMFBR) program to get $470 million of that amount. This would make it the most expensive energy research and development program in the budget. But the LMFBR project ran up politically embarrassing cost overruns. The total project is now expected to cost more than $10 billion, compared to a 1969 estimate of $3.3 billion.

There were other indications that nuclear energy is running into severe political opposition. The Joint Congressional Committee on Atomic Energy, which has consistently supported nuclear power programs, was weakened by the resignation from Congress of two of its most powerful members, Chet Holifield (D., Calif.) and Craig Hosmer (R., Calif.).

Four other committee members were defeated in congressional elections in November, 1974. At the same time, opponents of nuclear power in Congress became increasingly vocal over such issues as safety, the adequacy of safeguards against theft of nuclear ma-

terials, and the reliability of nuclear power plants.

Nonenergy support. The Ford Administration's budget proposed a massive spending increase for military research, which was slated to reach an estimated $10 billion in fiscal 1976, up from $8.6 billion actually spent in fiscal 1975. But the budget was relatively austere in other areas.

Ford proposed an increase of about 8 per cent for basic research, not even enough to keep pace with inflation. Federal research spending in colleges and universities was scheduled to increase by only about 2 per cent.

In keeping with Ford's policy of fighting inflation by holding down federal expenditures, no new federal projects were proposed in the budget in any area except for energy research and development. But the Administration approved the plan of the National Aeronautics and Space Administration (NASA) to build a third Earth Resources Technology Satellite and increased the NASA budget by $293 million to $3.4 billion for fiscal 1976.

Biomedical blues. The Administration's inflation-fighting strategies hit hardest in the area of biomedical research. On Jan. 30, 1975, President Ford asked Congress for authority to withhold $351 million of the $2 billion that had already been approved for the National Institutes of Health (NIH) for fiscal 1975. His proposed $1.75-billion NIH budget for fiscal 1976 would have reinstated only $72 million of that cut.

Even the cancer and the heart and lung research programs were slated for substantial cutbacks. But Congress refused to allow the cuts, and NIH supporters in Congress announced they would fight for increases in the Administration's proposed NIH budget.

The proposed cutbacks caused some consternation among biomedical scientists, but an event that took place a month earlier raised even louder protests and reopened a long-standing dispute between the Administration and the biomedical research community over how the NIH should be managed and how much freedom researchers should have. In December, 1974, Robert S. Stone, the director of NIH, was fired by Assistant Secretary for Health Charles C. Edwards. Edwards then resigned a week later. Although no reasons were given for the firing of Stone, many biomedical scientists assumed that he had been dismissed because he resisted Administration policies.

In protest, members of the Federation of American Scientists published a statement charging that Stone's removal was a political act representing unwarranted political interference in NIH activities. The statement's signers included three Nobel prizewinners — chemist Christian B. Anfinsen and physiologists Julius Axelrod and Marshall W. Nirenberg. It urged that scientists should be allowed to decide their own research priorities, and that biomedical research should not be run by managers in the Department of Health, Education, and Welfare (HEW).

Such protests have cropped up with alarming frequency in the past few years. Consequently, Congress passed a bill in September, 1974, establishing a seven-member panel to conduct a review of the Administration's biomedical research policies. The new panel was appointed on Jan. 31, 1975, with Franklin D. Murphy, former chancellor of the University of California at Los Angeles, as chairman.

In May, 1975, Donald S. Fredrickson, former president of the Institutes of Medicine of the National Academy of Sciences (NAS), became the new director of the NIH, and Theodore Cooper, former director of the National Heart and Lung Institute, was appointed assistant secretary for health in HEW. Biomedical scientists favored both appointments.

Ethical considerations. Researchers in some biosciences faced new restraints on their activities in 1974 and 1975. Ethical considerations arising from research involving human fetuses prompted Congress to pass a bill in June, 1974, imposing a temporary ban on federal support for such research. The bill also established a national commission to help determine whether and under what circumstances the ban should be lifted.

The commission reported in May, 1975, that such studies hold great promise for medical research, and it therefore recommended that the ban be lifted. In place of the ban, the commission proposed a set of strict controls to ensure that fetal research is carried out in an ethically acceptable manner.

It then turned its attention to drafting regulations governing research on children, prisoners, and the mentally ill. The commission's final report was to be issued in late 1975. It will then be replaced by a permanent National Advisory Council for the Protection of Subjects of Biomedical and Behavioral Research. Meanwhile, about 15 state and local governments passed laws making some types of research on human fetuses a crime.

Genetic engineering. In February, 1975, about 140 leading biological scientists met near Monterey, Calif., to discuss whether a self-imposed moratorium should be lifted. In July, 1974, a smaller group of scientists, acting under the auspices of the NAS, had appealed to their colleagues, suggesting that some experiments involving a revolutionary technique for transplanting genes from one organism into another should be temporarily halted. Although the technique promised to bring great advances in molecular biology, the scientists feared that it might produce virulent kinds of viruses and bacteria that could pose a public health hazard if they escaped from the laboratory. The appeal, published in *Science* and *Nature* magazines, apparently was universally respected.

At the February meeting, the scientists decided that most gene-transplant experiments could be carried out safely under stringent controls, including the stipulation that the experimental organisms be rendered incapable of living outside the laboratory environment. The scientists also recommended that other experiments be set aside until new safety precautions could be devised and that the most hazardous experiments should be banned entirely. See GENETICS; MICROBIOLOGY.

International problems. While U.S. scientists were troubled by lack of support for basic research, their colleagues in many other countries were suffering similar problems. For example, the British government cut research budgets early in 1974, and inflation took a savage bite out of what was left. Then,

in February, 1975, Britain announced a new policy of holding down expenditures on "big science" projects and using the limited funds to support as many scientists as possible.

A move toward closer collaboration among European scientists was made in November, 1974, when officers for a European Science Foundation were appointed. The foundation will provide a forum for discussions on specific scientific projects in which various European nations might participate.

The Soviet government continued to harass Russian Jewish scientists who applied for visas to emigrate to Israel. Many prominent Jewish scientists were fired from their jobs after indicating they wished to emigrate, but they attempted to keep up with their scientific work through secret seminars. In July, 1974, an international group of scientists organized such a seminar in the Moscow apartment of Russian physicist Aleksandr Voronel. Soviet authorities temporarily placed under house arrest or jailed the Russian scientists who had planned to attend.   [Colin Norman]

Space
Exploration

From Feb. 8, 1974, when the third and final Skylab crew returned to earth, through June, 1975, not a single U.S. astronaut ventured into space. Instead, the United States space effort focused on the July launchings of the joint Apollo-Soyuz Test Project (ASTP), in which U.S. astronauts Thomas P. Stafford, Vance D. Brand, and Donald K. Slayton were to rendezvous in orbit with Russian cosmonauts Alexei A. Leonov and Valery N. Kubasov.

The high point of the nearly 10-day space mission was to be two days during which the spacemen were scheduled to crawl back and forth between their coupled spacecraft, exchange greetings and gifts, share meals, and conduct joint scientific experiments.

This first international manned space mission was also the last flight of an Apollo vehicle. The next U.S. astronauts to go into space will be the first crew to orbit the space shuttle, scheduled to fly in 1979.

Russians in space. While U.S. astronauts were earthbound, about a dozen cosmonauts blasted off from Russian

launching pads. Cosmonauts aboard Soyuz 14 carried out a successful rendezvous and docking mission with the Salyut 3 space station in July, 1974. However, the August Soyuz 15 mission was cut short after only two days, apparently because of technical difficulties. The Soyuz 16 flight in December carried ASTP backup crewmen Anatoly V. Filipchenko and Nikolai N. Rukavishnikov and provided practice for U.S. and Russian tracking networks in preparation for the rendezvous. The spacecraft also carried the Russian half of the type of docking equipment to be used in the joint mission.

The longest Russian space mission yet flown was that of Soyuz 17. Alexei A. Gubarev and Georgi M. Grechko took off on Jan. 11, 1975, docked with the Salyut 4 space station, and did not come down until February 9. Salyut 4 was sent into orbit on Dec. 26, 1974. The Soyuz 17 mission carried out a greatly expanded program of scientific experiments, including X-ray studies of the sun, infrared temperature scans of earth's upper atmosphere, natural re-

Apollo-Soyuz crews dine together in a mock-up of the Russian Soyuz spacecraft while training at the Johnson Space Center near Houston. From left to right, the men are Donald K. Slayton, Alexei A. Leonov, Thomas P. Stafford, Valery N. Kubasov.

Space Exploration

Continued

sources investigations, and stellar astronomy experiments.

An April Soyuz flight was cut off when the launching rocket malfunctioned, and Russian officials declined even to give it a number. But the official Soyuz 18, launched on May 24, docked with the Salyut 4 station. Its two-man crew became the first Russians to successfully reoccupy a space station as had the second and third U.S. Skylab crews.

Mariner 10. Unmanned spacecraft, particularly the planetary probes, made some of the most significant achievements of the year. For sheer technological triumph, the hands-down winner was Mariner 10, which became the first man-made visitor to Mercury in March, 1974, and then flew by Mercury again in September, 1974, and March, 1975. Mariner 10's ultraviolet imagery also yielded the first close-ups of the cloud structure of Venus.

Even before it reached Venus on Feb. 5, 1974, Mariner was carefully aimed so that the cloudy planet's gravity would send the probe past Mercury at a precisely selected point. Mercury's gravity then steered the passing spacecraft into a sun-circling orbit exactly twice as long as Mercury's own path. As a result, while Mercury went around the sun once, Mariner went around two times, meeting the planet at the same spot as the initial encounter. This gave Mariner extra fly-bys. Mission planners envision long orbit and gravity boost techniques in future flights such as those to the moons of Jupiter.

During the first fly-by, when it discovered that Mercury has a weak but detectable magnetic field, Mariner 10 passed about 438 miles (705 kilometers) from the planet. Mariner 10 came around a second time on September 21 at a distance of 29,870 miles (48,071 kilometers) and took photographs. On its third and final encounter on March 16, 1975, the craft passed only about 200 miles (327 kilometers) from the planet, enabling researchers to determine that the magnetic field is the planet's own, not one produced by its interaction with the solar wind. See ASTRONOMY (Planetary Astronomy).

NASA Administrator James C. Fletcher congratulates Peter P. Purol, 17, of Baltimore. His emblem design for the Viking Mars mission, *above,* was judged best of the thousands of entries, including the nine runners-up, *below,* in a contest sponsored by NASA and the National Science Teachers Association.

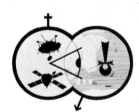

Mariner 10's accomplishments were all the more extraordinary considering that from the day of its launch on Nov. 3, 1973, it was beset with potentially crippling malfunctions, ranging from frozen cameras to erratic gyros. Yet controllers at the Jet Propulsion Laboratory in Pasadena, Calif., kept it going. And it became the first spacecraft to use a planet's gravity to guide it to another planet; the first to visit two planets; and the first to visit the same planet more than once.

Pioneer 11 visited Jupiter early in December, 1974, a year behind Pioneer 10. But instead of merely duplicating Pioneer 10's flight, which subjected the craft to a dose from Jupiter's radiation belts that would have killed a human being a thousand times over, Pioneer 11 flew much closer to the giant planet — 26,000 miles (42,000 kilometers) from the surface.

At scarcely a third of its predecessor's 81,000-mile (130,000-kilometer) distance, Pioneer 11 was subjected to a much greater amount of maximum radiation. But because it swooped in under the south pole of the planet and then northward, it spent less time than Pioneer 10 in the equatorial region where the radiation is most intense. Even so, some data were lost when radiation affected scientific instruments. Pioneer 11 provided reams of information on Jupiter's magnetic field, radiation belts, gravitation, temperature, atmospheric structure, and other characteristics. In addition, it gathered data on some of Jupiter's moons, notably the four largest — Callisto, Ganymede, Europa, and Io. See Solar Systems in Miniature.

Once past Jupiter, the spacecraft began speeding back in across the solar system for a 1979 rendezvous with the exotic ringed planet, Saturn. In March, 1975, the National Aeronautics and Space Administration (NASA) announced that a satellite had detected nonthermal radio emissions from Saturn, suggesting that the planet may have a magnetic field. This gave Pioneer 11 scientists a tantalizing preview of what they may find when the spacecraft gets there.

Russia launched several earth-orbiting satellites, and in October, 1974, they sent Luna 23 into an orbit around the moon. In June, 1975, they launched two probes toward Venus.

Satellites and astronomy. Satellites found a growing use in astronomical observations, largely because the earth's atmosphere distorts observations made from ground telescopes. The third NASA Small Astronomy Satellite, Explorer 52, was launched on May 7, 1975, to look at sources of X rays both inside and outside our Galaxy. Just five days before the satellite went into orbit, researchers from the Massachusetts Institute of Technology, Dartmouth College, and the University of Michigan made the first observations from the new McGraw-Hill Observatory at Kitt Peak, Arizona, established to look for visible-light objects in areas thought to be sources of invisible X rays.

Another X-ray watcher, the U.S.-British satellite Ariel 5, was launched on Oct. 15, 1974. It carried an all-sky X-ray monitor that can "see" the entire sky each time the satellite makes a complete turn on its axis.

The Netherlands joined the ranks of space-faring nations, launching its first satellite in August, 1974. The Dutch satellite was also designed to look for sources of X-ray and ultraviolet radiation. India was another newcomer to the space club. Its Aryabhata satellite, which also carried an X-ray detector, was launched by Russia in April, 1975.

West Germany launched a solar spacecraft called Helios on Dec. 10, 1974. It went closer to the sun than any other man-made object — within 28 million miles (45 million kilometers). Helios endured temperatures of 700°F. (300°C) — hot enough to melt lead — while it gathered data on the sun, its environment, and its influence.

Among the other satellites launched during the year was the sixth Applications Technology Satellite (ATS-6). It began the first satellite television broadcasts of educational courses to remote areas from Appalachia to Alaska on July 2. One of the most acclaimed satellites in NASA's history — the Earth Resources Technology Satellite — launched in 1972, was followed on Jan. 22, 1975, by a successor, Land-Sat 2. The new satellite continued the work of monitoring water supplies, crop conditions, and other phenomena on the earth below. [Jonathan Eberhart]

Transportation

Transportation service for many people was never better or more widespread than in 1975. Yet the disadvantages of the existing system have become increasingly apparent. Transportation continued to be a major cause of accidental death and serious injury and a major source of air pollution and urban noise, although recent technological advances and federal regulations have produced significant improvements. Transportation vehicles, especially automobiles, also are the primary users of scarce fuel. But in spite of the apparent need for better transportation systems, federal policymakers cut back sharply on support for new developments in favor of applying existing and often outdated technology.

Fuel consumption. About 25 per cent of all U.S. energy is used for transportation, and 60 per cent of that amount is supplied as gasoline for the nation's approximately 100 million cars and small trucks. Statistics compiled by the Department of Transportation (DOT) and the Environmental Protection Agency (EPA) show that reducing automotive energy consumption is one of the most important targets for petroleum conservation.

One way of increasing passenger-miles per gallon of fuel is by increasing the average passenger load per car from the present value of about 1.2 persons. A new approach to car pooling is the Commuter-Computer system in Boston, which uses computers to match riders who have similar departure and destination points. Nevertheless, even if half of all urban commuters used car pools, studies show that fuel consumption would be cut only 8 to 10 per cent.

Improving auto efficiency, however, could produce much greater fuel savings. The fuel efficiency of 1974 automobiles, averaged over all models, was 14 miles (22.5 kilometers) per gallon. This rose to 15.9 miles (25.6 kilometers) per gallon for 1975 models, due to more smaller cars and technological improvements. The DOT-EPA study suggested that cars could reach 19.6 miles (31.5 kilometers) per gallon by 1980, a 40 per cent improvement over 1974. Among the technological improve-

Man joins dummy in an impact machine to compare responses to a simulated auto crash, as part of a test of restraint systems.

Transportation

Continued

A linear induction motor research vehicle that tests high-speed performance went 255 miles (410 kilometers) per hour in August, 1974, a record speed for wheel-on-rail vehicles.

ments suggested to achieve this goal are streamlining to reduce aerodynamic drag and improving transmission and engine efficiency.

Soon to come is a tiny automotive computer, the microprocessor, an integrated circuit with hundreds of transistors on a single silicon chip. Installed under the hood of a car, the microprocessor can adjust engine performance to produce the most efficient fuel consumption. See ELECTRONICS.

Engine efficiency also can be improved by reducing the fuel-to-air ratio as much as possible, a process known as lean burning. The stratified-charge engine, in which the air-fuel mixture is rich near the spark plug, although lean on the average, can substantially boost efficiency. A controlled-combustion, stratified-charge engine developed by Texaco, Incorporated, showed up to 70 per cent improvement in fuel economy in road tests in early 1975. Ford Motor Company said its programmed-combustion, stratified-charge engine can improve fuel economy by 25 per cent in standard-sized autos, based on

tests in early 1975. The engines may be available in the mid-1980s.

Safety advances. Research continued on protecting passengers in car collisions. In late 1974, DOT's National Highway Traffic Safety Administration chose Calspan Corporation of Buffalo, N.Y., to conduct an 18-month program evaluating automotive restraint systems. For its tests, Calspan uses an impact machine to simulate auto crashes. Both human volunteers and lifelike dummies are used to determine the forces on humans during a collision.

Calspan is testing both seat belts and airbags, which inflate upon collision to cushion a passenger, even though interest in airbags waned in 1975. Data on seat belts suggested that they may be more effective than airbags in protecting passengers. Calspan reported that a study of 500 users of lap and shoulder belts involved in potentially deadly accidents since 1969 showed that not a single death occurred.

Computerized transit. Too much popularity spoiled an ambitious new public transit system, which was sharp-

A new winter tire, left, will grip ice without metal studs, thanks to rubber compounds first developed for the lunar rover tire, right.

Transportation

Continued

ly curtailed in May, 1975. The system, the first to combine fixed-route public transit with personalized dial-a-ride service, began operating over a 240-square-mile (622-square-kilometer) area in Santa Clara County, California, in November, 1974. The program used a fleet of low-pollution, propane-driven minibuses to serve 18 areas that were connected by arterial lines.

A customer called a toll-free number and a computer selected the nearest personalized-transit bus and estimated the time of pickup. Trips were assigned to buses by two-way radio. If the trip was outside a local service area, a bus took the rider to the nearest connecting arterial line. Unfortunately, the telephone-reservation and computer-scheduling systems could not handle the unexpected volume of requests, and the system was abolished for financial reasons, except in a small section.

Return of the dirigible? Air-transport researchers are taking a new look at lighter-than-air vehicles, including dirigibles and blimps, for freight transport. Continuing studies by the National Aeronautics and Space Administration indicate that new technological developments in structures, aerodynamics, and vehicle handling may revive this classic craft. The airship can move bulky, heavy cargo over long distances without the need for expensive runways and terminals. Its great capacity, combined with the ability to hover while using little energy, makes it a candidate for construction tasks and for aerial surveillance.

New materials developments may help spur the practical development of lighter-than-air vehicles. Du Pont Company has introduced a new synthetic fiber that, unlike nylon or Dacron, does not stretch. Made into rope, it is as strong as steel cable, but weighs only one-fifth as much, according to the company. ILC Industries in Dover, Del., has developed a way to weave strands of the new fiber in three directions instead of only two. A single layer of the resulting fabric is strong enough to serve as the skin of a dirigible, instead of the two layers normally required.   [Herbert H. Richardson]

Zoology

Zoologists discovered in 1974 that some apparently solitary animals may be more social than previously thought. Zoologists often categorize animals according to how much they associate with others of the same species. Bees and baboons would be at the higher end of the scale and raccoons and foxes would be considered nonsocial.

Secretly social. David P. Barash of the University of Washington at Seattle contended in August, 1974, that categorization depends upon the ability of observers to assess social characteristics in other species. Such assessment may be limited by the observers' own biases.

To prove his point, Barash studied recognition and social interaction among a number of foxes and raccoons. Some were captured in areas where they would have previously encountered each other, and others were captured far enough apart so that they were expected to be complete strangers.

Both types of animals showed a more aggressive response—growling or bristling—when meeting another member of the same species that had been trapped relatively far away. Fighting, the most intense aggressive level, was limited almost exclusively to animals captured several miles from each other. Those animals captured in the same neighborhood, however, behaved as if a definite pecking order had already been established.

Barash concluded that so-called solitary animals are not solitary at all. If they have met previously, they have interacted sufficiently to determine who is dominant. This type of behavior must be considered at least semisocial.

DDT ban pays off. David W. Johnston of the University of Florida at Gainesville reported in November, 1974, that federal restrictions on DDT and chemically related insecticides are beginning to reap benefits. He found decreasing quantities of DDT in fatty tissues collected from migratory songbirds killed accidentally between 1964 and 1973 when they flew into Florida television towers.

DDT builds up in the fatty tissues of certain birds that feed on insects, fish, and other prey that accumulate DDT.

Polar bears can barely be seen against snow in a conventional photo, *below*. But they appear black in an ultraviolet photo, *below right*. The technique will help in censuses of white arctic animals.

Zoology
Continued

Once applied, DDT may remain in the soil for at least 20 years, imposing obvious long-term dangers. One of the effects of DDT in birds is thinner eggshells. As a result, fewer eggs are hatched and the population declines.

Johnston found that fatty tissues from songbirds analyzed in 1973 – after a virtual total ban on DDT became effective in 1972 – contained only 10 per cent as much DDT as was found in 1969. The results suggest similar declines in DDT among other species that accumulate the pesticide.

Bats read the menu. Eight researchers at Washington University in St. Louis reported in December, 1974, that bats can use their sonar mechanism to discriminate among potential prey. Bats emit high-frequency sounds that bounce off nearby objects to help them locate prey. Potentially, a bat could waste a great deal of energy attempting to capture prey that would be too large or distasteful to eat, or too small to be worth the effort. The researchers showed, however, that bats can use echo cues to distinguish among targets.

In the experiment, the investigators presented bats with plexiglass squares drilled with holes to give off different arrays of echoes. Each square was just under 1 inch (2.2 centimeters) thick. The spectrum of echoes from each target was governed by drilling 24 holes partway through each square. The holes on any particular square were all the same depth, ranging from 0.24 inch (6 millimeters) to 0.32 inch (8 millimeters), and all the holes were the same diameter. The investigators trained the bats to discriminate among squares with different holes by rewarding the animals with food whenever they chose a square drilled with holes that were 0.32 inch deep.

The bats displayed an extremely acute ability to distinguish among the different targets. A bat determines the distance to an object by how long it takes an echo to return, but apparently the echo spectrum can also tell it something about an object's finer characteristics, such as size and shape. This enables the animal to judge whether an object is likely prey. The investigators

Drawing by Nurit; © 1974 The New Yorker Magazine, Inc.

An ant raises its gaster (part of the stomach), and releases a drop of a signaling chemical, *above,* to invite a nest mate to follow to a new food source. The nest mate touches the recruiting ant's gaster, *top right,* and hind legs, *middle right.* Then the recruiting ant leads the way to food, *bottom right.* The other ant follows, its antennae touching the recruiter.

Zoology

Continued

suggest that a bat may learn the echo response from different insects through experience – capture and tasting. It then uses the knowledge to discriminate among potential prey.

Early language training. Just as exposure to conversation during a child's earliest years helps the child learn a spoken language, learning sign language is also improved if the child is exposed to this type of communication early in life. Robert A. and Beatrice T. Gardner of the University of Nevada at Reno decided to see if early exposure would also help in teaching chimpanzees to use sign language. In February, 1975, they reported on their success with two chimps, Moja and Pili. Raised and cared for by deaf human beings from the time they were two days old, both quickly learned sign language. Within three months, both chimps spontaneously used signs meaning "come," "gimme," "go," "more," "drink," and "tickle." The study, part of an ongoing research effort, suggests that an infant, whether human or chimpanzee, learns a language more

quickly if it is exposed early in a favorable learning situation.

Instinct for survival. Although some degree of learning is involved in almost all types of animal behavior, there are times when depending on trial-and-error learning would be disastrous. For example, birds that prey on snakes that resemble others whose venom can kill in minutes cannot depend on learning which snakes are lethal and which are harmless. One error would quickly end the learning process. This suggests that the birds' ability to distinguish might be based on instinct.

Susan M. Smith of the University of Costa Rica reported an interesting test of this idea in February, 1975. For her experiment, Smith studied motmots, tropical birds related to the kingfisher, that prey upon snakes. She removed the motmots from their nests a few days after birth and raised them in the laboratory so they would have no experience in capturing snakes. Smith then showed them wooden snake models painted with various patterns of stripes and rings, including those characteris-

351

tic of deadly coral snakes, of snakes with less lethal poisons, and of harmless, nonvenomous varieties.

The motmots attacked the plain red, yellow, green, blue, or unpainted models, as well as the models with lengthwise stripes, but they avoided models painted with yellow and red rings, characteristic of the coral snake. Further testing showed that a ring pattern alone did not inhibit attack, but that the color of the rings was significant. The birds avoided the models most similar to the coral snake, and attacked those that resembled less harmful species. The results indicated that motmots have an inborn ability to recognize and avoid certain color patterns.

Romantic moths. Zoologists have long been interested in the guidance mechanism a male moth uses to locate a suitable mate over distances of several miles. The female, who remains still, secretes a luring odor called a pheromone into the air, and this is the major cue used by the male in its romantic pursuits. The main question is whether the pheromone alone guides the moth, or whether it simply alerts him to fly upwind, bringing him into the vicinity of the female.

A controversial 1973 report on how the male uses pheromone cues stimulated additional laboratory research in 1974. In the 1973 study, Stanley R. Farkas and Harry H. Shorey of the University of California at Riverside reported that male moths entering a chamber containing graduated amounts of pheromone flew toward the point of greatest concentration. But the suggestion that a moth could detect, monitor, and follow gradually increasing amounts of pheromone over long distances through woodlands seemed impossible to many zoologists.

J. S. Kennedy and D. Marsh of Imperial College Field Station in Ascot, England, reported in mid-1974 on an experiment to investigate more carefully how male moths locate females. If a moth flies upwind when it detects pheromone, the first question is how it determines the direction of the airstream. This may seem to be a simple-minded question, but when you are flying, the air, regardless of its direction of flow, is perceived merely as resistance from the direction in which you are

flying. It is like trying to determine wind direction by sticking your hand out the window of a moving car. The wind always seems to come from in front. To discover the wind direction, you would have to stop now and then to test the wind, unless you could make use of other information.

Kennedy and Marsh placed male moths in a wind tunnel and released pheromone into the air. The tunnel was carpeted with a pattern of stripes that could be kept stationary or moved either upwind or downwind in the tunnel. When the pattern was stationary, the males invariably flew upwind. But if the floor stripes were moved downwind, thus increasing the apparent ground speed of the fliers, the moths reduced their air speed accordingly and were carried downwind while still facing upwind. This maneuver prevented males from approaching the pheromone source. On the other hand, if the stripes were moved upwind, this decreased the moths' apparent ground speed. They would then speed up their flying and often overshoot the pheromone source.

The experiment indicated that moths apparently determine the wind direction by watching how the ground is moving. Kennedy and Marsh concluded that apparent wind movement, not pheromone strength, was the key factor in orienting the moth. Once a male draws near a female, perhaps he can locate her by the increasing strength of pheromone, as indicated by Farkas and Shorey. Finally, he comes close enough to see the female and make contact.

Discovery of the guidance system of the male moth may have important implications in controlling certain pest species. Many kinds of caterpillar – the immature larva stage of moth life – can wipe out acres and even entire forests of trees as they systematically strip the leaves. Conventional pesticides have not been able to curtail the annual spread of these hungry pests. The recent experiments increase our knowledge of how to use pheromones to eradicate moths. The pheromones, which are not harmful to people, animals, or plants, can be used to attract male moths to traps, preventing them from mating and thus reducing the population. [William J. Bell]

Rare desert animals, including many Biblical species, find a home in Israeli nature refuges. Seen in their natural habitat are young Nubian ibexes, *top left;* a Sahara oryx, *top right;* an addax antelope, *above;* a group of Somali wild asses, *above right;* and a Sinai leopard, once believed extinct, *right.*

Men and Women Of Science

More and more, scientists are attracting attention to their work at a younger age. This section, which recognizes outstanding scientists, describes two who have made their mark early in their career.

The Leakeys

By Roger Lewin

**In their 40 years of finding fossils, this family has
greatly changed our views about the origin of man**

The road to East Africa's Olduvai Gorge slithers down the western
slopes of Ngorongoro Crater, shell of a long-extinct volcano. The vast
Serengeti Plain sweeps out to the horizon; the land is hot and dry.

But 2 million years ago, the area looked entirely different. The
bright sun sparkled on a lacework of streams tripping down the volcano's slopes and emptying into a large lake in the valley below. A wide
variety of prehistoric animals, great and small, lived off the vegetation
surrounding the lake. And among the animals were at least two distinct types of manlike creatures. One type, with massive bones, including huge jaws for grinding vegetation, could walk upright much of the
time, but apparently had difficulty in traveling long distances. The
second type had a more humanlike jaw, was smaller and more agile,
and walked upright with an easy striding gait.

We can visualize this 2-million-year-old scene because of the many
exciting discoveries of a family of anthropologists named Leakey.

On the layered slopes of the gorge, which run for 25 miles (40
kilometers), Louis S. B. Leakey and his wife, Mary, later aided by
their sons Jonathan, Richard, and Philip, have patiently searched
since the 1930s for the fossil remains of early man and the tools he left
behind. Louis died in October, 1972, but family members still carry
on the work he began. Although Jonathan no longer helps in the

search for fossils, Mary, occasionally accompanied by Philip, continues to excavate at Olduvai Gorge, and Richard has established his own excavation site at Lake Rudolf in northern Kenya.

Driving the 320 miles (515 kilometers) from Nairobi, Kenya, to Olduvai Gorge in Tanzania takes a mere eight hours today, compared with the bone-jangling, seven-day treks the Leakeys had to make to reach the area in the 1930s. Mary relaxes visibly as we approach the gorge. To her, Olduvai Gorge is home.

"Isn't it beautiful?" she asks. And it is. But doesn't it seem rather isolated? "I like it that way," she answers, without hesitation.

Mary is quiet, shuns publicity, and is perfectly content with the occasional company of the African workers in the camp. Unlike her late husband, Louis, Mary will venture no opinions unless they are backed up with a wealth of data. Louis had a quick temperament and a confidence that often trapped him into making broad generalizations about evolution that later had to be rescinded. But in their work, Louis and Mary shared the invaluable characteristics of patience and sharp powers of observation which stood them in good stead during four decades of excavation.

Mary's Olduvai home is a collection of three-sided huts, or *bandas*. One thatched banda serves as a bedroom, another is a study and eating area. In the dining area, a wide sheet hangs from the ceiling above the table, because a swallow built its nest in the rafters.

Wouldn't it be simpler just to remove the nest? "If the swallow wants to nest there," she says, "it has every right to do so." This is typical of the Leakey respect for nature. Even though lions, rhinoceros, and other large wild animals roamed in and around the Olduvai Gorge excavation sites, the Leakeys never carried weapons. They kept alert for any sign of danger, and their Dalmatians, which have always accompanied the Leakeys on their expeditions, warned them of approaching wild animals and snakes. In 1938, Louis became a member of a committee to establish game reserves in Kenya, and he was a founding trustee of the national park system set up in 1946. Richard is a trustee of the East African Wildlife Society and, also, chairman of the Wildlife Clubs of Kenya.

At night, while oil lamps flare and sputter in the open banda, Mary describes the excitement she feels when looking for fossils. "Some days one is really keyed into it," she says. "The tension is right, and you know you will find something. Other times, it is just no use, and you might as well not even try."

This curious state of being tuned into the search is familiar to many anthropologists as intimate with their work as are the Leakeys. Just a short distance from where we talk, a concrete block topped by a white tablet marks the culmination of one of the special moments of tension. It is the site where on July 17, 1959, Mary made one of her most important discoveries, the fossil skull of *Zinjanthropus* (later renamed *Australopithecus boisei*), a relative of early man.

The author:
Roger Lewin is the science editor of *New Scientist* magazine.

In the sediment layers of Olduvai Gorge, *above,* the Leakey family found a fossil record dating 2 million years ago. In the early 1960s, Louis, Mary, and Philip, *left,* searched for fossils and artifacts at the campsite of a manlike being that lived 1.75 million years ago. Their everpresent dogs kept watch for snakes, lions, and other animals.

More than 500 miles to the north, Mary's son Richard has established his own base for archaeological exploration on the eastern shore of Lake Rudolf. In this area in 1972, Richard made a find that may again revolutionize our concept of how and when mankind evolved. He discovered a skull that he believes is an early form of man, a skull belonging to the genus *Homo*. It may be as much as 3 million years old. Richard feels about Rudolf as Mary feels about Olduvai. "This is my home," he says, looking over the blue waters of the massive lake, "or at least, one of them."

Like the climate at Olduvai, the climate around Lake Rudolf is hot. Temperatures often reach 115°F. (46°C) in the shade at midday, when the fossil hunters are forced to take a break from their work and cool down with a swim in the crocodile-infested lake. "They've never attacked anyone," says Richard, while crocodiles watch the swimmers from 30 yards (27 meters) away.

In addition to his work at Lake Rudolf, Richard is the director of the National Museums of Kenya. The museum job keeps him away from Lake Rudolf much of the time. But he, his wife Meave, and their two daughters fly up in his six-seater airplane about three times a month. The camp, which can accommodate about 60 scientists and workers, stands on a beautiful lakeside beach about midway between two major exploration sites. When Richard established the camp in 1968 and 1969, this was a good strategic position for the archaeologists to defend themselves against attacks by local hostile tribes. In fact, the first expeditions were accompanied by armed escorts. But Richard has

Mary pours coffee for U.S. Ambassador to Iran Richard M. Helms, a visitor at her home in Olduvai Gorge. The sheet hung above the table in the dining area protects it from a bird's nest up in the rafters.

since established friendly relations with the tribes, and there is no more trouble.

Unlike the vertical layered walls of Olduvai Gorge, the sediments at Rudolf are exposed in large flat areas. This is partly because the whole area tilted in a geological upheaval thousands of years ago. In many places, the deposits look like piles of rubble bulldozed into heaps. But this was caused by rain and wind erosion, which constantly exposes new fossils for Richard's team to spot and excavate.

The remarkable story of the Leakeys' odyssey into the past began with Louis, who was born on Aug. 7, 1903, in Kabete, Kenya, 8 miles (13 kilometers) from Nairobi. A group of Kikuyu tribal elders gathered to spit on the newborn white baby. They were showing their respect through an ancient tribal custom.

His parents were Church of England missionaries in East Africa, so young Louis grew up on the mission station. His playmates were Kikuyu children, and as a result, Louis became immersed in the tribal culture. "It never occurred to me to act anything other than Kikuyu," he recalled later. "In language and in mental outlook, I was more Kikuyu than English." He played and hunted like a Kikuyu boy and became a member of the tribe. He even underwent the Kikuyu initiation into manhood and never revealed the secrets, not even to Mary.

The Kikuyu upbringing taught Louis many things that he would never have learned growing up in England: the importance of patience and observation and an intimate feeling for the land—from the nature of its rocks and gullies to the habits of its animal life.

With one of her alert Dalmatians acting as back-seat driver, Mary heads for an excavation site in Olduvai Gorge.

Mary shows a tour group from the Denver Zoo a tablet marking the spot at Olduvai where she found the 1.75-million-year-old skull of *Zinjanthropus* in 1959, *top.* That discovery revolutionized scientific thought about the origins of mankind. An African fossil hunter, *above,* brings in some rock samples for Mary to inspect. At a promising new excavation site in the Olduvai area, *right,* she goes through the painstaking process of examining rock after rock for signs of fossils.

Young Louis found stone tools in the area near his home, and he learned by reading anthropology books that these were prehistoric implements. He also read Charles R. Darwin's *The Origin of Species* and believed Darwin's theory that man originally evolved in Africa. And so, when he was sent to school in England in 1919, Louis was already interested in the study of early man in East Africa. But he had not yet decided on anthropology as a career. In fact, he was planning to become a missionary like his parents.

Louis enrolled at Cambridge University and began playing rugby football. All went well until a fateful rugby match in 1923, when two kicks in the head left him with recurrent headaches. The headaches were so severe that a doctor suggested Louis take a year off to relax, a piece of advice that changed his life and, without doubt, the progress of physical anthropology.

The British Museum was sponsoring an expedition to Tanganyika (now Tanzania) to look for dinosaur fossils, and Louis went along. "Had it not been for my accident," he later commented, "I should certainly not have applied to go on this expedition, and I should therefore never have had the really practical training in methods of fossil collecting and preservation.... My luck had certainly changed in a most unexpected manner." And so it had. Indeed, so often did things seem to turn out well for Louis that, in the future, the phrase "Leakey's luck" would often be applied to his achievements.

After that expedition, Louis decided to make anthropology his career. In July, 1926, when he was almost 23, Louis and a fellow student set sail for Kenya on what he called the East African Archaeological Expedition. Louis' journey back through man's past was finally underway. It is a nice touch of historical irony that, just a little more than 40 years later, Richard was to set out as leader of his first major expedition (to the Omo Valley in Ethiopia) when he, too, was just approaching his 23rd birthday.

When Louis began combing the hills and valleys of Kenya for signs of man's origins, he was strongly advised not to waste his time there. Authorities said he would find nothing. At the time, scientists believed that through the process of evolution man had emerged several hundred thousand years ago. And they were certain that the site of man's origin was Asia, where the fossil Java man had been unearthed in the 1890s and Peking man in the 1920s.

Typically, Louis ignored their advice, and typically, he was right. Louis' first excavation was near Lake Nakuru, where he unearthed a Stone Age burial site. Over the next few years, Louis organized several more expeditions in Africa, each one yielding richer rewards—prehistoric skulls and skeletons, early animal fossils, and stone tools. Then he turned his attention to Olduvai Gorge.

The gorge was originally "discovered" by a German butterfly hunter in 1911. While he was hotly pursuing a prize specimen, he found a fossil bone which he took back to Berlin. German scientists

Richard Leakey travels to the Lake Rudolf camp in his plane. In his comfortable quarters at the Rudolf base, Richard makes dinner for himself and an African colleague.

launched an expedition to Olduvai in 1913 and found many fascinating animal fossils, but no sign of man. World War I put an end to further exploration there, until Louis organized a second expedition in 1931. Within eight hours of their arrival, Louis and his colleagues had found stone tools used by our prehistoric ancestors. And the gorge came to dominate Louis' attention for the rest of his life and provided major fossil finds that have reshaped our knowledge about early man.

This fossil-rich area began forming 2 million years ago as the water level of the ancient lake rose and fell, picking up and depositing sediments. It built up a layer cake of time, with the oldest layer buried deep and the most recent at the surface. Finally, the lake dried up, leaving an arid basin. But then a small, seasonal river began to etch its way from west to east across the basin and, over a period of thousands of years, it sliced down through the layered deposits–through time itself. Today, a 350-foot (107-meter)-deep gash exposes 2 million years of prehistory in neat layers, beautifully arranged for archaeologists and anthropologists to study.

During the more than 40 years since the Leakeys came to the gorge, the world heard about their discoveries through the publicity that was enthusiastically generated by Louis. But many of the finds were made by Mary, who preferred to remain in the background. Mary actually spent more time out in the field excavating than Louis, because he was often busy with the affairs of the national museum in Nairobi. From 1945 to 1961, Louis was the director of the museum, the position that Richard now holds.

Louis met Mary while on a visit to England. He was writing *Adam's Ancestors* (1935), an account of early man, and was looking for someone to illustrate it. Mary Douglas Nicol had the best qualifications for the job: She had both artistic talent and prehistory in her blood. Her father was landscape painter Erskine E. Nicol, and her great-great-grandfather was John Frere, the first to link stone tools to early man. Mary's own interest in prehistory was sparked when she helped excavate some prehistoric cave paintings in France.

When Louis saw Mary's drawings of prehistoric artifacts, he described them as "the best representations of stone tools I had ever seen." She got the job, and their relationship became not only professional but personal as well. They married on Christmas Eve, 1936. Jonathan was born in 1941; Richard, in 1944; and Philip, in 1949.

Mary and Louis surveyed Olduvai Gorge for about 20 years before launching a major excavation there. They were hampered by a lack of funds and then by a lack of time after 1945, when Louis began working full time for the museum. It was not until 1959 that Mary uncovered the now-famous *Zinjanthropus*. That fateful July morning, Mary spotted a piece of bone and two fossilized teeth embedded in a rocky slope. She immediately recognized them as being humanlike. Using dental picks and fine brushes, the Leakeys uncovered 400 fragments of the creature's skull—missing only the lower jaw. One U.S. anthropologist described the discovery of the ancient skull fragments as "the event that opened the present modern era of truly scientific study of the evolution of man." Patience certainly had its reward.

By 1959, scientists working in South Africa had discovered evidence of ancient manlike apes—the small *Australopithecus africanus* and the larger *Australopithecus robustus*. At first, they believed that *Australopithecus africanus*, which was found by Raymond A. Dart in 1924, had no connection with man's origins. But by the mid-1940s, that view had changed. Many anthropologists had come to regard these small Australopithecines as being on the direct evolutionary route to modern man, *Homo sapiens sapiens*. Mary's discovery of *Zinjanthropus*, which means *East Africa man*, was a major event. Even though Zinj, as they called it, was similar to *Australopithecus robustus*, it was sufficiently different for Louis to proclaim it closer to man. After all, Louis reasoned, it had been found along with stone tools, which, he assumed, were manufactured by *Zinjanthropus*. The tools and their supposed maker were dated at a staggering 1.75 million years old.

Soon the world of anthropology was to be turned on its head once again. And again, the Leakeys were behind it all. In 1960, they had received a grant from the National Geographic Society, which enabled them to devote more time to their research. Between 1960 and 1964, they excavated fossils that Louis was certain were of an even closer relative of man. And they were the same age as Zinj. This new creature was smaller than *Zinjanthropus*, had smaller teeth, a bigger brain, and appeared to have walked more upright than Zinj.

These characteristics were convincing enough for Louis to propose that the creature was a direct ancestor of modern man. He called it *Homo habilis*, meaning skillful man. Louis proposed that the tools found near *Zinjanthropus* had clearly been fashioned by the sophisticated hands of *Homo habilis* and not by Zinj. However, Louis' proposal did not go unchallenged. A number of anthropologists argued that this creature was simply a relative of the small *Australopithecus*, not *Homo*. But Louis stuck to his position.

Instead of there having been one steady progression of manlike apes, evolving from a common stock toward man, it appeared clear to Louis that nature had made several attempts at producing manlike creatures. The lines such as *Australopithecus africanus* and *Zinjanthropus*, he believed, went up evolutionary blind alleys and eventually became extinct. But, in Louis' opinion, *Homo habilis* appeared to have soldiered onward to become modern man.

Geologically, Olduvai Gorge is a youngster. Two million years is a mere blink in evolutionary time. So the Leakeys looked in other areas for evidence of early ancestors of man, and they found some. For example, the Leakeys' senior African assistant, Heselon Mukiri, unearthed a 14-million-year-old fossil in 1962 at Fort Ternan, near Lake Victoria. Although the fossil was much like an ape, Louis believed it showed sufficient signs of ancient human ancestry to indicate it had already taken the evolutionary turn toward man. And there are signs that this creature, known as *Kenyapithecus wickeri*, used lumps of lava to break open animal bones to get at the nutritious marrow—an example of very early tool use.

At a Rudolf excavation area, Richard spots a potentially interesting fossil, *below,* and then pauses to examine it closely, *below right.*

The Fort Ternan discovery put man's origins back to at least 14-million years, and placed them in Africa. This ancient creature, Louis suggested, evolved from a more basic stock which he termed *Kenyapithecus africanus*. Such a creature, called *Ramapithecus*, was found in India in 1936, and Louis proposed that it is merely a migrant from the African stock, a view also held now by Richard. So until evidence to the contrary comes to light, Richard believes, as his father did, that man was "born" in Africa.

Although Louis won fame by formulating broad theories about the finds, a great number of people labored on the Leakey projects. All the excavations were expertly manned by skilled and dedicated African workers. And for many years, three small boys also lent a hand—Jonathan, Richard, and Philip Leakey. Their parents carried them around the excavation sites even when they were tiny infants. As they grew older, the Leakey children crawled around with Louis and Mary and either joined in the delicate digging of fossil bones, or wandered off to look for animals and butterflies.

Both Jonathan and Philip made important contributions. Jonathan unearthed the first signs of *Homo habilis* at Olduvai in 1960. In addition to digging, Philip helped establish new excavation sites. And when Philip was only 12 years old, Louis described him as "the best Land Rover driver in Africa."

Jonathan, the quietest and most reserved of the three, is a herpetologist and now has completely given up searching for fossils. He keeps a snake farm and grows melons at Lake Baringo, in a remote part of northern Kenya. This, he admits, is a deliberate escape from the broad umbrella of the Leakey legend.

But Philip still hovers on the edge of a career in anthropology. He has dabbled in business and, recently, in politics. In fact, he was almost elected as the first white member of Kenya's National Assembly in 1974. But he occasionally helps at Olduvai, and Mary encourages his inclinations toward anthropology. "He has everything one needs in a fossil hunter," she says, "observation, drive, and the right state of mind." Philip's wife, Valerie, works in the office Mary maintains in the museum in Nairobi.

Ironically, Richard was the least interested of the three in fossil hunting as a youngster. As a boy, he recalls, "The idea of collecting fossils struck me as totally boring." More than anything else, he wanted to be independent. "I did not want to do what my father and mother were doing," he says. So at an age when most boys are still in high school, he was learning how to trap animals for research in breeding programs, and gathering up animals that had died accidentally or of natural causes so that he could sell their skeletons. And then he ran a successful safari business for nature lovers. He became financially independent, but he considered his parents' house in Langata, a suburb of Nairobi, home. "I was seldom there," he says, "and my parents were away more often than not. We saw very little of each other."

Then, in 1963, he went along on an archaeological expedition to Lake Natron in Tanzania, just out of curiosity, and the team found the first known lower jaw of *Zinjanthropus*. "This was very exciting for a young lad," Richard says.

He returned to the dig the following year and was soon so interested in anthropology that he began to think that attending the university "was not such a bad idea after all." So, with an income from the safari business to support him, Richard set off for London in 1965 to prepare for his university entrance examinations. He finished the preparations in six months but had to return to Kenya because his safari business ran into financial trouble.

After returning to Kenya, Richard went on several more expeditions, and by the end of 1966, his interest in actual fossil hunting had grown so great that he decided not to go to the university. It would be too time-consuming. He was also involved in helping run the museum in Nairobi. Two years later, when he was 23, he became director of the National Museums, a post that involves responsibility for all museums and archaeological sites in Kenya.

While preparing for his 1967 Omo Valley expedition, Richard had flown over Lake Rudolf, a 150-mile (240-kilometer) stretch of jade-colored water that once joined with the Nile River. Richard noticed mile upon mile of exposed sediments on the lake's eastern side and thought, "That looks like a good fossil site. I'll try that." As it turned out, Rudolf is even richer in fossils than Olduvai.

"There was no specially good reason why Rudolf should have been fossil rich," he admits, "but I was very naïve then." After a moment's reflection, he adds, "I hope I never stop being naïve. As long as you are, you will try anything, and sometimes you'll do the unexpected."

The first Rudolf expedition was backed by the National Geographic Society, which has supported much of the Leakey family's work. The society gave 22-year-old Richard a $25,000 grant—against the advice of his father. "The fact that my father was against it," Richard says, "made me even more determined to do it. Maybe he was against it because he knew I would react that way. I don't know." At any rate, National Geographic's faith was justified, and the east Rudolf region began to yield its ancient treasures to the young Leakey.

Because father and son had such similar, powerful personalities, and because they were achieving in the same scientific area, a clash between them was probably inevitable. "Relations between us were rather strained for a long time," Richard says, "but it gradually changed, and we had become great friends by the time he died. I was pleased that we had."

On the first major exploration of Rudolf's deposits in 1969, Richard came across an incredible prize, an anthropologist's dream—the intact skull of a creature resembling the Olduvai *Zinjanthropus* just waiting to be picked up from the ground. Richard rushed back to camp to show the find to his mother. Almost exactly 10 years previously, she had

Richard takes a break from studying skulls of modern animals at the Nairobi museum to chat with Mary over coffee.

spent weeks carefully sifting out of the gorge the skull fragments of the first *Zinjanthropus*. Then she spent months painstakingly reconstructing the shattered skull. And her son simply had to bend down and pick up an intact skull. Was Mary envious? "No," she says, "I was just very, very pleased." Richard's skull turned out to be about the same age as the one that Mary had found at Olduvai.

Meave G. Epps, who later married Richard, went along on that expedition. She had been recruited by Louis to run the National Primate Research Center in Kenya, and she recalls that her first contact with Richard was less than pleasant. "I had been running the center in Louis' absence for less than a month," she says, "when I got a rocket from Richard for spending too much money."

Because Meave is an expert in the comparative anatomy of monkeys, Richard suggested she accompany him to Rudolf and study fossil monkeys. This she gladly did, and soon she became interested in the fossils of early man.

For the next few years, the Rudolf dig yielded dozens of fossil fragments of the same type as Richard's first find. Richard's team also found some that were tantalizingly akin to genuine early man, to the genus *Homo*, but nothing of which he could be certain. Then, in September, 1972, one of Richard's colleagues, Bernard Ngeneo, made the discovery that was to reshape ideas on man's origins once again.

Ngeneo discovered the fragments of a skull that appeared undeniably to belong to the genus *Homo*. It had a big brain cavity—more than half the size of modern man's—and delicate, manlike features. According to some dating estimates, it was 3 million years old. It was at least a million years older than Olduvai's *Homo habilis*, yet it appeared to be further down the evolutionary road toward modern man. The Rudolf team later discovered leg bones of the creature which showed that, like *Homo habilis*, it had walked upright. Richard named this Rudolf find 1470 man, because 1470 was its catalog number.

Meave undertook the job of reconstructing the skull from its fragments. She could do little else at the time because she was caring for their 3-month-old daughter. Meave spent weeks at Lake Rudolf and also at the museum in Nairobi patiently piecing 1470 together. The task was somewhat like working a three-dimensional jigsaw puzzle, but with no picture to use as a guide and with some of the pieces missing. "It was tremendously exciting," she recalls, "because we really didn't know what it would look like."

According to Richard, the brain size and general development of 1470 man implies that the evolution of the genus *Homo* as a separate line may have begun about 5 million years ago, much earlier than anyone would have guessed. No one knows why it started then, but it coincides with a period of evolutionary blossoming in many animal families. Perhaps a powerful burst of cosmic radiation caused a rash of mutations, or there may have been a dramatic, widespread change in the environment. Richard suggests that *Homo*'s cousins, *Australopithecus africanus* and *Zinjanthropus*, were born at this time also. But they were destined to become extinct.

Richard's wife, Meave, a comparative-anatomy expert, discusses finds with museum workers. Philip and Valerie lunch with Mary, Richard, and Meave at the museum.

Anthropologists are now searching for even older forms of *Homo*, and some may be found at current excavation sites in Ethiopia. Does Richard mind if someone else discovers man's earliest ancestor, some 5-million years old? "No. I used to be intensely competitive, but I'm not anymore. We found 1470 man and that gave the important prediction about man's origins. Now I don't mind who makes an earlier find."

Like his father, whose intimacy with the land and the fossils led him to be rather impatient with academic anthropologists, Richard holds much the same feelings. He disdainfully dismisses those who question his proposals because he lacks academic training. Both Louis and Richard had the advantage of being broadly based in areas such as anthropology, archaeology, geology, anatomy, and geochronology, without necessarily being specialists in them.

The Leakeys devised innovative ways to check the validity of their theories. For example, they made stone tools and used them as *Homo habilis* probably did. They were able to consider a number of aspects about the problem of understanding primitive manlike creatures, such as how they lived, what they ate, and whether they were hunters or scavengers. Pulling all the aspects together in probing such a large problem is something most people are unable to do. This, as much as anything, is the secret behind the Leakey legend.

Also like his father, Richard drives himself very hard. He works a 12-hour day at the museum and then spends more time on research at home. His many activities in running the museum, his involvement in wildlife organizations and foreign lecture tours—plus visits to the Rudolf excavations—produce a full schedule. Again like his father, Richard strongly identifies with Kenya and the African people. He refuses to visit South Africa, where there are some important fossils, because of that country's policy of separating the races. Richard will only go when his Kenyan colleagues can also visit there on an equal basis.

The Leakey family has always shared the concerns of the black nationalists in Kenya. Louis was desperately upset in the early 1950s by the tactics of the Kikuyu-led freedom fighters in the Mau Mau uprising. Louis identified totally with the Mau Mau aims of independence for the people of Kenya. But he deplored and publicly opposed their violent methods. The Kikuyu were Louis' blood brothers, and the situation caused him intense distress, though he had predicted a bloody conflict in a book he wrote some years before. He had seen that if independence from Great Britain did not come soon enough and in the right way, fiery passions would overflow.

Richard's own close identification with Kenya and its people has prompted many to speculate about a political career for him. Having achieved so much so young, Richard is constantly asked what he plans to do in the future. But he delights in remaining tantalizingly mysterious about his plans. He repeatedly denies the possibility of a political career—but with an inscrutable smile. Who knows, then, what direction the Leakey legend will take in the future?

James Gunn

By John F. Henahan

**Theorist and observer, this young astronomer
creates controversy with simple answers
to nature's most complex cosmic questions**

It is a cold November night and starlight bathes the silvery dome that houses the 200-inch (5-meter) Hale telescope on Palomar Mountain in southern California. A small room at the base of the telescope is dark except for the flickering green light of an oscilloscope. The room is unheated to ensure the accuracy of sensitive equipment that is crammed together as tightly as that in an astronaut's space capsule. To keep warm during the night, one of the men in the room wears a U.S. Air Force "hot suit," which is connected to an electric outlet. He is James E. Gunn, professor of astronomy at the California Institute of Technology (Caltech) in Pasadena and staff member of the Hale Observatories. Although he does not rocket away from the earth in his work, the far-reaching eye of the big telescope takes him deeper into space than any astronaut will ever go.

James Gunn is only 36 years old, but he has already earned a reputation as one of the most brilliant, energetic, and dedicated astronomers at work today. A short man with moderately long hair and a ready smile that flashes periodically in the eerie glow of the oscilloscope, Gunn watches intently as an electronic system that he designed transforms data from the incoming light of a distant star into a pattern on the screen. He pushes a button, and the system records the data on the magnetic memory of a computer.

But James Gunn is much more than an innovative observational astronomer, willing to spend nearly 50 nights a year in this cold room. "Jim is a brilliant theoretician as well as an excellent observer, two qualities that you don't often find combined in one astronomer," says J. Beverley Oke, associate director of the Hale Observatories. "He's the sort of person that can do what you can do, but he can do it better. He's just an incredible combination of so many things."

In a field where scarce and imprecise data produce almost as many different theories as there are astronomers, Gunn has often been at the center of controversy. However, unlike other young astronomers who make their mark by leaping to far-fetched conclusions, Gunn patiently pieces together his solutions to cosmological puzzles within the framework of known and well-established physics. This approach makes him a conservative in a field dominated by colorful and flamboyant liberals who are quite willing to invoke as-yet-unseen matter or unknown forces to explain maverick observations.

For example, in December, 1974, Gunn and three other astronomers published a paper that shook the astronomy community. In it, they argued forcefully for the open, or ever-expanding, model of the universe. Pulling together seemingly disconnected measurements and deductions, they maintained that the expanding universe contains only about one-tenth of the matter that would be needed for it to slow down, stop expanding, and eventually collapse to a superdense clump of matter under the pull of gravity. Within weeks, news magazines widely circulated their conclusions.

It was a move totally characteristic of James Gunn, who admitted that the paper was "basically a biased review." Many other astronomers believe strongly that the missing matter might be hidden in unseen places in the universe. But to Gunn, this is an unnecessary complication. "All our findings are highly suggestive, but not watertight evidence, that the universe is open," he says. "For that matter, they could very well be all wrong. But if you say they're wrong, you almost have to devise strategies that complicate what is basically a very simple picture. I love simple things and, so far, nature has always indicated that she also prefers simple things."

In search of simple theories, Gunn willingly spends long hours in the small, cold room on Palomar Mountain painstakingly gathering more data. Each time he presses that button, another clue joins all the bits and pieces that he and his colleagues have been gathering from light that has been traveling toward the earth for millions of years. Gunn hopes that the clues will help him answer the questions that have obsessed him for most of his professional life: Where is the universe going, and what will it do when it gets there?

James Gunn's trek to the edge of the universe began in Beeville, Tex., a small town near the Gulf Coast that he says is noted primarily for "oil, cattle, and the Navy." His father, a geophysicist, encouraged him to study science. As a high school student, Gunn toyed with the

The author:
John F. Henahan
is a free-lance science
writer. He wrote
"The Biologists Bite
Back" for the 1975
edition of *Science Year*.

In the Hale telescope's Cassegrainian focal cage, *above,* James Gunn prepares for a nightlong vigil. Then he descends to the main data room, *below.*

Gunn monitors a TV display, *above,* linking him to the scope. He then checks data in the computer, *below.*

A passion for building tools with which to probe the heavens began when, as a teen-ager, Gunn built his own telescope, *right*. He now spends long hours improving various telescope attachments, *below* and *below right*, at the Hale Observatories.

idea of becoming a chemist, but he soon became more interested in the distant bodies that shone nightly in the Texas skies.

During his high school years in Beeville, Gunn first showed his penchant for observation by building a small, but sophisticated, telescope. He patiently ground the telescope's 8-inch (20-centimeter) reflecting mirror and later added an electronic drive and camera attachment. The telescope as well as pictures that Gunn had taken of the Orion Nebula and the Andromeda Spiral were so impressive that they were prominently featured some years later on the pages of two issues of *Sky and Telescope* magazine.

While James Gunn the observer was perfecting his telescope, James Gunn the theoretician began to emerge at Rice University in Houston. In 1961, he earned a bachelor's degree in mathematics and physics, the tools that he would later need to build his astrophysical theories. Even then, Gunn's abilities were obvious. He was elected to Phi Beta Kappa, and won the Miller fellowship for being the highest-ranking undergraduate. His fellow students coined the phrase "to Gunn an exam," meaning to do exceptionally well on a difficult test.

Only after leaving Rice did Gunn decide on a career in astronomy. "When I was in physics, I was looking for answers at the ultimately small level," he says. "A move into astronomy meant a shift to the ultimately large level of the universe. At the time, I was beginning to realize that astronomy might be my first love, but I still didn't have a very strong idea of what professional astronomers did for a living. In retrospect, I took a kind of shot in the dark."

In choosing a graduate school, Gunn was torn between his love for observation and his new-found love of theoretical physics. The choice became one of geography. The West Coast schools boasted first-rate telescopes and viewing conditions, whereas the Eastern universities had a reputation for producing first-rate theorists. He found the best of both worlds at Caltech, where he earned his Ph.D. in 1965 after completing a thesis on clusters of galaxies and how they might be used conveniently to measure the expansion of the universe. Within a year, however, his career was at least slightly sidetracked by military service. A Reserve Officers Training Corps graduate, the young astronomer spent two years in the U.S. Army Corps of Engineers, and he eventually reached the rank of captain. Through an arrangement with the National Aeronautics and Space Administration, he was assigned as a senior space scientist to the Jet Propulsion Laboratories (JPL) at Caltech. "Considering the alternatives, it was a very nice tour of duty," Gunn recalls. "I was able to work on theoretical studies of planetary atmospheres, and I really got a good handle on the electronic side of astronomy."

Gunn's electronics expertise might well have won acclaim even if he had not become a brilliant astronomer. A standard observational technique calls for passing light from a star through a spectrometer. This prismlike device breaks the light into a pattern of colored lines that uniquely identifies the chemical elements of the light source. Astronomers also use these spectral patterns to calculate how fast the source is moving toward or away from them by measuring where the recognizable patterns appear in the spectrometer relative to patterns produced by similar elements on the earth. For objects that are speeding away, the lines are shifted toward the end of the spectrometer where red light normally appears. The greater this red shift, the faster the source is speeding away.

Gunn and his colleague Roger Griffin designed a system for the Palomar Observatory telescope that displays a spectral pattern instantly on the cathode-ray tube of an oscilloscope. At the push of a button, the lined pattern becomes a retrievable mark on the magnetic tape of a computer memory. Their new system dramatically speeds up the process of observation and cataloging by eliminating the long wait for photographic plates to develop—the standard, time-consuming method for recording spectrographic lines.

"Sometimes I wonder whether I'm a scientist or an electronics engineer," Gunn says with a laugh. He explains that building his own hi-fi while in high school is what got him interested in electronics in the first place. Listening to music is one of the few forms of relaxation for both Gunn and his wife, Rosemary, whom he met in Beeville and dated at Rice. Over the years, their tastes have shifted from the lush music of the Russian romantic composers to the intricate compositions of Johann Sebastian Bach and other baroque composers.

Earlier in his career, Gunn had more time to build other things. Shortly after he and Rosemary married in 1961, he built furniture for the small house they own in a wooded section of Altadena, Calif. He also built an addition to the house that doubled its size. "I liked doing it...once," Gunn says wryly.

Gunn also likes mountain climbing, cross-country skiing, and reading science fiction. But he is most relaxed at home with Rosemary—framing a "not too modern" etching or lithograph for their collection, or cooking apple pancakes for a Sunday brunch with their white cat, Morgan le Fay. Like her husband, Rosemary is outwardly low keyed but inwardly intense. Bemused by his preoccupation with astronomy, she is nevertheless impressed by the fame he has earned so quickly. She expresses her strong political feelings by serving as president of the Pasadena Area League of Women Voters.

Gunn shares his wife's political leanings, but confesses to a "vague and not overwhelming" interest in these activities. But that is to be expected of a man who spends so much time at work, especially since he became a full professor at Caltech in 1972. In addition to the 50 nights he spends on Palomar Mountain each year and the countless hours he spends theorizing, he teaches, acts as adviser to a number of graduate students, and serves on the faculty board and graduate study committee. He also spends some time adding electronic improvements to the various Hale Observatories telescopes.

At home, Jim and Rosemary Gunn frame art prints. Later, they cook and eat breakfast while their cat, Morgan le Fay, waits impatiently for a tasty tidbit.

The Gunns can pick
lemons in their backyard,
thanks to the southern
California climate.
Inside, Gunn pursues
one of his hobbies —
designing and building
furniture for his home.

It is often difficult for students to see Gunn because of his heavy schedule. "He ought to give out numbers like they do in the supermarket," one former student suggested in mild exasperation.

"But even though he is extremely busy, he is always willing to spend as much time with you as you need," says another student. "He pays attention to you and seems to know what you're thinking so well that he finishes your sentences for you." "He's really a remarkable guy and, in his own way, a tremendous challenge," says Gus Oemler, a former student. "He thinks fast and he moves fast. Jim tends to overestimate people, perhaps. He tells you 'you can do that,' but what he really means is 'I can do it.' "

But beyond his considerable prowess, a certain measure of serendipity—or blind luck—has also helped propel this modest man to the forefront of astronomy at such a tender age. In July, 1967, a group of astronomers at Cambridge University in England happened to aim their new radio telescope at a seemingly uninteresting constellation. Unexpectedly, they began receiving pulses of radio-frequency energy precisely every 1.33730109 seconds. Other observers soon found more of these pulsars. One that turned on and off as often as 30 times per second was found in October, 1968, in the Crab Nebula—a gaseous blob in our Galaxy, the Milky Way. Most astronomers believe the Crab is the remains of a star that Oriental observers watched explode in broad daylight in the year 1054. Each pulse from the Crab pulsar sent 10 billion times the electric power generated on the earth hurtling into space. Theoreticians raced to explain the energetic pulses.

As fate would have it, James Gunn was working among the East Coast theorists in 1968. He had gone to Princeton University to teach after completing his military service at JPL. To Gunn, the answer to the Crab pulsar puzzle was obvious. "People had been looking for a neutron star in the Crab Nebula for more than 30 years, and now, just where you might have expected to find one, we found the pulsar," he recalls. The connection seemed too strong to be pure coincidence.

Neutron stars are balls of condensed nuclear matter, each comparable to having squeezed all the sun's mass into a sphere about 10 miles (16 kilometers) in diameter. Because of their tiny size and great distance, astronomers had given up hope of ever seeing light from neutron stars, if indeed they existed outside the fertile minds of astrophysicists. "But the best chance for finding neutron stars lay in the gaseous remains of once-collapsed stars like the Crab," explains Gunn, "where the forces might have provided just the right conditions."

To explain the pulsar's punctual behavior, Gunn and his Princeton colleague Jeremiah P. Ostriker sided with those astronomers who believed that the pulsar spun rapidly about its axis instead of actually blinking on and off. The effect would be like that of the rotating beacons that sit atop police cars and emergency vehicles. They worked out the detailed picture of the spinning neutron star model and made measurable predictions that were later confirmed by observations.

"The repeating radio signal is produced when the pulsar, with an enormous magnetic field, rotates rapidly," Gunn explains. "Every time the pole of the magnet sweeps by, a burst of radiation is carried along the lines of magnetic force and hurled outward toward the observer. In effect, we have a great flywheel in the sky."

In a paper published in 1969, Gunn and Ostriker explained why pulsars were very gradually slowing down. They proposed that the flywheellike neutron stars were spinning down by throwing energy into space. "We're now pretty sure that the radiation of very low-frequency radio energy by the spinning magnet slows down the pulsar," Gunn says. Although his pulsar theory is a front-runner after seven years, it is only one of many. And because new observations of pulsars have outpaced theoretical explanations for their behavior, he admits that "we seem to know less and less about pulsars as time goes on." Gunn expresses mild frustration over the fact that astrophysicists cannot quickly measure their success with a single experiment, as chemists or biologists frequently can. "Quite often, it's not known until some time after you've advanced an explanation whether or not it is correct," Gunn says. "Even then, some details aren't fully explained. When you're looking at objects millions of light-years away, you just can't get inside them to see how they work."

When such frustrations seem particularly burdensome, the young scientist looks forward to summer, when he spends as much time as he can backpacking and climbing in the mountains of California, Colorado, or Washington. Even then, Gunn says, the chief value of his summer trips is "to get out from under the immediate, niggling demands that have to be taken care of the rest of the year, and to get myself back into thinking about theoretical astronomy." Gunn was honored in 1972 with an Alfred P. Sloan fellowship that let him do any research he wanted for two years with a $20,000 grant. He used part of it—"at a trickly rate"—to pay for summers in Aspen, Colo., and Cambridge, England, where the main fare was astronomy.

It may have been on one of his climbs in the Rockies that Gunn hardened his position on another controversy that has yet to be decided—the mysterious behavior surrounding quasi-stellar radio sources, *quasars*. They were found in the early 1960s as faint, starlike points on a photograph of a portion of the heavens from which strong radio signals had been detected. Astronomers were astounded when they examined the red shifts of these objects. The spectral patterns suggested that the sources are among the fastest moving and, hence, most distant objects in the universe. To appear as bright as they do from such a distance, quasars must be radiating so much energy that even the prodigious pulsars seem puny by comparison.

Gunn first became involved with quasars after he left Princeton in 1970 to return to Caltech and the Hale Observatories. There, he analyzed the brightness and red shifts of quasars and galaxies seen through the 48-inch (122-centimeter) Schmidt telescope on Palomar

The astronomer-teacher combines blackboard equations with lively gestures to make a point to his graduate class in astrophysics.

Mountain. Gunn found that in many cases there was a correspondence between the red shifts of quasars and the galaxies with which they seemed to be associated. "If you believe that the red shifts of galaxies are reliable indicators of their speed and distance," he argues, "and if the red shifts of a quasar and the galaxy it is associated with agree, then that is strong evidence that they are both receding at the same speed. By now, the statistics are overwhelming that some quasars are as far away as their surrounding galaxies appear to be–that is, some are near the edge of the visible universe."

Not so, say Geoffrey R. Burbidge of the University of California at San Diego, and Gunn's Caltech colleague Halton C. Arp. Burbidge and Arp believe that quasars are just too bright to be so far away. They may indeed be moving at fantastic speeds as their red shifts suggest, but perhaps that is because they are being rapidly expelled from relatively nearby galaxies.

Gunn believes that the expansion of the universe is a perfectly sound reason for the quasars' large red shifts and he is confident that a reason for their brightness will be found. "My own view is that although we don't have a handy explanation for exactly how quasars are triggered

With his colleagues
Barbara Zimmerman
and J. Beverley Oke,
Gunn works out a
computer program.
Alone, he ponders the
cosmic puzzles that have
become his lifework.

in distant galactic cores," Gunn says, "the answer will be found within the physical framework that has been built so far. I just don't think there is any overwhelming evidence to the contrary."

On April 1, 1974, Gunn and Oke published the results of a four-year study that they believe answers the key question: Are quasars associated with galaxies, or are they a separate class of object? With the help of engineer Earle Emery, they used a series of tiny disks to block out the central bright light from the source named BL Lacertae as its light passed through the Hale telescope. With the central portion obliterated, the astronomers measured the red shift of the fuzzy light surrounding BL Lacertae, which was at one time thought to be a variable-brightness star in the Milky Way. The spectral pattern told them that the fuzz was probably a typical galaxy filled with older stars and located about a billion light-years away.

Independent measurements from radio telescopes showed that the brilliant central core was a quasar. "Thus, BL Lacertae is not a variable star, but is evidently a large galaxy with a quasar in it," Gunn concludes. "Though very far away, it is the nearest known quasar."

Within a few months of the Gunn-Oke report, Margaret Burbidge, an observational astronomer who is married to Geoffrey Burbidge, announced that she was unable to confirm Gunn's observations. She used the 120-inch (305-centimeter) telescope at the University of California's Lick Observatory near San Jose, Calif. She suggested strongly that her analyzer was better than Gunn's, and implied that Gunn and Oke had measured the red shift of background light from other stars, not the light from the fuzz around BL Lacertae. Meanwhile, Gunn maintains that observations he made in 1975 reinforce his original claim.

"There is no doubt that Jim is very bright and that you have to take his observations seriously," Geoffrey Burbidge says. "He seems to want to believe that the red shifts of quasars are the result of the expanding universe. On the other hand, he must realize that this is a question that will ultimately be resolved by observers and it cannot be done until others see what he says he is seeing."

Gunn insists that his apparent talent for inserting himself into controversial areas is due more to the nature of modern astronomy than to any obstreperous streak of his own. "Interesting problems are always controversial and that's why they interest me," Gunn says.

So, for James Gunn, a shot in the dark has become a skyrocketing career in astronomy that has taken him to the center of controversy about the ultimate fate of the universe. But whether or not the cosmic questions that obsess him are resolved during his lifetime, he will no doubt continue to leave his conservative mark wherever he seeks the answers. For, as Geoffrey Burbidge says, "Jim is a believer." Gunn disagrees with Burbidge on many points, but he probably would agree on that one. "When you've arrived at the correct theory, you know it," Gunn says. "It has the ring of rightness."

Awards
And Prizes

A listing and description of major science awards and prizes, the men and women who won them, and their accomplishments

Earth and Physical Sciences

Chemistry. Major awards in the field of chemistry included:

Nobel Prize. Paul J. Flory, professor of chemistry at Stanford University, received the 1974 Nobel prize in chemistry for his research, "both theoretical and experimental, in the physical chemistry of macromolecules."

Flory pioneered research on the chemical and physical properties of giant natural and synthetic molecules, including rubbers, fibers, plastics, proteins, and nucleic acids (DNA and RNA). He was a member of the Du Pont research team that created nylon, one of the earliest synthetic polymers to be widely applied.

A polymer is a molecule made up of many single molecules, or monomers, linked together in a long chain. Until Flory developed his ideas and methods, chemists were unable to study polymers in their various states or determine their usefulness for various applications.

Flory, a native of Sterling, Ill., studied at Manchester College in Indiana and at Ohio State University. He was a research chemist for Du Pont, Standard Oil Company of New Jersey, and Goodyear Tire & Rubber Company and was executive director of research at the Mellon Institute in Pittsburgh. He served as professor of chemistry at Cornell University from 1948 to 1956. Flory received the American Chemical Society's Priestley Medal in 1974.

Perkin Medal. Carl Djerassi, professor of chemistry at Stanford University and chief executive of Zoecon Corporation of Palo Alto, Calif., received the 1975 Perkin Medal. Djerassi was honored for his work in synthesizing norethisterone, which led to development of the first oral contraceptives. The award, highest given in the United States for applied chemistry, is by the American Section of the Society of Chemical Industry.

Djerassi was born in Vienna, Austria, in 1923 and came to the U.S. at the age of 16. He attended Kenyon College in Gambier, Ohio, and the University of Wisconsin. Djerassi and his co-workers synthesized norethisterone while working in Mexico City. He received the National Medal of Science in 1973.

Paul J. Flory won the Nobel chemistry prize for work on macromolecules.

Earth and Physical Sciences

Continued

Priestley Medal. Henry Eyring, distinguished professor of chemistry and metallurgy at the University of Utah, received the 1975 Priestley Medal, highest award given by the American Chemical Society. Eyring's absolute rate theory (ART), formulated in 1929, became a broad guide to how chemical reactions take place.

Eyring was born in Mexico and studied at the University of Arizona and the University of California, Berkeley. He taught at Princeton University from 1931 to 1946, then served as dean of the graduate school at the University of Utah for 20 years.

Physics. Awards recognizing major work in physics included:

Nobel Prize. Two British astronomers shared the 1974 Nobel prize in physics for their studies of distant parts of the universe. Both Sir Martin Ryle, professor of radio astronomy, and Antony Hewish, lecturer in radio astronomy, are at Cambridge University. This was the first Nobel prize in physics awarded for work in astronomy.

Ryle was honored for developing the aperture synthesis technique, in which several small radio telescopes are aligned to produce the resolving power of a much larger telescope. This technique is credited with much of the progress in recent radio-astronomical observation.

Hewish was honored for his discovery of pulsars, those mysterious sources of pulsed radio waves that originate far out in space.

Ryle, director of the Mullard Radio Astronomy Observatory at Cambridge, received the Catherine Wolfe Bruce Medal in 1974. Hewish has worked on Ryle's research team since World War II except for a year spent as visiting professor at Yale University.

Buckley Prize. Albert W. Overhauser, Stuart Professor of Physics at Purdue University, received the 1975 Oliver E. Buckley Solid State Physics Prize. He was honored by the American Physical Society "for his invention of dynamic nuclear polarization and for the stimulation provided by his studies of instabilities of the metallic state."

387

Overhauser received his A.B. and Ph.D. degrees from the University of California at Berkeley, and worked in the Ford Motor Company's research laboratories for 15 years before going to Purdue. He also was a research associate at the University of Illinois and taught at Cornell University.

Franklin Medal. Russian physicist and mathematician Nikolai N. Bogolyubov received the 1974 Franklin Medal, highest honor given by the Franklin Institute of Philadelphia. Academician Bogolyubov was honored for "his powerful mathematical methods in nonlinear mechanics."

The Franklin Medal is awarded annually to the person working in physical science or technology whose efforts are deemed to have done the most to advance knowledge of physical science or its application.

Bogolyubov is recognized internationally as a theoretical physicist and teacher. He has held top academic positions at Kiev State University, Moscow State University, and the Academy of Science of the U.S.S.R.

Michelson Medal. Peter P. Sorokin, a researcher for International Business Machines at the Thomas J. Watson Research Center in Yorktown Heights, N.Y., was awarded the Albert A. Michelson Medal in 1974. The Franklin Institute award was for outstanding contributions to quantum optics, particularly his discovery of the organic-dye laser.

Sorokin and his associates used a giant pulse ruby laser in 1966 to produce the first stimulated light emissions from organic molecules.

Geosciences. Awards for important work in the geosciences included: ·

Carty Medal. J. Tuzo Wilson, professor of geophysics at the University of Toronto, Canada, received the 1975 John J. Carty Medal for the Advancement of Science. The award, by the National Academy of Sciences (NAS), includes a $3,000 honorarium and is granted for distinguished accomplishment in any field of science.

Wilson pioneered the plate tectonic, or sea-floor spreading, theory and ap-

British astronomers Antony Hewish, left, and Sir Martin Ryle shared the 1974 Nobel prize in physics, the first to be awarded for work in astronomy.

Earth and Physical Sciences

Continued

Nikolai N. Bogolyubov

plied the concept to many aspects of marine geology. Recent deep-sea drilling studies tend to bear out his theories. He used his knowledge of the ancient continents and modern sea floor to reconstruct the history of the Atlantic.

Day Medal. Alfred E. Ringwood, a noted Australian geoscientist, won the Arthur L. Day Medal given by the Geological Society of America (GSA).

Ringwood led in the development and operation of a geophysics laboratory at the Australian National University in Canberra. This lab has become one of the world's leaders in studies of phase changes in the earth's mantle.

Ringwood was born in Melbourne and attended Melbourne University and Harvard University. He became a senior research fellow at the Australian National University in 1959.

Penrose Medal. W. Maurice Ewing, a pioneer researcher in the earth sciences, was awarded the Penrose Medal posthumously by the GSA. Ewing died on May 4, 1974, at the age of 67.

Ewing founded the Lamont-Doherty Geological Observatory at Columbia

University and was its director from 1948 to 1972. Since 1972, he had served as chief of the Earth and Planetary Sciences Division at the University of Texas Marine Biomedical Institute in Galveston.

Ewing was the first scientist to use seismic waves to study the ocean floor. He also developed a seismograph that has become a standard in measuring the strength of earth tremors.

Vetlesen Prize. Chaim Leib Pekeris, professor at the Weizmann Institute of Science in Israel, was awarded Columbia University's 1974 Vetlesen Prize. The award, a gold medal and $25,000, recognizes "achievement in the sciences resulting in a clearer understanding of the earth, its history, or its relation to the universe."

Pekeris, a mathematician, has studied the earth's modes of free oscillation, convection within the earth, propagation of sound in layered media, and the theory of the earth's magnetic field. He and his Weizmann colleagues designed and built three computers which he has used to compute tides on a global scale.

Life Sciences

G. Wesley Hatfield

Biology. Among the awards presented in biology were the following:

Horwitz Prize. Boris Ephrussi, 73, French geneticist, received the 1974 Louisa Gross Horwitz Prize. The $25,000 prize is awarded annually for outstanding research in biology.

Ephrussi discovered the "petite mutants" of yeast, from which developed the genetics of mitochondria, minute bodies found within cells that help in cell metabolism. This discovery "permitted the development of this whole new field of science."

Lilly Award. G. Wesley Hatfield, microbiologist at the University of California at Irvine, received the 1975 Eli Lilly and Company Award in Microbiology and Immunology. The medal and $1,000 were for research on the autoregulation of gene expression.

Hatfield studied at Santa Barbara, Purdue, and Duke universities.

Tyler Ecology Award. Ecologist Ruth Patrick, chairman of the board of the Academy of Natural Sciences in Philadelphia, won the $150,000 John and Alice Tyler Ecology Award for her

studies of polluted streams. The award, largest in the world for scientific achievement, is given to the individual or group judged to have made the year's greatest contribution to ecology.

U.S. Steel Foundation Award. Bruce M. Alberts of the Princeton University Department of Biochemical Sciences received the $5,000 U.S. Steel Foundation Award in molecular biology. He was honored particularly for his discovery of a protein, referred to as gene-number-32 protein, in phage T4. This protein is required for genetic recombination and is involved in deoxyribonucleic acid (DNA) replication.

Medicine. Major awards in medical sciences included the following:

Nobel Prize. Three pioneers in cell biology shared the 1974 Nobel prize in physiology and medicine. They are Albert Claude, director of the Jules Bordet Institute in Brussels, Belgium; George E. Palade of Yale University; and Christian de Duve of Rockefeller University and the University of Louvain in Belgium.

The men were honored for work they did about 30 years ago at the Rockefeller Institute for Medical Research (now Rockefeller University).

Claude was the first to use the electron microscope to study cells. He also pioneered in using centrifuges to separate cell components for analysis. Palade and De Duve refined Claude's methods to obtain more sophisticated analyses. They discovered ribosomes and lysosomes and determined that cells use ribosomes as conveyor belts in manufacturing proteins. Lysosomes, they found, break down large chunks of food so the cell can digest them.

Gairdner Awards. Seven medical scientists received Gairdner Foundation International Awards in 1974. They were honored by the Gairdner Foundation of Toronto, Canada, for research.

Judah H. Quastel of the University of British Columbia received $10,000 for his biochemical research.

Hans J. Müller-Eberhard from the Scripps Clinic and Research Foundation in San Diego, Calif., won $10,000 for expanding understanding of the molecular basis of the complement system in man.

Hector F. DeLuca of the University of Wisconsin won $10,000 for his explanation of the metabolism of vitamin D.

Roger Guillemin of the Salk Institute in San Diego and Andrew V. Schally of Tulane University received $5,000 each for their work in the identification, synthesis, and clinical application of hypothalamic-releasing hormones.

David Baltimore of the Massachusetts Institute of Technology in Cambridge and Howard M. Temin of the University of Wisconsin received $5,000 each for their research on how viruses act in tumor production.

Kittay Award. Harry F. Harlow, 70, former director of the Primate Research Center at the University of Wisconsin, won the $25,000 Kittay International Award in 1975. The award, given annually by the Kittay Scientific Foundation, was for his studies of infant monkeys deprived of their mothers, which helped psychologists understand the nature of the bond between mother and infant.

Biologist George E. Palade shared the 1974 Nobel prize in medicine for his part in pioneering studies on the inner structure and workings of the cell.

Space Sciences

Aerospace. The highest awards granted in the aerospace sciences included:

Collier Trophy. John F. Clark, director of the National Aeronautics and Space Administration (NASA) Goddard Space Flight Center in Greenbelt, Md., and Daniel J. Fink, vice-president of General Electric's Space Division, received the 1974 Robert J. Collier Trophy. The National Aeronautic Association decided on LANDSAT, the Earth Resources Technology Satellite (ERTS) program, as 1974's outstanding aerospace event.

Clark and Fink were honored as leaders of the NASA-industry team that proved the value of remote sensing from space. ERTS-1, the program's first satellite, scanned the earth from 570 miles (917 kilometers) up, and its remote-sensing system provided an astonishing mass of data on the earth's resources.

Goddard Award. Gordon E. Holbrook and George Rosen shared the 1975 Goddard Award. Holbrook is director of engineering, aircraft gas turbines, Detroit Diesel Allison Division, General Motors Corporation. Rosen is chief of propeller research for the Hamilton Standard Division of United Aircraft Corporation.

The men were honored for their "joint leadership and contributions in pioneering the development and production of widely used turbopropeller propulsion systems."

Hill Space Transportation Award. Rocco A. Petrone, associate administrator of NASA, received the 1974 Hill Space Transportation Award, presented by the American Institute of Aeronautics and Astronautics.

Petrone was honored for his contributions to space-flight technology in directing the last six successful Apollo missions to the moon. The award includes a certificate and a $5,000 cash honorarium.

Astronomy. Important contributions in astronomy were honored with the following awards:

Arctowski Medal. Jacques M. Beckers, solar observer at the Sacramento Peak Observatory in Sunspot, N. Mex., received the Henryk Arctowski Medal in 1975. The award is made by the NAS for studies of solar activity changes and their effects upon the ionosphere and terrestrial atmosphere. It includes a gold medal and a $5,000 honorarium.

Beckers was honored for his ingenuity in developing new instrumentation to observe and discover new aspects of solar activity. Instruments he invented are now being used by other astronomers.

Bruce Medal. Allan R. Sandage of the Hale Observatories received the 1975 Catherine Wolfe Bruce Medal. The gold medal is presented by the Astronomical Society of the Pacific.

Sandage is noted for his discovery of quasars, or quasi-stellar objects, in space. He found the first quasar in 1960 and the first X-ray star in 1966.

He published a series of papers in 1972 summarizing more than 40 years of work on the relationship between red shifts and the distances of remote galaxies. This work resulted in new values for the expansion of the universe and supports the "big bang" theory — that the universe originated in an explosion of hydrogen.

Gould Prize. Astrophysicist Lodewyk Woltjer was awarded the Benjamin Apthorp Gould Prize for 1975. The award, which includes a $5,000 honorarium, cited Woltjer's observational and theoretical studies, which advanced understanding of the physics, dynamics, and astrophysics of interstellar matter.

The award, made by NAS, also cited his work as editor of *The Astronomical Journal* from 1967 to 1974. Woltjer was recently named director-general of the European Southern Observatory.

Watson Medal. Gerald M. Clemence, an astronomer with the U.S. Naval Observatory and the Yale Observatory, was posthumously awarded the James Craig Watson Gold Medal for 1975. Clemence died on Nov. 22, 1974.

Clemence was honored for his distinguished career in dynamical astronomy and his independent determination of a new and more accurate theory of the motion of stars, which he completed in 1961.

Clemence served on the staff of the Naval Observatory from 1930 to 1940, in the Nautical Almanac Office from 1942 to 1958, and was scientific director of the observatory from 1958 to 1963. Then he joined the Yale Observatory.

John F. Clark

Daniel J. Fink

Allan R. Sandage was awarded the 1975 Bruce Medal for outstanding research in astronomy.

General Awards

Science and Man. Awards for outstanding contributions to science and mankind included the following:

Founders Medal. James B. Fisk, former chairman of the board of Bell Laboratories, received the 1975 Founders Medal for "his leadership in the advancement of communications technology for the benefit of society." The award is given by the National Academy of Engineering (NAE).

Fisk joined Bell Labs in 1939 as a research engineer. He became president in 1959 and board chairman in 1973, then retired in 1974. He received a Presidential Certificate of Merit for directing the development during World War II of the microwave magnetron, a device used to generate the pulses in high-frequency radar.

Oersted Medal. Robert Resnick, professor of physics at Rensselaer Polytechnic Institute in Troy, N.Y., received the 1975 Oersted Medal. The award, the highest honor given by the American Association of Physics Teachers (AAPT), was for his "notable contributions to the teaching of physics."

Resnick is the author or coauthor of several textbooks. One of these, *Physics,* has been translated into 14 languages. He has also been active in several national projects on physics instruction. He was a consultant to Harvard Project Physics and contributed materials to it. He was co-director of the AAPT project which prepared the two-volume *Physics Demonstration Experiments.*

Resnick received his Ph.D. degree in theoretical nuclear physics from Johns Hopkins University in 1949.

Zworykin Award. Jack S. Kilby, a consultant who invented the monolithic integrated circuit, won the 1975 Vladimir K. Zworykin Award. The $5,000 award is presented annually by the NAE.

Kilby hit upon the idea of forming electronic components in a single block of semiconductor material in 1958, the year he joined Texas Instruments, Incorporated, in Dallas. His invention made possible such devices as the electronic calculator and electronic watch. He received the National Medal of Science in 1969. [Joseph P. Spohn]

Major Awards and Prizes

Award winners treated more fully in the first portion of this section are indicated by an asterisk (*)

Adams Award (organic chemistry): Rolf Huisgen

American Physical Society High-Polymer Physics Prize: Walter H. Stockmayer

*Arctowski Medal (astronomy): Jacques M. Beckers

Arthur L. Day Prize (physics): Drummond H. Matthews, Fred J. Vine

Bonner Prize (nuclear physics): Chien Shiung-wu

Bowie Medal (geophysics): Sir Edward Bullard

*Bruce Medal (astronomy): Allan R. Sandage

*Buckley Solid State Physics Prize: Albert W. Overhauser

Carski Foundation Award (teaching): Matthew C. Dodd

*Carty Medal: J. Tuzo Wilson

Centennial of Chemistry Award: Sir Derek Barton

Clowes Prize (cancer research): Elwood V. Jensen

*Collier Trophy (astronautics): John F. Clark, Daniel J. Fink

Davisson-Germer Prize (optics): Homer D. Hagstrum, James J. Lander

*Day Medal (geology): Alfred E. Ringwood

Debye Award (physical chemistry): Herbert S. Gutowsky

Fields Medal (mathematics): David B. Mumford

Fleming Award (geophysics): Carl E. McIlwain

*Founders Medal (engineering): James B. Fisk

*Franklin Medal (physics): Nikolai Bogolyubov

*Gairdner Awards (medicine): Dr. David Baltimore, Dr. Hector F. DeLuca, Dr. Roger Guillemin, Dr. Hans J. Müller-Eberhard, Dr. Judah H. Quastel, Dr. Andrew V. Schally, Dr. Howard M. Temin

Garvan Medal (chemistry): Marjorie C. Caserio

Gibbs Medal (chemistry): Herman F. Mark

*Goddard Award (aerospace): Gordon E. Holbrook, George Rosen

*Gould Prize (astronomy): Lodewyk Woltjer

Guggenheim Award (aerospace): Floyd L. Thompson

Haley Astronautics Award: Paul J. Weitz, Charles Conrad, Jr., Joseph P. Kerwin

Heineman Prize (mathematical physics): Ludwig Faddeev

*Hill Space Transportation Award (astronautics): Rocco A. Petrone

*Horwitz Prize (biology): Boris Ephrussi

Hubbard Medal (geography): Alexander Wetmore

Ives Medal (optics): David L. MacAdam

*Kittay Award (psychiatry): Harry F. Harlow

Klumpke-Roberts Prize (astronomy): Isaac Asimov

Langmuir Prize (chemical physics): Robert H. Cole

Lasker Awards (medical research): Dr. John Charnley, Dr. Ludwik Gross, Dr. Howard E. Skipper, Dr. Sol Spiegelman, Dr. Howard M. Temin

*Lilly Award (microbiology): G. Wesley Hatfield

*Michelson Medal (astronomy): Peter P. Sorokin

NAS Award for Environmental Quality: John T. Middleton

*Nobel Prize: chemistry, Paul J. Flory; physics, Sir Martin Ryle and Antony Hewish; physiology and medicine, Albert Claude, George E. Palade, and Christian de Duve

*Oersted Medal (teaching): Robert Resnick

Oppenheimer Memorial Prize (physics): Nicholas J. Kemmer

Pendray Award (aerospace): William R. Sears

*Penrose Medal (geology): W. Maurice Ewing (posthumous)

*Perkin Medal (chemistry): Carl Djerassi

*Priestley Medal (chemistry): Henry Eyring

Reed Award (aerospace): Antonio Ferri

Royal Astronomical Society Gold Medal: Jesse L. Greenstein

Trumpler Award (astronomy): J. R. Gott III

*Tyler Ecology Award: Ruth Patrick

*U.S. Steel Foundation Award (molecular biology): Bruce M. Alberts

*Vetlesen Prize (geophysics): Chaim Leib Pekeris

*Watson Medal (astronomy): Gerald M. Clemence (posthumous)

*Zworykin Award (engineering): Jack S. Kilby

Notable scientists who died between June 1, 1974, and June 1, 1975, include those listed below. An asterisk (*) indicates that a biography of the person appears in *The World Book Encyclopedia.*

Alexanderson, Ernst F. W. (1878-May 14, 1975), Swedish-born electronics engineer and inventor whose high-frequency alternator made possible the first voice radiobroadcast in 1906.

Apgar, Virginia (1909-Aug. 7, 1974), physician, developed the Apgar Score, an internationally recognized test that measures five body functions and helps to determine 60 seconds after birth if a baby needs help to sustain life.

Baker, James A. (1910-April 14, 1975), animal virologist and educator, led research at Cornell University in developing vaccines for puppies.

Bergmann, Ernst D. (1903-April 6, 1975), German-born Israeli organic chemist and educator, was chairman of Israel Atomic Energy Commission.

Blackett, Lord Patrick M. S. (1897-July 13, 1974), British physicist, won the 1948 Nobel prize in physics for cosmic-ray studies that confirmed the existence of the positron. A leader in research during World War II, he was made a life peer by Queen Elizabeth II in 1969.

Blagonravov, Anatoly A. (1894-Feb. 4, 1975), Russian scientist who played a key role in the Sputnik program.

Brode, Wallace R. (1900-Aug. 10, 1974), chemist, was associate director of the National Bureau of Standards from 1947 to 1957. He won the Priestley Medal in 1960, highest award given by the American Chemical Society.

Bronowski, Jacob (1908-Aug. 22, 1974), Polish-born British mathematician and writer, was a fellow of the Salk Institute for Biological Studies in San Diego. He filmed the prizewinning BBC-TV series "The Ascent of Man," and wrote many books and two plays.

*****Bush, Vannevar** (1890-June 28, 1974), electrical engineer, president of the Carnegie Institution of Washington from 1939 to 1955. He was director of the Office of Scientific Research and Development during World War II. In the 1930s, at the Massachusetts Institute of Technology (M.I.T.), he developed a differential analyzer, the forerunner of the electronic analogue computer.

Carbonara, Victor E. (1895-Feb. 23, 1975), Italian-born mechanical engineer and former president of Kollsman Instrument Company, helped to prepare charts and navigational data for Charles A. Lindbergh, Richard E. Byrd, and other early transatlantic pilots.

Carpenter, Clarence R. (1905-March 1, 1975), psychologist and anthropologist at Pennsylvania State University from 1940 to 1969, studied monkeys and apes for clues to human behavior.

*****Chadwick, Sir James** (1891-July 24, 1974), British physicist, won the 1935 Nobel prize in physics for his discovery of the neutron. He headed the British team that worked on the Manhattan Project to develop the atomic bomb.

Coolidge, William D. (1873-Feb. 3, 1975), physical chemist and electrical engineer, discovered in 1908 the method for drawing out of tungsten the hair-thin filaments used in incandescent light bulbs. In 1913, he developed the "Coolidge tube" which remains the basis of modern X-ray units.

Craig, Lyman C. (1906-July 7, 1974), chemist, invented countercurrent distribution process used for isolating and studying synthetic antimalarials, antibiotics, hormones, and proteins.

Drickey, Darrell J. (1934-Dec. 10, 1974), experimental physicist, led the joint Russian-U.S. team in Serpukhov, Russia, that in 1970 made the first measurement of the pi meson.

Fajans, Kasimir (1887-May 18, 1975), Polish-born professor of chemistry at the University of Michigan from 1936 to 1957, established Fajans' absorption indicators for volume analysis.

Fok, Vladimir A. (1898-Dec. 27, 1974), Russian theoretical physicist, won the 1946 Stalin Prize and the 1960 Lenin Prize, established the scalar relativistic wave equation for a particle with no spin in an electromagnetic field.

Goldsmith, Alfred N. (1887-July 2, 1974), prolific inventor and electronics engineer, created the first commercial radio phonograph.

Gordon, Antoinette K. (1891-March 24, 1975), anthropologist and authority on Tibetan religion and art, translated ancient stories of wisdom and folly from the Tibetan in *Tibetan Tales* (1953).

Heffner, Hubert (1924-April 1, 1975), professor of applied physics at

Lord Patrick M. S. Blackett

Jacob Bronowski

Vannevar Bush

Deaths of Notable Scientists

Continued

Sir Julian S. Huxley

Percy L. Julian

Sir Robert Robinson

Stanford University from 1954 to 1975, was deputy director of the White House Office of Science and Technology from 1969 to 1971.

*Huxley, Sir Julian S. (1887-Feb. 14, 1975), British biologist, author, and humanist, served as the first director-general of the United Nations Educational, Scientific, and Cultural Organization from 1946 to 1948. He did research in ornithology and the experimental analysis of development. He was knighted in 1958 for his contributions to science.

*Julian, Percy L. (1899-April 19, 1975), grandson of slaves, overcame a deficient early education to become a renowned organic chemist. He discovered a low-cost synthesis for cortisone and a way to mass-produce physostigmine, which is used to treat glaucoma.

Kartveli, Alexander (1896-July 20, 1974), Russian-born aeronautical engineer, designed the P-47 fighter plane of World War II and the F-105 fighter-bomber of the Vietnam War.

Kirk, John E. (1905-April 7, 1975), Danish-born physician and educator, cited in 1958 by the National Gerontological Research Foundation for his outstanding contributions to the study of the aging process.

Letov, Aleksandr (1912-Sept. 29, 1974), Russian mathematician and expert on control systems, won the 1972 Soviet State Prize for his contribution to control theory.

Lewis, Warren K. (1882-March 9, 1975), chemical engineer, was a leader in researching chemical engineering principles in the production and refining of petroleum. He was on the M.I.T. faculty from 1910 to 1948 and won the 1947 Priestley Medal.

Lindemann, Erich (1900-Nov. 16, 1974), German-born psychiatrist and psychologist, was a professor at Harvard University from 1935 to 1965 and pioneered in applying social sciences approaches to psychiatric problems.

Michels, Walter C. (1906-Feb. 27, 1975), professor of physics at Bryn Mawr College from 1932 to 1972, won the 1963 Oersted Medal for notable contributions to physics teaching.

Mirsky, Alfred E. (1900-June 19, 1974), early molecular biologist, found methods nearly 30 years ago to isolate chromatin for study.

Mudd, Stuart (1893-May 6, 1975), microbiologist and educator, helped develop a process for freeze-drying blood plasma during World War II.

Perey, Marguerite (1909-May 13, 1975), French nuclear chemist and educator, co-worker of Marie Curie, discovered element 87, a radioactive substance now called francium.

*Robinson, Sir Robert (1886-Feb. 8, 1975), British organic chemist, won the 1947 Nobel prize for chemistry, and the 1953 Priestley Medal. He determined the structure and behavior of many important plant substances, including penicillin. He was a professor of organic chemistry in Great Britain and Australia and was knighted in 1939.

Serin, Bernard (1922-June 18, 1974), physicist, discovered in 1950 the isotope effect that demonstrated the essential role of the electron-phonon interaction in superconductivity.

Sober, Henry A. (1918-Nov. 26, 1974), biochemist, co-winner with Elbert Peterson of the 1970 Hillebrand Award for development of the modified cellulose ion exchangers.

Swingle, Wilbur W. (1891-May 20, 1975), physiologist and professor of biology at Princeton University from 1929 to 1959, developed a drug for treating Addison's disease in 1929.

Vance, John E. (1905-March 19, 1975), professor emeritus of chemistry at New York University who worked on processing ores and the preparation of pure uranium compounds for the Manhattan Project, which developed the atomic bomb.

Vaurie, Charles (1906-May 13, 1975), French-born curator of the Department of Ornithology of the American Museum of Natural History from 1967 to 1972.

Young, Rodney S. (1907-Oct. 25, 1974), University of Pennsylvania archaeologist, directed excavations from 1950 to 1974 in Gordium, Turkey, where he unearthed a palace believed to have been used by King Midas of the legendary golden touch.

Zmuda, Alfred J. (1921-July 14, 1974), physicist at Johns Hopkins University since 1951, whose studies with research satellites led to a better understanding of the earth's magnetic field and the causes of the bands of light of the aurora borealis.   [Irene B. Keller]

American Science: The First 200 Years

By A. Hunter Dupree

As with other forces that contributed to the development of the United States, science and technology have shaped and been shaped by the character of the people

Americans have always thought of themselves as a practical and ingenious people and, since World War II, they have considered the United States the world's leading scientific nation. The atomic bomb, the flights to the moon, and the mechanical and electrical networks that make up most of the material culture in the United States have all come from American science and technology.

Most Americans support science, including the billions of dollars the federal government spends each year on research and development. Some people have developed reservations about science because of the atomic bomb and the peacetime nuclear power industry, which produces deadly poisonous wastes as a threat both to man and to the environment. But both friends and foes agree on one thing—science is powerful. It can do more good, and more harm, than any other force in American life.

Such was not the case 200 years ago when the nation was born. In revolutionary America the fruits of science—mainly medicine and agriculture—came from European vines. Yet, even then, a few major figures had begun to emerge who, with the mixture of curiosity and practicality peculiar to Americans, were to build a foundation of science that would have a major influence in the nation's development.

One of the early giants was also one of the Founding Fathers–
Benjamin Franklin (1709-1790). Born in Boston, Franklin made his
fortune as a printer in Philadelphia, and by the time he was 40 he had
retired from business to devote himself to public service and to science.
His electrical experiments spread his fame to Europe. With his famous
kite demonstration, he showed that lightning is a form of electricity.
He developed the theory that electricity is a single "fluid" which gives
matter a positive or negative charge. He introduced the technical
words "plus," "minus," "charge," and "battery."

Late in 1776, Franklin went to France to negotiate an alliance, and
he stayed on in Paris as ambassador until 1785. Much of his effectiveness there stemmed from his scientific reputation, for here was a man
who combined the political ideals of the American Revolution and the
rationality of the Age of Reason.

Equally as important as scientists were the patrons of science.
Thomas Jefferson (1743-1826) was both. His contributions to science
ranged from national measuring systems, such as our decimal mone-
tary system, to paleontology. His *Notes on Virginia* (1784) tied his
natural and political philosophies together to describe his native state
as a favorable human environment.

As the third President of the United States (1801-1809), Jefferson
believed national institutions to be as necessary to the developing

The inventive genius of men like Benjamin Franklin,
left, and the leadership and encouragement of men
like Thomas Jefferson, *below,* gave a direction to
American science that soon made it as independent
from the Old World as the new nation itself.

nation as roads and canals. The United States Military Academy at West Point, N.Y., which was the nation's first advanced engineering school, and the Coast and Geodetic Survey, forerunner of the National Ocean Survey, were begun during Jefferson's Administration. He also set the pattern for scientific exploration for most of the 19th century when he instituted and obtained the money for one of the most famous expeditions of all time.

Under the command of Meriwether Lewis (1774-1809) and William Clark (1770-1838), an expedition set out in 1804 to explore the new Louisiana Territory. The group traveled up the Missouri River to its source, found a way through the Rocky Mountains and down the Columbia River to the Pacific Ocean. The journey was more than a geographical exploration. Jefferson had sent Lewis to Philadelphia before his departure to learn how to make scientific collections. As a result, the explorers brought back a variety of dried plant specimens, animal skins, and voluminous journals containing observations on the peoples and objects they encountered. The data not only contributed to American science, but much of it also ultimately reached European science centers.

The Lewis and Clark Expedition was followed by many others. Matthew F. Maury (1806-1873), director of the U.S. Navy Department's Depot of Charts and Instruments, cut many days off the passage around Cape Horn by his research on the physical geography of the sea. Army expeditions, under the command of a special West Point-educated Corps of Topographical Engineers, crisscrossed vast areas west of the Mississippi, finding and mapping wagon roads and railroad routes and developing a comprehensive picture of the plants, animals, and geology of the area.

American scientific research extended beyond its shores in 1838, when the U.S. Navy sent out a United States Exploring Expedition under the command of Captain Charles Wilkes (1798-1877). Carrying a corps of scientists, Wilkes's ship ranged from the Antarctic to Puget Sound and across the Pacific. Expedition mineralogist and geologist James D. Dana (1813-1895) made important studies of the growth of coral reefs, and naturalist Charles Pickering (1805-1878) pioneered in observations of the geographical distribution of man in the Pacific area. The worldwide collection of plants, animals, anthropological observations, and astronomical and hydrographic information flooded the U.S. institutions of the time, stimulating support for more museums and observatories, and the publication of scientific books.

Before the Civil War, colleges and universities played minimal roles in science, although by 1840, institutions such as Harvard and Yale had begun scientific research. At Yale, chemist Benjamin Silliman (1779-1864) and Dana, veteran of the Wilkes Expedition, trained a large number of professional scientists.

Harvard also began to reflect the demand for knowledge of science. Mathematician Benjamin Peirce (1809-1880) was the moving spirit

The author:
A. Hunter Dupree is George L. Littlefield Professor of History at Brown University. He wrote *Science in the Federal Government* and *Asa Gray*.

behind the Harvard Observatory and brought the national *Nautical Almanac* to Cambridge. Botanist Asa Gray (1810-1888) studied and incorporated into his herbarium the plants collected west of the Mississippi on the Wilkes Expedition, and on the American expeditions to open Japan in the 1850s. Jeffries Wyman (1810-1874) established an anatomical museum and later studied anthropology. Eben N. Horsford (1818-1893) brought German analytical chemistry to Cambridge.

Harvard's scientific atmosphere began to generate sparks in 1846 with the arrival from Switzerland of the flamboyant Louis Agassiz (1807-1873). He was both a zoologist and a geologist, with the whole of the animal kingdom, living and fossil, at his command. He worked with the Coast and Geodetic Survey and established a Museum of Comparative Zoology at Harvard, for which he raised huge sums of money. He introduced a new generation of students to the thrill of original research. Agassiz would give a new student a preserved fish and, without accompanying books or explanation, tell him to describe it. After hours of staring at the fish, the student would begin to notice some interesting features. Two or three days later, Agassiz would come back, praise the student, and begin to instruct him. The student seldom forgot the lesson of the value of close observation of nature, independent of theory.

By the 1850s, American university scientists were competent enough in theory and technique to participate in the major revolution in scientific thinking begun by Charles R. Darwin (1809-1882). Darwin corresponded with Gray and encouraged him to study the geographical

Lewis and Clark met and studied Indian tribes as part of their charter to explore and report on the Louisiana Territory.

Antarctica's chilly shore was the site of one series of scientific investigations by geologists on expeditions led by Charles Wilkes.

Abraham Lincoln signed a bill establishing the National Academy of Sciences in 1863. The academy's founders included, from left, Peirce, Bache, Henry, Agassiz, Lincoln, Wilson, Davis, and Gould.

John Wesley Powell championed the Indians whose land he studied.

distribution of plants in North America. Gray demonstrated, even before Darwin's *Origin of Species* was published in November, 1859, that certain plants—dried specimens of which were collected in the southern Appalachians and on the American expeditions to Japan—were unique in that they appeared only in the two areas. Although similar, they were sufficiently distinct to be considered separate species. Gray speculated in a paper published in early 1859 that, before the period of glaciers, the plants, now separated in eastern Asia and eastern North America, had been part of one *flora* across the top of Asia and North America, including a then-dry Bering Strait. The plants had been pushed south by the glaciers, varying enough through many generations to become separate species.

This analysis agreed with Darwin's ideas. Interestingly, while Gray became the champion of a fair hearing for Darwin, Agassiz opposed Darwin's theory of evolution by natural selection to the end of his life.

While science and scientists of both North and South played a part in the development of weapons during the Civil War, the real significance of the period from 1861 to 1865 was the legislative climate in which the National Academy of Sciences (NAS), the land-grant college system, and the U.S. Department of Agriculture were created. The federal government's support of science had begun.

After the Civil War, the United States began moving toward its position as one of the richest, most populous, and most energy-intensive of the world's industrialized nations. The number of scientists began to increase and they were better organized, better supported, and more professionally competent.

The old geographic-scientific expeditions were now giving way to government-supported surveys. A survey in the 1870s produced one of the most creative scientific statesmen in American history. School-

Simple but serviceable, the Powell camps were a reflection of the rugged outdoors that was the laboratory of explorers and geologists.

teacher John Wesley Powell (1834-1902), a one-armed veteran of the Civil War, began the explorations which culminated in the great adventure of running the Colorado River in 1869. Starting at Green River, Wyo., on May 24, 1869, with a party of 11 men in four boats, Powell ran the rapids of the Colorado, emerging below the Grand Canyon on August 29. It was a journey of nearly 900 miles. Powell returned repeatedly to the Colorado Basin because he saw it not just as a place for heroics, but also as a geological museum. He saw that the canyons were formed by the erosion of rivers in a plateau that was experiencing gradual elevation. In the 1880's, other surveys, consolidated under Powell, were organized under the name of the United States Geological Survey, with eminent scientists such as geologists Grove K. Gilbert (1843-1918) and Clarence E. Dutton (1841-1912). Gilbert traced the shores of the extinct Lake Bonneville in Utah and Nevada. Dutton studied in detail the geology of the Grand Canyon.

Powell saw the American West as more than just a scientific laboratory. He saw a landscape dominated by a lack of water interacting with successive populations—the Indians, Spanish, Mormons, and the settlers who poured in after the Civil War. He envisioned that scientific knowledge could control this process of settlement rather than leave the rapidly filling continent to the blind exploitation of settlers moved only by short-range economic goals. It was not to happen.

In 1888, legislation was passed that allowed Powell to tie the classification and distribution of all public lands in the West to topographic and hydrographic surveys that would determine the best sites for dams and lands most suitable for irrigation. But by 1890, the political supporters of rapid development in the West had won. Nevertheless, Powell showed the American people that science could be geared to the public interest and the long-run demands of nature.

401

In 1876, American science began its second century with the dedication of the Johns Hopkins University in Baltimore. The English advocate of Darwinism, zoologist Thomas H. Huxley (1825-1895), was the speaker, and his presence meant that the first modern graduate school in America would emphasize science and evolution. All of the new universities were beginning to emphasize basic research, specialization, and rigorous advanced training. By 1900, graduating American Ph.D.s in many fields were better educated, better equipped, and had many more research opportunities than their elders, who had had to get their scientific education in Europe.

The career of physicist Albert A. Michelson (1852-1931) showed the transition from the old pattern to the new. Brought to the near-frontier West from Strelno, Germany, by his parents at the age of 2, he was educated at the U.S. Naval Academy and learned his science at the U.S. Naval Observatory. As a professor at Case School of Applied Science in Cleveland, he and chemist Edward W. Morley (1838-1923) made history when their experiments with a sensitive instrument they designed—the interferometer—tested the relative motion of the earth and disproved the existence of a medium in space called ether. This work helped reshape the basic concepts of physics. Michelson did not pursue this theoretical line, but continued to improve his measuring apparatus. In 1907, he became the first American to win a Nobel prize—in physics.

In the period before World War I, while universities were becoming comfortable homes of basic research in all the major branches of the physical and biological sciences, other institutions were developing science in its applications to agriculture, engineering, and medicine.

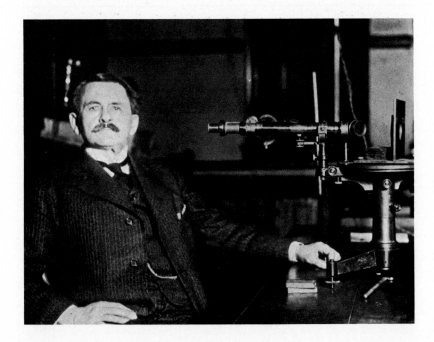

Albert A. Michelson, a craftsman of measuring instruments, used his skills to determine the speed of light.

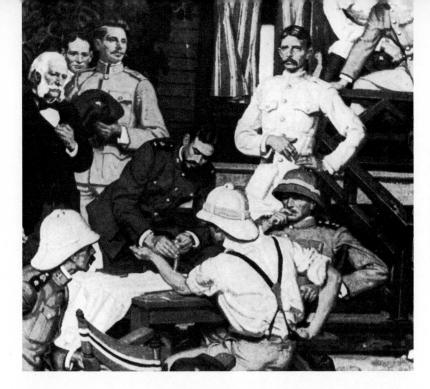

Inoculation against yellow fever following the discovery of its cause was a scientific triumph for Dr. Walter Reed and his colleagues and signaled the beginning of the science of public health.

The land-grant colleges had developed agriculture and engineering to a unique degree by the early 1900s. The agricultural-experimental stations took the controlled conditions of the laboratory outdoors to experimental plots and, by 1915, were sending the results directly to farmers through agricultural extension services.

Since the beginning of the republic, many scientists had been medical doctors, but only in the late 1800s did the physician-scientist begin to apply organized science to his profession. The new era began when William H. Welch (1850-1934) established a pathology laboratory similar to those in Europe at the Johns Hopkins Medical School. Theobald Smith (1859-1934), a medical doctor who had studied at Johns Hopkins, Toronto, and Cornell universities, became director of the Pathological Laboratory of the Bureau of Animal Industry of the Department of Agriculture in 1884. A bacteriologist working with a team that included entomologists, Smith proved that the organism that caused Texas fever in cattle was carried by a secondary host, a tick. The conquest of Texas fever opened the way for the attack on many human diseases, including yellow fever and malaria.

By 1900, the new scientific medicine paid off handsomely by identifying the mosquito as the link between man and yellow fever. Dr. Walter Reed (1851-1902) and his colleagues, working in Cuba after the Spanish-American War, not only identified the insect vector but also pointed the way to practical elimination of the disease by public health measures leading to mosquito control.

The methods at Johns Hopkins were adopted at all major medical schools after educator Abraham Flexner (1866-1959) recommended in 1910 that many of the existing medical schools be closed down and

Best known for his many inventions, Thomas A. Edison was also the creator of the world's first industrial lab.

resources be concentrated on those research-oriented institutions which were connected to universities and hospitals.

Meanwhile, industry had begun to realize the value of research and development. Thanks to Thomas A. Edison (1847-1931), scientists and engineers had reached a commanding position in the electrical industry by 1914. Edison did not consider himself a scientist, but rather an inventor. Among his first inventions were modifications of the telegraph. At age 23 he received $40,000 for a stock-ticker system to be used in the gold market. He used his profits to set up a laboratory devoted solely to invention. Edison was at his best in identifying a need and then developing a system to fulfill it. For example, when he developed the incandescent light bulb in 1879, it had characteristics that fit into an electric illumination system capable of lighting entire cities.

By the early 1890s, Edison had gone on to other interests and the General Electric Company (GE), which took over the Edison electrical manufacturing interests, began to develop a laboratory that was a prototype in that it joined basic and applied research. Charles P. Steinmetz (1865-1923), for example, used advanced mathematics to develop the many components needed for alternating-current generating and distribution systems. GE's Irving Langmuir (1881-1957) had an astonishingly fruitful career–working with chemical reactions at high and low temperatures, atomic structure, and atmospheric science. He received the Nobel prize in 1932 for work on surface chemistry and the electrical theory of matter. But his early practical work at GE – developing a tungsten filament in a globe filled with inert gas, which was the basis of a new electric-lamp industry—more than repaid the company for supporting his basic research.

Thus basic research, which was becoming as sophisticated and as well supported in American universities as that in Europe, found its way into industry. At the same time, applied research continued to grow. Nevertheless, World War I was a nasty shock to American technology. The cutting off of German supplies of chemicals and optical glass and Chilean nitrates revealed serious gaps in a number of disciplines in the United States.

Also, the threat of American involvement and the horror caused by "unusual" weapons—the submarine, airplane, and poison gas—demanded that scientists bring their skills to the weapons field. In 1915, the U.S. Navy created a Navy Consulting Board and the next year President Woodrow Wilson (1856-1924) set up the National Research Council (NRC) as an arm of the NAS. Physicist Robert A. Millikan (1868-1953) and astronomer George E. Hale (1868-1938) were the moving spirits of the new NRC. They brought together physicists to work on underwater listening devices and chemists to work on poison gas and develop countermeasures against it.

The postwar years brought many challenging problems. Young American physicists thrilled at the intellectual vistas opened by the

Edison's contemporary Charles Steinmetz, right, was a leading scientist at the General Electric Research Lab that evolved from the Edison organization.

With equipment such as the 200-inch Hale telescope on Palomar Mountain in southern California, American astronomers began to make many important contributions to the science of the stars.

LIGHT PATH TO PRIME FOCUS, f 3.3
" CASSEGRAIN = f 16.
" COUDE = f 30

APPROXIMATE SCALE

R. W. PORTER. '38.

THE TWO HVNDRED INCH TELESCOPE~

theory of relativity and quantum mechanics. Astronomers were making brilliant use of the great telescopes on Mount Wilson and Palomar Mountain in California. Harlow Shapley (1885-1972) and Edwin P. Hubble (1889-1953) vigorously argued the shape and size of the universe on the basis of a rapid increase in available data. In biology, genetics became a rigorous laboratory science, Thomas H. Morgan (1866-1945) of Columbia University and the California Institute of Technology leading the way with his studies of the fruit fly. He received the Nobel prize in 1933 for his work on the functions of chromosomes in heredity. Chemistry was developing so rapidly that hundreds of corporations established research laboratories, offering many opportunities to scientists.

The Great Depression of the 1930s brought hardship to students and shrunken budgets to many university, industrial, and governmental laboratories, but research continued. Ernest O. Lawrence (1901-1958) had his cyclotron going at the University of California at Berkeley by 1930. The Lawrence Radiation Laboratory steadily grew as a place for experimental physics and launched the careers of many talented physicists and chemists, including several who would follow Lawrence as winners of a Nobel prize.

Even as unlikely a field as soil microbiology at an agricultural experimental station was leading Selman A. Waksman (1888-1974) and René J. Dubos (1901-   ) in the direction of antibiotics in the late 1930s.

By 1938, the pattern of science in the United States was one of good organization and equipment and well-trained people. A government committee concerned with planning and natural resources issued a report that year, *Research—A National Resource.* It pointed up the process of science as a part of America's strength in conservation, agriculture, and industry. Some critics felt that the one thing lacking in the busy, organization-oriented lives of American scientists was that occasional soaring intellect who could set the direction for the work of others. However, thanks to the misfortunes of Europe, a great windfall of superlative scientific talent was already descending on the United States.

The Nazi take-over in 1933 cast hundreds of the world's greatest scholars and scientists out of Germany. They went to Denmark, to Great Britain, to Russia, to Turkey, and to America. But only the United States had enough laboratories, equipment, and professorships to allow the large numbers of refugees to continue with their work.

Albert Einstein (1879-1955), who settled at the Institute for Advanced Study in Princeton, N.J. in 1933, was a symbol of the intellectual migration. Yet the sheer weight of brainpower carried by the obscure as well as the famous wrought a sudden change in the quality of science in the United States. Here was the opportunity to make America the research center of the world. However, the message the refugees brought to Americans was not one of congratulation. Rather, it was a call to wake up and save Western civilization from totalitarianism.

When word reached the Danish physicist Niels Bohr (1885-1962), attending a conference in the United States in January, 1939, that Otto Hahn (1879-1968) and Lise Meitner (1878-1968) in Germany had achieved fission of the nucleus of a uranium atom, he realized that the way to a chain reaction releasing undreamed-of amounts of energy had opened up. Hungarian refugees Leo Szilard (1896-1964), Edward Teller (1908-   ), and Eugene Wigner (1902-   ) were so concerned about the possibility of a fission bomb in the hands of Germany that they enlisted Einstein to write of their fears to President Franklin D. Roosevelt in 1939. The letter helped to set in motion research that led to the atomic bomb.

By early 1940, it was clear that American science badly needed a uniting leadership if it were to provide research that would influence the outcome of an inevitable war. As France fell in June, 1940, four leaders met, determined to do just that. They were James B. Conant (1893-   ), president of Harvard University and a distinguished chemist; Karl T. Compton (1887-1949), president of Massachusetts Institute of Technology; Frank B. Jewett (1879-1949) of Bell Telephone Laboratories and president of the NAS; and Vannevar Bush (1890-

Albert Einstein was one of the many scientific giants who fled Europe in the 1930s and found sanctuary in American laboratories.

The National Defense Research Committee (NDRC) was the key to American scientific readiness at the beginning of World War II. Seated from left, Strong, Conant, Bush, Tolman, Jewett; and standing, Compton, Stewart, and Bowen.

A giant mushroom cloud announced the achievement of one goal of the NDRC.

1974), president of the Carnegie Institute of Washington and chairman of the National Advisory Committee on Aeronautics. Bush went to President Roosevelt with a proposal to bring the scientific community into the war effort, specifically to work on weapons.

Thus the Office of Scientific Research and Development (OSRD) was born. Bush and his colleagues considered the OSRD a temporary emergency agency and concentrated on pulling in the scientists and laboratories needed from universities and industry to work on specific projects. They developed the idea of research contracts which quickly put government funds into private laboratories. By Dec. 7, 1941, when the U.S. was drawn officially into World War II, the machinery for bringing scientists into the war effort had been set up. Even more striking, decisions about the direction of research had also been made. The OSRD would concentrate on radar and electronics, and, more daringly, would start on the road to an atomic bomb.

After the war, the concepts of science organized to do big jobs were carried on in different forms. In 1948, Bush published a report— *Science, the Endless Frontier* —which proposed a comprehensive National Science Foundation (NSF) to support basic research and education permanently with government contracts to private institutions in the same way that OSRD had functioned during the war.

The comprehensive program never became reality. The NSF finally established in 1950 was only a fragment of the original conception. But, after 1945, the Department of Defense, the NSF, and the National Institutes of Health became the sources of government funds. Research contracts and grant programs became the legal instruments which regulated this flow of government money.

Thus American science entered a golden age. The scale of experimentation grew in many fields, and "big science" became the norm. The abundant, costly equipment—from electron microscopes to particle accelerators—exceeded the scientists' wildest dreams.

The results were as spectacular as the scale. The science of the golden age extended mankind's vision in a wide spectrum. At one end of the scale, the interior of the nucleus opened up to a bewildering array of elementary particles. At the other end, radio telescopes opened the borders of the universe. In between came a new way of defining information in mathematical terms that hastened the growth of electronic computers. These devices, in turn, increased information capacity, and information theory became a great industrial product.

One of the most spectacular results of combining the new capabilities was the penetration of space. Vannevar Bush thought in 1948 that rockets would never become strategic weapons because they could not be aimed accurately. But inserting the rocket into an electronic envelope for guidance, communication, and control made the intercontinental ballistic missile and the artificial research satellite possible by the mid-1950s, and led to man's landing on the moon. It was electronics, not just rockets, that made a spaceship into a skylab.

Biology also thrived in the new environment. Borrowing techniques from nuclear chemistry and crystallography, James D. Watson (1928-) and Francis H. C. Crick (1916-) worked out the molecular structure of the genetic material deoxyribonucleic acid (DNA) in 1952. Within a decade, Marshall W. Nirenberg (1927-), using information theory, deciphered the genetic code by which DNA transmits messages to the proteins, the building blocks of organisms.

The era also saw a new relationship between science and the White House. Russia launched Sputnik I in 1957, and this prompted the creation of a complex set of advisers to the President of the United States. The post of the President's Science Adviser and the President's Science Advisory Committee were established in 1957. The Federal Council for Science and Technology and the Office of Science and Technology followed soon afterward.

By the mid-1960s, most people assumed that the postwar system of American science had become as permanent as the Constitution itself. The few critics usually merely regretted that research was unduly concentrated at a few universities. The leaders of the scientific community worried about the implications of the huge nuclear stockpile. But they felt that they could best serve the causes of science and peace by working within the Department of Defense, confident that their advice and research would contribute to security and act as a brake on the excesses of officials.

But American science and technology has suffered rude shocks since 1965. The complicated weapons systems used in the Vietnam War seemed less and less appropriate to the tasks given them, and the scientists and engineers who created them less enlisted on the side of

A leader of the new biology is James D. Watson who shared a Nobel prize for his work in determining the structure of DNA.

Putting men on the moon
was the culmination
of years of cooperative
effort by American
scientists and engineers
from many disciplines.

humanity than had been their counterparts in World War II. The waste of energy and the scarring of the environment posed severe problems in the United States. Since scientists had been so prominent in shaping the post-World War II economic and social systems, they could not escape a share of the blame. Meanwhile, scientists were losing their influence with government. The low point came in early 1973 when President Richard M. Nixon abolished the post of science adviser, the President's Science Advisory Committee, and the Office of Science and Technology.

The history of 200 years of American science has not been one of unbroken and inevitable progress. There have always been serious problems that the wisdom of the times could not solve. The scientific community often did not have the answers either, but it usually had the flexibility to change itself for a better attack on the problems.

As the United States enters its third century, its scientists and engineers must exercise this flexibility and, more important, find a strong sense of direction that now seems lost. A complicated new mix of disciplines and institutions, including biology and the social sciences along with the physical sciences and engineering, is arising. American scientists must recognize this and begin to define new goals if American science is to continue to serve the nation and its people.

Index

This index covers the contents of the 1974, 1975, and 1976 editions of *Science Year,* The World Book Science Annual.

Each index entry is followed by the edition year in *italics* and the page numbers:

Cosmic rays, *76*-40, *74*-256, 340

This means that information about Cosmic rays begins on the pages indicated for each of the editions.

An index entry that is the title of an article appearing in *Science Year* is printed in boldface italic letters: ***Archaeology.*** An entry that is not an article title, but a subject discussed in an article of some other title, is printed: **Ultrahigh sound.**

The various "See" and "See also" cross references in the index are to other entries within the index. Clue words or phrases are used when the entry needs further definition or when two or more references to the same subject appear in *Science Year.* These make it easy to locate the material on the page.

Uranium: dynamics, *Close-Up, 76*-258; environment, *75*-290; nuclear, *75*-336. See also headings beginning **Uranium. . . .**

The indication *"il."* means that the reference is to an illustration only, as:

Sea urchins, *il., 76*-271

Index

Index

Index

Index

Jonsson, J. Eric, *75-402*
Jorpes, J. Erik, *75-405*
Josephson, Brian D.: awards and prizes, *75-398*; *Close-Up, 75-284*; *il., 75-396*
Josephson tunneling: *Close-Up, 75-284*
J-particle: *Special Report, 76-112*
Julian, Percy L., *76-395*
Jupiter: astronomy, *74-253*; *Close-Up, 75-249*; gravity and comets, *ils., 75-22, 23*; Pioneer 10, *75-352*; planetary astronomy, *76-237, 75-246*; planetary satellites, *Special Report, 76-210*; space exploration, *76-345*
Juvenile hormones: *Close-Up, 74-246*; pest control, *75-195*

K

Kalahari Desert, *76-181*
Kartveli, Alexander, *76-395*
Kates, Joseph R., *75-399*
Kellermann, Kenneth I.: Gould Prize, *74-410*
Kelp: energy, *76-278*
Kendall, Henry W., *75-406*
Kenya, *76-358*
Kenyapithecus: Men and Women of Science, 76-366
Kerwin, Joseph P., *75-401, 74-357*
Kety, Seymour: neurology, *75-323*
Kidney: artificial, *74-314*; zoology, *74-368*
Kies, Constance: nutrition, *75-326*
Kikuyu, *76-361*
Kilbourne, Edwin D., *75-32*
Kimberlite pipes: geophysics, *75-299*
Kirk, John E., *76-395*
Kistiakowsky, George B., *74-407*
Kitt Peak National Observatory: planetary astronomy, *75-248*; space exploration, *75-351*
Kittay Award, *76-390, 75-400, 74-409*
Klebsiella pneumoniae: nitrogen, *Special Report, 76-101*
Knuckle: artificial, *il., 75-313*
Koburger, John A.: nutrition, *75-326*
Kohoutek: *ils., 75-12, 15, 19, 20, 21, 22, 23, 24*; organic matter, *75-296*; *Special Report, 75-13*
Kohoutek, Lubos, *il., 75-15*
Kow Swamp: anthropology, *74-248*
Kubasov, Valery N.: Apollo-Soyuz, *76-342, 75-351*
Kuffler, Stephen W.: Horwitz Prize, *74-408*
Kuiper, Gerard P., *75-405*
!Kung San: *Special Report, 76-181*
Kuru, *76-56, 75-319*

L

Labeling standards: drugs, *76-268*
Labrador rocks, *76-287*
Lac operator: biochemistry, *75-257*
Lactase: nutrition, *74-329*

Lactose: biochemistry, *76-248*; microbiology, *75-320*
Lake Rudolf: *Men and Women of Science, 76-358*
Lana: anthropology, *il., 76-229*
Land, Edwin H., *75-398*
Land-Sat 2, *76-345*
Language: anthropology, *76-229*; artificial intelligence, *Special Report, 75-76*; zoology, *76-351*
Large-scale integration (LSI), *75-281*
Laroxin: drugs, *76-268*
Laser cane, *76-72, il., 74-204*
Laser enrichment, *76-258, 260*
Laser-induced fluorescence: chemical dynamics, *75-268*
Laser ranger, *74-204*
Laser-scan photography, *il., 74-205*
Laserphoto: electronics, *74-289*
Lasers: atomic and molecular physics, *76-321*; chemical dynamics, *76-257, 75-268*; dynamics, *Close-Up, 76-258*; electronics, *74-287*; fusion, *75-288*; pellet fusion, *76-328, 75-338*; *Special Report, 74-199*; ultraviolet, *74-337*
Lava, *75-294, il., 297*; FAMOUS, *Special Report, 76-140*; geochemistry, *76-288*. See also **Volcanoes.**
Lawrence Berkeley Laboratory (LBL), *76-105*
Lawson criterion, *76-327*
Lead-acid battery, *76-126, 255*
Lead pollution: botany, *74-269*; environment, *74-297*
Leakey, Louis S. B., *76-357, 74-413*
Leakey, Mary, *76-357*
Leakey, Richard E., *76-357, 75-241, 74-247*
Lean-burn system: electronics, *76-273*
Learning: biochemistry, *74-264*; neurology, *74-327*; psychology, *76-332, 74-348*. See also **Education; Psychology.**
Lederberg, Joshua, *il., 75-387*
Lefschetz, Solomon, *74-414*
Left-handedness, *76-335*
Legume, *76-96*
Lehrman, Daniel S., *74-414*
Leonov, Alexei A., *76-342, 75-351*
Leontief, Wassily, *il., 75-396*
Lepton: *Special Report, 76-107*
Lesotho: geophysics, *75-299*
Letov, Aleksandr, *76-395*
Leukemia, *76-300*
Leukocytes: immunology, *75-305*
Lewis, Warren K., *76-395, 74-411*
LH: hormones, *Special Report, 76-85*
Lidar: lasers, *Special Report, 74-204*
Life expectancy, *76-335*
Life span: aging, *Special Report, 74-120*
Ligands, *76-261*
Ligase: genetics, *76-285*; microbiology, *76-310*
Light: blind, *Special Report, 76-77*; infrared, *74-255*; optical communications, *75-275*

Light Amplification by Stimulated Emission of Radiation. See **Lasers.**
Lighter-than-air vehicles, *76-348*
Lilly Award (microbiology), *76-389, 75-399, 74-408*
Lindemann, Erich, *76-395*
Lipid molecules: cell, *Special Report, 75-112*
Lipofuscin: age pigment, *74-127*
Liquid fuel, *76-255*
Liquid metal-cooled fast-breeder reactor (LMFBR): environment, *76-281, 75-290*; science support, *76-340*
Liquids, magnetic: chemical technology, *75-267*
Lithium-sulfur battery, *76-127*
Little Butser: archaeology, *76-232*
Liver: alcoholic cirrhosis, *75-308*; cancer, *75-291*; public health, *76-336*; surgery, *76-318*
Livestock: agriculture, *75-239, 74-245*
Lizards, *74-284*
Lobsters, *74-367*
LoBuglio, Albert: immunology, *75-305*
Logic circuits: electronics, *75-281*
Logs, *il., 75-236*
Lorenz, Konrad, *75-399, il., 75-396*
Los Angeles, *il., 75-123*
Lousma, Jack R.: awards and prizes, *75-401*; space exploration, *75-351*
Lunar climate: *Special Report, 75-150*
Lunar highlands, *74-299*
Lunar rocks: geochemical research, *75-295*
Lunar science: geophysics, *75-295*
Lunar soil: agriculture, *76-228*; geochemistry, *74-300*; space colonies, *76-34*
Lung: *Special Report, 76-42*; zoology, *74-368*
Lunokhod 2: space exploration, *74-361*
Lurgi process: coal, *Special Report, 75-226*
Lust, Reimar: Guggenheim Award, *74-410*
Luteinizing Hormone (LH), *76-85, 74-410*
Lymphocytes: B cells, *75-306*; immunology, *76-295, 75-304*; *Special Report, 74-108*; T cells, *75-306*
Lymphoma: internal medicine, *75-308*
Lynch, Gary: neurology, *75-324*
Lysine: agriculture, *76-277, 75-236*

M

MacDowell, Edwin C., *75-405*
Macrophages: biochemistry, *74-265*; dentistry, *76-297*; immunology, *76-296*, *Special Report, 74-112*
Magnesium, *76-326*
Magnesium oxide: conversion from olivine, *75-297*
Magnet: *Special Report, 76-121*
Magnetic field: astronomy, *76-235*;

421

Index

Index

Index

Index

V

Vaccines: dentistry, *76*-297; influenza, *74*-314, *Special Report*, *75*-28; Jonas Salk, *74*-389; public health, *il.*, *76*-336
Vaiana, Giuseppe S.: stellar astronomy, *75*-251
Vale, Brenda, *75*-208
Van de Kamp, Peter, *75*-251
Van Flandern, Thomas C.: *Close-Up*, *76*-245
Vanda: Antarctica, *il.*, *75*-98
Vegetation: ecology, *75*-281
Vela: gamma rays, *Special Report*, *76*-196; high-energy astronomy, *75*-252; X-ray emission, *74*-72
Vela X.: stellar astronomy, *75*-250
Velikovsky, Immanuel: *Close-Up*, *75*-249
Venezuela: archaeology, *74*-253
Venus: *Close-Up*, *75*-249; geochemistry, *74*-301; meteorology, *76*-307; planetary astronomy, *75*-248; space exploration, *76*-343; space probe, *74*-254; *Special Report*, *75*-149. See also *Space Exploration.*
Venus, Mount: FAMOUS, *Special Report*, *76*-142
Veterinary medicine, *76*-54, *74*-160
Vetlesen Prize, *76*-389, *74*-407
Video record player, *76*-275
Vietnam refugees, *76*-337
Vietnam War: cloud seeding, *74*-320
Viking program: climate, *Special Report*, *75*-159; space exploration, *75*-353
Vindolanda, *75*-244
Vinland map: *Close-Up*, *75*-264
Vinyl chloride: environment, *75*-291; public health, *76*-336
Viroids: *Close-Up*, *74*-326
Virology: *Special Report*, *76*-57
Virus: A-type, *il.*, *75*-33; *Books of Science*, *76*-260; C-type, *76*-300; cancer, *Close-Up*, *76*-300; *Close-Up*, *75*-319; flu, *ils.*, *75*-33, 34, 36, 37, 38; immunology, *75*-305; influenza, *Special Report*, *75*-26; internal medicine, *75*-308; London flu, *il.*, *75*-33; medicine, *74*-311; research, Jonas Salk, *74*-389; slow, *Special Report*, *76*-54; viroids, *74*-326. See also **Cancer; Leukocytes;** *Microbiology;* **Vaccines.**
Virtanen, Artturi, *75*-405
Vishniac, Wolf V., *75*-405
Vision: artificial intelligence, *Special Report*, *75*-78
Vitamin A: drugs, *76*-268, *75*-276
Vitamin B12: biochemistry, *74*-262; chemical synthesis, *74*-278; nutrition, *Close-Up*, *76*-315
Vitamin C: nutrition, *76*-315, *74*-329
Vitamin D: drugs, *76*-268, *75*-276
Vitamins: *Close-Up*, *76*-315; drugs, *76*-268; nutrition, *76*-314

Volcanoes: geophysics, *76*-292; lava, *il.*, *75*-297; oceanography, *76*-317
Vortex liners, *76*-329
Vostok: Antarctica, *il.*, *75*-98

W

Waksman, Selman A., *75*-405
Wall, Patrick D.: pain, *Special Report*, *75*-65
Wallpaper, *76*-257
Wankel engine: environment, *74*-295
Waste: alternative technology, *Special Report*, *75*-209; chemical technology, *75*-263; industrial, *74*-297; ocean disposal, *74*-333; radioactive materials, *75*-414. See also **Air pollution;** *Ecology;* **Ocean pollution; Pollution.**
Water: alternative technology, *Special Report*, *75*-210; astronomy, *74*-254; *Books of Science*, *76*-251; botany, *76*-253; energy source, *76*-119, 277, *74*-52; environment, *75*-307; fluoridation, *75*-307; geology, *76*-289; irrigation, *74*-244; on Mars, *75*-294; planetary astronomy, *76*-237; recycling, *75*-265; solid state physics, *76*-331; space colonies, *76*-38
Water pollution: nitrogen, *Special Report*, *76*-101. See also *Ecology;* **Ocean pollution; Pollution; Thermal pollution.**
Waterways: *Special Report*, *76*-14
Watson, James D., *76*-283, 409
Watson Medal, *76*-391
Watson-Watt, Sir Robert, *75*-405
Weather: Antarctica, *Special Report*, *75*-96; atmospheric models, *75*-317; *Books of Science*, *75*-259; city weather, *75*-317; *Close-Up*, *76*-308; communications, *75*-274; disaster, *Special Report*, *76*-168; ecology, *75*-278; forecasting, *74*-238; modification, *74*-320, *Special Report*, *74*-220; space colonies, *76*-38; *Special Report*, *76*-182. See also **Climate;** *Meteorology.*
Weed control, *76*-12, *74*-244
West Germany: space exploration, *76*-345
Westar 1: satellite communication system, *75*-274
Whales: *Books of Science*, *76*-251; oceanography, *74*-335
Wheeler, Willis B.: nutrition, *75*-326
Whey, *76*-254
Whipple, Fred L.: comets, *75*-19
Whirlpools, *76*-329
White, Paul Dudley, *75*-405
White blood cells: cancer, *Close-Up*, *76*-300; immunology, *75*-305
White dwarf stars, *76*-206
Wilcox, John M.: meteorology, *75*-315
Wildt, Rupert: *Close-Up*, *75*-249
Wilkinson, Geoffrey, *75*-397
Wilson, Robert: awards and prizes,

75-402; biography (*A Man of Science*), *74*-372
Wind patterns: meteorology, *75*-315
Windmills: energy, *76*-118
Winston, Roland, *75*-288
Windstorms: disaster, *76*-168
Wise, C. David: neurology, *75*-323
Wolf spider, *il.*, *75*-359
Wolff, Sheldon, *75*-292
Women: scientists, *Special Report*, *74*-24
Wood, *76*-257
Wood lice, *76*-270
Woodward, Robert B.: Cope Award, *74*-404
Work: psychology, *76*-335
World Soy Protein Congress: nutrition, *75*-325

X

X-ray astronomy, *76*-199, *74*-65
X-ray crystallography: biochemistry, *75*-257; structural chemistry, *76*-260, *75*-270
X-ray diffraction: geoscience, *75*-298
X-ray emission: solar, stellar astronomy, *75*-251
X-ray lasers, *76*-321
X rays: astronomy, *74*-259; geoscience, *75*-298; medical electronics, *75*-283; space exploration, *76*-345; *Special Report*, *74*-65
Xenon: chemical synthesis, *75*-272

Y

Yang: acupuncture, *75*-58
Yellow fever: *Essay*, *76*-403; mosquito, *Special Report*, *75*-194
Yellowstone National Park: *Books of Science*, *76*-250; ecology, *75*-279
Yerkes Observatory: stellar astronomy, *75*-251
Yin: acupuncture, *75*-58
Young, John W., *75*-366, *il.*, 369

Z

Zabriskie, John R.: immunology, *75*-305
Zhirmunskii, A. V., *75*-45
Ziegler, Karl, *75*-405
Zinc-air battery, *76*-256
Zinjanthropus, *76*-358
Zoology, *76*-349, *75*-358, *74*-366; animal locomotion, *Special Report*, *75*-170; Antarctica, *Special Report*, *75*-105; *Books of Science*, *76*-251, *75*-260, *74*-268; chimps, *Special Report*, *74*-34; diving sea mammals, *74*-335; dogs, *Special Report*, *74*-160; *Special Reports*, *76*-13, 54; whales, *74*-335. See also *Ecology.*
Zwicky, Fritz, *75*-405
Zworykin Award, *76*-392, *75*-402, *74*-411

Acknowledgments

The publishers of *Science Year* gratefully acknowledge the courtesy of the following artists, photographers, publishers, institutions, agencies, and corporations for the illustrations in this volume. Credits should be read from left to right, top to bottom, on their respective pages. All entries marked with an asterisk (*) denote illustrations created exclusively for *Science Year.* All maps were created by the *World Book* Cartographic Staff.

Cover
James A. Powell, Jr.

Special Reports

10   Woods Hole Oceanographic Institution; Mel Konner, Anthro-Photo; Bobbye Cochran*; Paul R. Alexander*; James A. Powell, Jr.
12   James A. Powell, Jr.
14   Anne B. Vietmeyer
15   Ron Church, Tom Stack & Associates
16-17   Noel D. Vietmeyer; Noel D. Vietmeyer; James A. Powell, Jr.
19   James Teason*; James A. Powell, Jr.
20-21   Noel D. Vietmeyer
22   James A. Powell, Jr.
23   Central & Southern Florida Flood Control District
24   James A. Powell, Jr.
26-27   Paul R. Alexander*
28   Princeton University
29   Mas Nakagawa*
30-31   Paul R. Alexander*
32   Mas Nakagawa*
33   Paul R. Alexander*
35   Mas Nakagawa*
36-39   Paul R. Alexander*
43   James P. Gerner
44   William Notman & Son, Montreal
45   Bobbye Cochran*; Ontario Research Foundation; Ontario Research Foundation; Technical aid, The Hospital for Sick Children, Toronto
46   Bobbye Cochran*
47-48   Werner A. Meier, M.D., Rush-Presbyterian St. Luke's Medical Center
51   Bobbye Cochran*
52   Werner A. Meier, M.D., Rush-Presbyterian St. Luke's Medical Center
55   John Freas*
56   Wilma Stevens*
57   Richard T. Johnson, M.D., The Johns Hopkins University School of Medicine
58   John Freas*; Gabriele M. Zu Rhein, M.D., University of Wisconsin; Gabriele M. Zu Rhein, M.D., University of Wisconsin
60   Monique Dubois-Daley, M.D., Department of Health, Education, and Welfare; John Freas*; Robert Herndon, M.D., The Johns Hopkins University School of Medicine
61   Diane Griffin, The Johns Hopkins University School of Medicine
62   Robert Herndon, M.D., The Johns Hopkins University School of Medicine
64-65   John Freas*
68-69   Joseph A. Erhardt*
71-77   Veterans Administration Hospital, Blind Rehabilitation Center, Hines, Ill. (Joseph A. Erhardt)*
78   Lou Bory Associates*; Dobelle, Mladejovsky and Girvin, from *Science.* Copyright 1974 by the American Association for the Advancement of Science; Dobelle, Mladejovsky and Girvin, from *Science.* Copyright 1974 by the American Association for the Advancement of Science

80   Lynette Morningstar*
82   Michael Reed*
84-89   Lynette Morningstar*
90   Maurice Dubois, Station de Physiologie de la Reproduction, Nouzilly, France; R. Unger, Dallas; R. Assan, Paris; W. Gepts, Brussels, Belgium
93   Jim White, Black Star; Fred Ward, Black Star
94   Joseph A. Erhardt*
95   Terry Wickart*
96   Winston J. Brill and Stanley Carlson*
97   Terry Wickart*
98   Joseph A. Erhardt*
100   Terry Wickart*
101   Joseph A. Erhardt*
102   Terry Wickart*
104   Lawrence Berkeley Laboratory, University of California
106   Fred C. Eckhardt, Jr.*
107   Stanford University
109   Vera Lüth; Vera Lüth; Vera Lüth; Stanford University
114   Samuel C. C. Ting, Massachusetts Institute of Technology
115   Brookhaven National Laboratory
116-117   Consumers Power Company
120   Jackson-Zender*
122   Jackson-Zender*; Los Alamos Scientific Laboratory, University of California
123-124   Jackson-Zender*
126   Jackson-Zender*; Argonne National Laboratory*; Argonne National Laboratory*
128   Ontario Research Foundation (Roger Chapman)*
130-131   Ontario Research Foundation (Roger Chapman)*; Technical aid, Cambridge Instrument Company
132   Bobbye Cochran*; Ontario Research Foundation (Roger Chapman)*; Technical aid, Cambridge Instrument Company
133-137   Ontario Research Foundation (Roger Chapman)*
134   Technical aid, Marion R. Weimann, Jr., University of Chicago; Stanley Carlson, University of Wisconsin, Madison
135   Technical aid, The Hospital for Sick Children, Toronto
136   Technical aid, Cambridge Instrument Company
138   Woods Hole Oceanographic Institution; Woods Hole Oceanographic Institution; Woods Hole Oceanographic Institution; Centre National pour L'Exploitation des Océans; Centre National pour L'Exploitation des Océans
141   Copyright Editions Pierre Charron-Tanguy de Remur, Paris
142   Woods Hole Oceanographic Institution
143   Woods Hole Oceanographic Institution; Centre National pour L'Exploitation des Océans; Woods Hole Oceanographic Institution
144   Woods Hole Oceanographic Institution
145   Woods Hole Oceanographic Institution; Woods Hole Oceanographic Institution; Woods Hole Oceanographic Institution; Centre National pour L'Exploitation des Océans
147-150   Woods Hole Oceanographic Institution
152   J. R. Eyerman
156   U.S. Geological Survey; U.S. Geological Survey; J. R. Eyerman

159   Jackson-Zender*; Cooperative Institute for Research in Environmental Sciences; Cooperative Institute for Research in Environmental Sciences; Eve S. Sprunt and W. F. Brace, Massachusetts Institute of Technology
160   Don Ivers, California Institute of Technology; Harry Cullum*; Rockwell International
161   J. R. Eyerman
163   WORLD BOOK photo
164   Sovfoto; Carl Kisslinger, Cooperative Institute for Research in Environmental Sciences
166-167   Henri Bureau, Sygma
168   Michael J. Novia
169   Harry Redl, Black Star; Alain Nogues, Sygma
173   Flip Schulke, Black Star
174   Don Goode, Black Star; United Press Int.
175   United Press Int.; Don Goode, Black Star
176   Don McCoy, Black Star
177   Gerald Brimcombee, Black Star; Edward Grant, Miller Services Ltd.
178   Richard L. Grossman
180   Mel Konner, Anthro-Photo
182   Richard B. Lee
184   Irven DeVore, Anthro-Photo; Richard B. Lee; Irven DeVore, Anthro-Photo
185   Richard B. Lee
186-187   Laurence K. Marshall; Mel Konner, Anthro-Photo
188   Irven DeVore, Anthro-Photo
189   Richard B. Lee
191   Irven DeVore, Anthro-Photo; Mel Konner, Anthro-Photo; Richard B. Lee
193   Richard B. Lee
194   Richard B. Lee; Irven DeVore, Anthro-Photo
197   Frank Rakoncey*
198   Los Alamos Scientific Laboratory, University of California
199   Frank Rakoncey*
200   TRW Inc.; Los Alamos Scientific Laboratory, University of California
201   Frank Rakoncey*
202   Mount Hopkins Observatory
204-207   Frank Rakoncey*
211   Alex Ebel*
212   Cornell University
214-217   Alex Ebel*
219   NASA
220-221   Alex Ebel*

Science File

224   National Bureau of Standards; D.C. Basolo; © Leiden Observatory; H. D. MacGinite, University of California at Berkeley; NASA
226   D. C. Basolo
227   Pennwalt Corporation
229   Charles E. Oxnard, University of Chicago
230   Cleveland Museum of Natural History
231   D. Markovic
232   © Times Newspapers Ltd., 1974
233   University of Michigan
234   U.S. Forest Service from *Science*. Copyright 1974 by the American Association for the Advancement of Science
236   NASA; Uwe Fink, University of Arizona; NASA (Arthur Grebetz*)
239   Kitt Peak National Observatory
240   T. R. Gull, Kitt Peak National Observatory
241   © Leiden Observatory
243   National Astronomy and Ionosphere Center, Cornell University
246   Arthur Grebetz*
247   D. Papahadjopoulos, E. Mayhew, G. Poste, and S. Smith, Roswell Park Memorial Institute, Buffalo; W. J. Vail, University of Guelph, Canada
248   John Brandt, Fred V. Lucas, Sr., Arlene Martin, and Marie L. Vorbeck, University of Missouri School of Medicine, Reprinted with permission from *Biochemical and Biophysical Research Communications.* © Academic Press, Inc.
252   Leo J. Hickey, Smithsonian Institution
254   Drawing by D. Fradon; © 1975 The New Yorker Magazine, Inc.
256   National Bureau of Standards
258   Lawrence Livermore Laboratory, University of California; Arthur Grebetz*
261   Argonne National Laboratory
262   International Business Machines Corporation
263   Lawrence Berkeley Laboratory, University of California
265   Bell Telephone Laboratories
266   Westinghouse Research Laboratories
267   Melvin H. Van Woert, M.D., Mount Sinai School of Medicine, New York City
269   Arthur Grebetz*
271   Norman S. Smith, University of Arizona
272   University of Utah
273   Hughes Aircraft Company
274   Drawing by Chon Day; © 1975 The New Yorker Magazine, Inc.
276-277   Lawrence Livermore Laboratory, University of California
279   Jan van Wessum
282   U.S. Environmental Protection Agency
283   Arthur Grebetz*
284   National Academy of Science
286   Roy Stephen Lewis, University of Chicago
287   University of Massachusetts
289-290   U.S. Geological Survey
291   Theodore L. Sullivan, *The New York Times*
293   Texas Memorial Museum
294   From *Trilobites, A Photographic Atlas* by Ricardo Levi-Setti, University of Chicago Press, 1975
295   Peck-Sun Lin, Donald F. Hoelzl Wallach, Tufts-New England Medical Center, from *Science*. Copyright 1974 by the American Association for the Advancement of Science
297   Israel Kleinberg, M.D., State University of New York at Stony Brook; Ontario Research Foundation (Roger Chapman)*
298   Joseph A. Erhardt*
299   Wide World
302   Charles Moore, Black Star
303   C. Everett Koop, M.D., The Children's Hospital of Philadelphia
305   Arthur Grebetz*; Alton Ochsner Medical Foundation, New Orleans; Alton Ochsner Medical Foundation, New Orleans
306   Arthur Grebetz*
307   U.S. Geological Survey
311   Isama Kondo, M.D., and Noriko Hasegawa, The Jikei University School of Medicine, Tokyo
313   Pasko Rakic, Edward P. Sayre and Richard L. Sidman, Harvard Medical School; Larry J. Stenas, University of Utah
316   Drawing by Lorenz; © 1975 The New Yorker Magazine, Inc.
318   U.S. Geological Survey
320   National Oceanic and Atmospheric Administration
322   L. S. Bartell and C. L. Ritz, University of Michigan
324   David M. Baurac, California State University at Northridge
326   Sandia Laboratories
328   KMS Fusion, Inc.
329   Lawrence Berkeley Laboratory, University of California
330   Drawing by Porges; © 1974 The New Yorker Magazine, Inc.

331 Lawrence Berkeley Laboratory, University of California
333 David Premack, University of California at Santa Barbara, from *Science*. Copyright 1975 by The American Association for the Advancement of Science
334 Drawing by Lorenz; © 1974 The New Yorker Magazine, Inc.
336 Seattle King County Health Department
337 Bell Telephone Laboratories
343-344 NASA
346 Calspan Corporation
347 U.S. Department of Transportation
348 Goodyear
349 D. M. Lavigne, University of Guelph, Canada
350 Drawing by Nurit; © 1974 The New Yorker Magazine, Inc.
351 M. Möglich, University of Frankfurt
353 Holy Land Conservation Fund, Inc.

Men and Women of Science
354 Gene Trindl, Globe*; John Reader*; John Reader*; John Reader*; Gene Trindl, Globe*
356 John Reader*
359 Baron Hugo van Lawick, © National Geographic Society; Robert F. Sisson, © National Geographic Society
360-370 John Reader*
372-384 Gene Trindl, Globe*

387 Stanford University
388 Cambridge University
389 The Franklin Institute, Philadelphia; University of California at Irvine
390 William B. Carter, Yale University
391 NASA; Fabian Bachrach
392 Orlando, Globe*
394 Pictorial Parade; Salk Institute for Biological Studies; Wide World
395 Pictorial Parade; Julian Research Institute; Wide World
397 Philadelphia Museum of Art; New York Historical Society
398 Brown University
399 Detail from *Into The Unknown* by J.K. Ralston, Jefferson National Expansion Memorial, St. Louis; Yale University
400 National Academy of Sciences; U.S. Geological Survey
401 U.S. Geological Survey
402-403 Bettmann Archive
404 Edison National Historic Site
405 General Electric Company
406 Hale Observatories
407 Wide World
408 *Science News Letter*, July 13, 1940; U.S. Department of Defense
409 Wide World
410 NASA

Typography

Display—Univers and Baskerville
Monsen Typographers, Inc., Chicago
Text—Baskerville Linofilm
Total Typography, Inc., Chicago
Text—Baskerville Linotron
Black Dot Computer Typesetting Corporation, Chicago

Electronic Color Separations

Printing Developments, Inc., Chicago

Offset Positives

Jahn & Ollier Engraving Co., Chicago
Process Color Plate Company, Chicago

Printing

Kingsport Press, Inc., Kingsport, Tenn.

Binding

Kingsport Press, Inc., Kingsport, Tenn.

Paper

Text
Childcraft Text, Web Offset (basis 60 pound)
Mead, Escanaba, Mich.

Cover Material

Flax Lexotone
White Offset Blubak
Holliston Mills, Inc., Kingsport, Tenn.